Reverse Osmosis

Reverse Osmosis

Membrane Technology, Water Chemistry, and Industrial Applications

Edited by
Zahid Amjad, Ph.D.

CHAPMAN & HALL

I(T)P® International Thomson Publishing

New York • Albany • Bonn • Boston • Cincinnati • Detroit • London • Madrid • Melbourne
Mexico City • Pacific Grove • Paris • San Francisco • Singapore • Tokyo • Toronto • Washington

This edition published by Chapman & Hall, New York, NY

Printed in the United States of America

For more information contact:

Chapman & Hall
115 Fifth Avenue
New York, NY 10003

Chapman & Hall
2-6 Boundary Row
London SE1 8HN
England

Thomas Nelson Australia
102 Dodds Street
South Melbourne, 3205
Victoria, Australia

Chapman & Hall GmbH
Postfach 100 263
D-69442 Weinheim
Germany

International Thomson Editores
Campos Eliseos 385, Piso 7
Col. Polanco
11560 Mexico D.F.
Mexico

International Thomson Publishing - Japan
Hirakawacho-cho Kyowa Building, 3F
1-2-1 Hirakawacho-cho
Chiyoda-ku, 102 Tokyo
Japan

International Thomson Publishing Asia
221 Henderson Road #05-10
Henderson Building
Singapore 0315

2 3 4 5 6 7 8 9 XXX 01 00 99 98 97 96

Library of Congress Cataloging-in-Publication Data

Amjad, Zahid.
    Reverse osmosis: membrane technology, water chamistry & industrial applications / by Zahid Amjad.
        p.   cm.
    Includes bibliographical references and index.
    ISBN 0-442-23964-5
    1. Reverse osmosis. I. Title.
    TP156.07A45   1992   92-40463
    660'.2842--dc20   CIP

Visit Chapman & Hall on the Internet http://www.chaphall.com/chaphall.html

To order this or any other Chapman & Hall book, please contact **International Thomson Publishing, 7625 Empire Drive, Florence, KY 41042.** Phone (606) 525-6600 or 1-800-842-3636. Fax: (606) 525-7778. E-mail: order@chaphall.com.

For a complete listing of Chapman & Hall titles, send your request to **Chapman & Hall, Dept. BC, 115 Fifth Avenue, New York, NY 10003.**

To Rukhsana and Naureen

Contents

About the Editor   xi
Contributors   xiii
Preface   xv

1. **Reverse Osmosis Membranes State of the Art**   1
   Donald C. Brandt, Gordon F. Leitner, and Wendy E. Leitner

   Introduction and Overview   1
   Reverse Osmosis Theory   2
   Membrane Materials   6
   Membrane Devices and Configurations   8
   Membrane Device Performance   14
   RO System Design Considerations   16
   Applications of Reverse Osmosis   24
   Recent Developments   30
   Reverse Osmosis Membrane Manufacturers—A Worldwide
      Perspective   31
   Summary   32

2. **Future Trends in Reverse Osmosis Membrane Research
   and Technology**   37
   Takeshi Matsuura

   Introduction   37
   Recent Development of Membranes and Membrane Materials   37
   Development of Materials for High Chlorine Resistant Membranes   44
   Origin of the Membrane Pore   46
   Membrane Transport   50
   Pore Size Control of Reverse Osmosis Membrane   54
   Membrane Fouling   58
   Module Design   60
   Membrane Applications   64
   Conclusion   71

3. **The Economics of Desalination Processes   76**
    Gregory A. Pittner

    Introduction   76
    Methods of Financial Analysis   77
    Energy Consumption in Desalination   80
    Overview of Desalination Processes   81
    Typical Capital and Operating Costs   87
    Comparison of RO and Ion Exchange for Production of High
        Purity Water   95
    Summary   102

4. **Design Considerations for Reverse Osmosis Systems   104**
    Robert Bradley

    Introduction   104
    Water Supplies and Characteristics   104
    Materials of Construction   108
    Pretreatment   108
    Membrane Selection   119
    System Design Parameters   125
    Instrumentation and Record Keeping   137

5. **The Importance of Water Analysis for Reverse Osmosis Design
    and Operation   139**
    Joseph P. Hooley, Gregory A. Pittner, and Zahid Amjad

    Introduction   139
    Water Source Investigation   140
    Sampling   141
    Types of Analyses   143
    Cost of Water Analysis   159
    Who Benefits from Water Analysis   159
    Summary   161

6. **Mechanistic Aspects of Reverse Osmosis Membrane Biofouling
    and Prevention   163**
    Hans-Curt Flemming

    Introduction   163
    Biofouling in RO Water Systems   164
    Mechanistic Aspects of Membrane Fouling   170
    Control and Prevention of Biofouling   186
    Do We Have to Live with Biofouling?   199

7. **Considerations in Membrane Cleaning   210**
    Zahid Amjad, Kenneth R. Workman, and Donald R. Castete

    Introduction   210
    Deposit Removal Techniques   214
    Membrane Cleaner Selection Criteria   217
    System Troubleshooting   219
    Rejuvenation and Sterilization   224
    Case Histories   228

8. **Analytical Techniques for Identifying Reverse
    Osmosis Foulants   237**
    James D. Isner and Robert C. Williams

    Introduction   237
    Analytical Techniques   238
    Summary   273

9. **RO Application in Brackish Water Desalination and in the
    Treatment of Industrial Effluents   275**
    C. A. Buckley, C. J. Brouckaert, and C. A. Kerr

    Brackish Water Desalination   275
    Treatment of Industrial Effluents   282
    Examples of Reverse Osmosis Effluent Treatment Processes   291

10. **Applications of Reverse Osmosis Technology in the
    Food Industry   300**
    S. S. Köseoḡlu and G. J. Guzman

    Introduction   300
    Applications   301

11. **High Purity Water Production Using Reverse
    Osmosis Technology   334**
    Gregory A. Pittner

    Introduction   334
    High Purity Water Applications and Quality Requirements   335
    Water Purification with Reverse Osmosis   340
    Problems in High Purity Water Systems   342
    Double-Pass Reverse Osmosis   347
    Pretreatment and Posttreatment   349
    Field Experience   354
    Market Size and Growth Projections   361
    Conclusion   362

**12. An Overview of U.S. Military Applications of
        Reverse Osmosis   364**

S. A. Choudhury

Introduction   364
Background   364
Design Considerations   365
Military Water Treatment   366
600-GPH ROWPU   367
3,000-GPH ROWPU   369
150,000-GPD ROWPU   372
Reverse Osmosis Elements   376
Chemicals   377
Ongoing and Future Efforts   377

13. An Overview of RO Concentrate Disposal Methods   379

Laura S. Andrews and Gerhardt M. Witt

Introduction   379
Regulatory Background   380
Surface Water Disposal   381
Deep Well Injection   383
Spray Irrigation/Land Application   387
Other Disposal Methods   388

Index   390

About the Editor

Zahid Amjad is Research and Development Fellow for the Speciality Polymers and Chemicals Division of The BFGoodrich Company in Brecksville, Ohio, where he has served since 1982. He received an M.S. degree in chemistry (1967) from Punjab University (Pakistan), and a Ph.D. degree in physical chemistry (1976) from Glasgow University (Scotland). Dr. Amjad was assistant research professor at the State University of New York at Buffalo from 1977 until 1979, when he joined the Calgon Corporation. His research interests include membrane separation technologies for water purification, membrane fouling and prevention, and investigations on the physico-chemical events at the solid-liquid interfaces. Dr. Amjad has published more than 60 papers in various technical journals and is a holder of 23 patents. He is listed in *American Men and Women of Science, Who's Who in Technology,* and *Who's Who of American Inventors.* Dr. Amjad is a certified professional chemist and a member of several professional organizations.

Contributors

Zahid Amjad, The BFGoodrich Company, Brecksville Research Center, 9921 Brecksville Road, Brecksville, OH.

Laura S. Andrews, Camp Dresser & McKee, Inc., 201 Montgomery Avenue, Sarasota, FL.

Robert Bradley, Arrowhead Industrial Water, Inc., A BFGoodrich Company, 300 Tristate International, Suite 320, Lincolnshire, IL.

Donald C. Brandt, 908 Fairthorne Avenue, Wilmington, DE.

C. J. Brouckaert, Chemical Engineering, Pollution Research Group, University of Natal, King George V Avenue, Durban 4001, South Africa.

C. A. Buckley, Pollution Research Group, University of Natal, King George V Avenue, Durban 4001, South Africa.

Donald R. Castete, Arrowhead Industrial Water, Inc. A BFGoodrich Company, 999 North F.M. 3083, Conroe, TX.

S. A. Choudhury, U.S. Army Belvoir Research, Development, and Engineering Center, Fuel and Water Supply Division, Fort Belvoir, VA.

Hans-Curt Flemming, Institute für Siedlungswasserbau, Universität Stuttgart, Bandtäle 1, D-7000 Stuttgart 80, Germany.

G. J. Guzman, Food Protein Research and Development Center, Engineering Biosciences Research Center, Texas A&M University Systems, College Station, TX.

Joseph P. Hooley, Arrowhead Industrial Water, Inc., A BFGoodrich Company, 10321 Brecksville Road, Brecksville, OH.

James D. Isner, The BFGoodrich Company, Avon Lake Technical Center, P.O. Box 122, Avon Lake, OH.

C. A. Kerr, Pollution Research Group, University of Natal, King George V Avenue, Durban 4001, South Africa.

S. S. Köseoglu, Food Protein Research and Development Center, Engineering Biosciences Research Center, Texas A&M University Systems, College Station, TX.

Gordon F. Leitner, Leitner & Associates, Inc., Consulting Engineers, 815 Shadow Lawn Drive, Elm Grove, WI.

Wendy E. Leitner, Leitner & Associates, Inc., Consulting Engineers, 815 Shadow Lawn Drive, Elm Grove, WI.

Takeshi Matsuura, Faculty of Engineering, University of Ottawa, 161 Louis Pasteur, Ottawa, Canada.

Gregory A. Pittner, Arrowhead Industrial Water, Inc., A BFGoodrich Company, 1441 East Washington, Los Angeles, CA.

Robert C. Williams, The BFGoodrich Company, Avon Lake Technical Center, P.O. Box 122, Avon Lake, OH.

Kenneth R. Workman, Arrowhead Industrial Water, Inc., A BFGoodrich Company, 300 Tristate International, Suite 320, Lincolnshire, IL.

Gerhardt Witt, Parson Brinckerhoff Gore & Storrie, West Palm Beach, FL.

Preface

Reverse osmosis (RO) is a valuable membrane separation unit operation. The uses of RO encompass a variety of industries, including those engaged in the desalination of water (i.e., producing potable water from brackish water and seawater), the purification of ancillary process streams (i.e., for cooling water and boiler water), the production of high purity rinse waters (i.e., for paint electrodeposition and semiconductor wafer baths), contact water for chemical manufacture (i.e., pharmaceutical or cosmetic), food and beverage production (i.e., concentration of fruit juices and cheese making), and cleaning industrial effluents (i.e., before discharge or reclaim). As the technology matures, innovative uses are found for the process.

This book was designed to provide a comprehensive discussion of RO membrane technology, applications of the process, limitations and pitfalls inherent to its operation, and trends in research and technology. It should prove useful to process and design engineers, operations personnel (plant managers and technicians), and scientists in various research institutes. Chapters 1 through 3, cover membrane technology. Chapters 4 through 8, present a discussion on system design, water chemistry, and membrane foulants and cleaners. Chapters 9 through 13 address some of the applications of membrane separations.

The book starts with an overview of membrane technology, RO processes, membrane types and performance expectations, the state of the membrane industry, and a brief introduction to membrane applications. This is followed by a detailed presentation of trends in the industry for the near future. There is special emphasis on materials of construction and mechanical properties. A chemical analysis of the membranes allows projections of operating characteristics. Economic considerations explain why membrane processes are rapidly supplanting more traditional operations.

The next chapters explain the importance of construction material, membrane selection, instrumentation, pretreatment, and posttreatment on the process design. Other considerations are the intended use of the process and the effect of water chemistry and the importance of chemical analyses to optimize an operation. Microbiological fouling is a frequent cause of performance deterioration in membrane-based systems and is rarely understood. Therefore, a comprehensive discussion of bacterial species commonly found in water supplies, the mechanisms of biofouling, and methods of prevention is given. Other foulants common to RO and cases of failures resulting from such foulants are described, along with methods for their removal or prevention. The various analytic techniques used to identify RO foulants and the causes of mechanical degradation are described in detail.

The final chapters begin with an account of various applications, from brackish water desalination to the treatment of industrial effluents to the processing of food products with membranes and membrane-based systems. In the food production industry alone, the variety of uses for membrane separations is impressive. The separations process allows the blending of milk and cheese products, the concentration of fruit juices, the removal of alcohol from wine and beer, and the recovery of cereal proteins. The production of high and ultrahigh purity water and the problems encountered in such systems are discussed. An overview of the use of RO technology by the U.S. military is also presented. The military must be able to produce pure water from any available source for deployed land-based forces. An overview of methods for the disposal of the concentrate generated from the RO process itself concludes the book. This overview includes such techniques as deep well injection, spray irrigation, and surface and tidal water disposal.

I thank all the contributors for their hard work and cooperation in preparing these chapters and producing them in a timely fashion, which made my job of editing much easier. I am indebted to Robert W. Zuhl, Jeff Pugh, and the reviewers who have read parts or all of the manuscripts and have made valuable comments and helpful suggestions.

I thank Ralph F. Koebel for his encouragement and continued support, and am grateful to The BFGoodrich Company for providing support and an excellent environment in which most of the work was done.

Thanks are also extended to Pam Jensen and Helena Edding for their excellent secretarial assistance. Furthermore, it is a pleasure to acknowledge the help and cooperation of the Van Nostrand Reinhold staff during the writing and production of this book.

Finally, I extend my sincere appreciation to my wife, Rukhsana, whose continued encouragement, moral support, and invaluable help have made the editing of this book an enjoyable experience.

1

Reverse Osmosis Membranes State of the Art

Donald C. Brandt

Donald C. Brandt, P.E.

Gordon F. Leitner and Wendy E. Leitner

Leitner and Associates, Inc.

INTRODUCTION AND OVERVIEW

The phenomenon of osmosis was discovered by Abbè Nollet more than 200 years ago when he observed the transport of water across a pig bladder covering the mouth of a jar containing "spirits of wine" (Lonsdale 1982). The first synthetic membrane was prepared by Moritz Taube in 1867 and is discussed in a recent review (Mason 1991). Modern-day reverse osmosis (RO) processes are an outgrowth of research with cellulose acetate (CA) membranes in the late 1950s at the University of Florida (Reid and Breton 1959) and at the University of California, Los Angeles (Loeb and Sourirajan 1963). Loeb and Sourirajan are credited with making the first high-performance membranes by creating an asymmetric cellulose acetate structure with improved salt rejection and water flux. The history and background of these developments are well documented elsewhere (Merten 1966; Sourirajan 1970a; Loeb 1981).

The commercial era for RO began in the late 1960s by Gulf General Atomics (later known as the Fluid Systems Division of Universal Oil Products and now of Allied-Signal Corporation) and Aerojet General, using Loeb-Sourirajan cellulose acetate membranes in a spiral wound configuration developed through funding by the U.S. Department of Interior, Office of Saline Water. Hollow fiber membranes and devices were developed and commercialized by Du Pont in the late 1960s. In 1971, Du Pont introduced the Permasep® B-9 permeator containing millions of asymmetric aromatic polyamide (aramid) hollow fine fibers (Chemical Engineering 1971). In late 1973, it introduced the Permasep® B-10 permeator, also using asymmetric aramid fibers, capable of producing potable water from seawater in a single pass.

In the mid-1970s, cellulose triacetate hollow fiber permeators were introduced by Dow Chemical Company, followed by Toyobo of Japan, and spiral
wound polyamide thin film composite membranes were introduced by the Fluid
Systems Division and by FilmTec (now a subsidiary of Dow Chemical Company).

Throughout the 1980s, improvements were made to these membranes to
increase water flux and salt rejection with both brackish water and seawater.
Today the predominate membrane materials are still aramids, polyamides, and
cellulose acetate and triacetate in spiral wound and hollow fiber configurations.
They are being used for numerous applications including potable water production, waste recovery, food applications, kidney dialysis, high-purity water for
boiler feed, and ultrapure water electronics applications. In 1990, RO technology
was used to treat more than one billion gallons of water per day, and this market
is expected to grow by about 18 percent a year for the rest of the century.

REVERSE OSMOSIS THEORY

Reverse osmosis, as its name implies, is a process whereby the natural phenomenon of osmosis is reversed by the application of pressure to a concentrated
solution in contact with a semipermeable membrane. If the applied pressure is in
excess of the solution's natural osmotic pressure, the solvent will flow through
the membrane to form a dilute solution on the opposite side and a more
concentrated solution on the side to which pressure is applied. If the applied
pressure is equal to the solution's natural osmotic pressure, no flow will occur; if
the applied pressure is less than its natural osmotic pressure, there will be flow
from the dilute solution to the concentrated solution.

The rate of water transport across the membrane depends on the membrane
properties, the solution temperature, and the difference in applied pressure, across
the membrane, less the difference in osmotic pressure between the concentrated and
dilute solutions. Osmotic pressure is proportional to the solution concentration and
temperature and depends on the type of ionic species present. For solutions of
predominantly sodium chloride at ambient temperatures, a rule of thumb is that
osmotic pressure is 10 psi (0.7 atm) per 1000 mg/l concentration.

Basic Equations

Water Transport
The water transport (at constant temperature) through a semipermeable membrane is described by the following equation:

$$Q_w = K_w (\Delta P - \Delta \pi) A / \tau \tag{1}$$

Where

Q_w = water flow rate through the membrane
K_w = membrane permeability coefficient for water
ΔP = hydraulic pressure differential across membrane
$\Delta \pi$ = osmotic pressure differential across membrane
A = membrane surface area
τ = membrane thickness

Osmotic Pressure

Osmotic pressure depends on solute concentration, solution temperature, and the type of ions present. For dilute solutions, osmotic pressure is approximated using the van't Hoff equation (Lonsdale 1986):

$$\pi = v_i \, c_i \, RT \qquad (2)$$

Where

π = osmotic pressure
c_i = molar concentration of the solute
v_i = number of ions formed if the solute dissociates (e.g., for NaCl, $v_i = 2$; for BaCl$_2$, $v_i = 3$)
R = gas constant
T = absolute temperature

For more concentrated solutions, this equation (2) is multiplied by an osmotic coefficient that is estimated from vapor pressure data or from freezing-point depression of the solution involved (Hwang and Kammermeyer 1984). For practical purposes, the van't Hoff equation can be used for brackish water, and experimental osmotic pressure data obtained from sea salt solutions (Sourirajan 1970b) can be used for seawater.

Salt Transport

The salt transport across a membrane is proportional to the concentration or chemical potential difference across the membrane and is described by the following equation:

$$Q_s = K_s \, (\Delta C) A / \tau \qquad (3)$$

Where

Q_s = salt flow through membrane
K_s = membrane permeability coefficient for salt
ΔC = salt concentration differential across membrane
A = membrane surface area
τ = membrane thickness

In this equation salt transport across the membrane is dependent only on the concentration differential and is independent of applied pressure.

The transport of salt across a membrane is commonly expressed as *salt passage* or *salt rejection*. Salt passage is the percentage of the salt in the feed that passes through the membrane into the permeate, and is calculated as follows:

$$SP = C_p/C_f \times 100 \qquad\qquad (4)$$

Where

SP = % salt passage
C_p = salt concentration in product stream
C_f = salt concentration in feed stream

The percentage of salt rejection is 100 percent minus the percentage of salt passage.

$$C_p = Q_s/Q_w \qquad\qquad (5)$$

Factors Affecting Membrane Performance

Recovery/Conversion
Recovery or conversion is commonly used to define the percentage of the feed water that is converted to permeate and is calculated as follows:

$$Y = Q_p/Q_f \times 100 \qquad\qquad (6)$$

Where

Y = % recovery or conversion
Q_p = product stream flow rate
Q_f = feed stream flow rate

At 75 percent conversion, 100 gpm of feed water is converted to 75 gpm of product water (permeate) and to 25 gpm of concentrated brine. The brine will contain most of the dissolved solids in the feed and will be approximately four times more concentrated than the feed. To conserve energy, it is desirable to operate at as high a recovery rate as possible to minimize upstream capital costs (e.g., pretreatment equipment and pumps, and high-pressure pump electricity costs). Excessively high recovery can create high brine concentrations, which reduce permeate flow and increase salt passage. This may lead to membrane fouling or scaling resulting from the precipitation of sparingly soluble salts from the concentrated brine.

Temperature
Temperature changes affect both osmotic pressure and water flux (K_w). The effect on osmotic pressure can be seen in equation 2. K_w is also directly proportional to temperature and is usually proportional to the variation in the water viscosity from one temperature to another. A rule of thumb is that membrane capacity increases about 3 percent per degree Celsius increase in water temperature.

Pressure
For a given set of feed conditions, increasing pressure results in increased water flow per unit of membrane area. Although the transport of salt (Q_s) across a membrane is not affected by pressure, increased water flow with pressure dilutes the salt passing through the membrane, which results in a lower permeate salt concentration (a reduction in salt passage).

Compaction
The water transport, or flux, through a clean membrane can decrease with time as a result of membrane compaction. Compaction is caused by creep deformation of polymeric membranes over time and is dependent on the membrane material, applied pressure, and temperature. As pressure and temperature increase, the tendency to creep is greater. This tightens the membrane's rejecting layer and reduces its water transport as a log function of time. A log-log plot of water flux versus time at a given temperature and pressure produces a straight line. This effect is more pronounced in asymmetric homogeneous membranes. These data, available from membrane manufacturers, are used to predict performance at a future time to provide the design basis for RO system capacity. Initially, an RO system has excess membrane capacity, which is offset by lower operating pressures that are gradually increased to the design pressure over time.

Concentration Polarization
Concentration polarization results from the buildup of a boundary layer of more highly concentrated solute on the membrane surface than in the bulk liquid. This

occurs because water permeation at the membrane surface leaves the more concentrated solute layer which must diffuse back into the bulk liquid. Owing to the higher flux rates, spiral membranes have a greater tendency toward concentration polarization than hollow fiber membranes. Proper membrane device design to maintain high fluid flow parallel to the membrane surface and promote mixing at the membrane surface is important to minimize the boundary layer thickness.

Concentration polarization increases the osmotic pressure at the membrane surface, which causes a reduction in water flux and an increase in salt transport across the membrane. If the concentration of sparingly soluble salts in the boundary layer exceeds their solubility limits, precipitation or scaling will occur on the membrane surface. At such higher concentrations, colloidal materials become less stable and may agglomerate and foul the membrane surface.

MEMBRANE MATERIALS

The ideal membrane has the following characteristics:

- High water flux rates
- High salt rejection
- Tolerant to chlorine and other oxidants
- Resistant to biological attack
- Resistant to fouling by colloidal and suspended material
- Inexpensive
- Easy to form into thin films or hollow fibers
- Mechanically strong, e.g., tolerates high pressures
- Chemically stable
- Able to withstand high temperatures

Cellulose Acetate

Cellulose acetate (CA) membranes are cast from cellulose diacetate and cellulose triacetate formulations and blends of the two. With increasing acetyl content, salt rejection and chemical stability increase and flux decreases. The Loeb-Sourirajan asymmetric structure is developed by casting the CA solution from an alcohol or ether with a "doctor blade" onto a porous substrate (e.g., sailcloth), where a thin skin is created by air drying the surface. This dense skin comprises a layer of about 0.2 μm on the more porous layer, which has an overall thickness of about 100 μm. This technique can also be used when casting membranes into tubular and hollow fiber forms.

CA membranes have poor chemical stability and tend to hydrolyze over time at a rate dependent on a combination of temperature and pH conditions. They can operate continuously at temperatures in the range of 0° to 30°C and with pH in the range 4.0 to 6.5. They are also subject to biological attack, but this can be offset by their ability to withstand continuous exposure to low levels of chlorine. As a result of their poor stability, CA membranes tend to decrease in salt rejection with time. The popularity of these materials, however, is due to their availability through numerous sources and low cost.

Aromatic Polyamide (Aramid)

Asymmetric aromatic polyamide (aramid) membranes (Richter and Hoehn 1971) were pioneered by Du Pont in the hollow fine fiber form. These fibers are produced by solution spinning. A dense skin, approximately 0.1 to 1.0 μm thick, forms on their outer surface from controlled evaporation of the spinning solvent. The remaining fiber structure is a porous supporting structure approximately 26 μm thick. Salt rejection takes place on this dense layer, which is further enhanced by posttreating the fibers with polyvinyl methyl ether (PT-A) for brackish applications, and with PT-A and tannic acid (PT-B) for seawater applications.

Aramid membranes are characterized by excellent chemical stability, as compared with cellulosic membranes. They can operate continuously at temperatures in the range of 0°C to 35°C and pH in the range 4 to 11 and are not susceptible to biological attack. Aramids are, however, susceptible to chlorine attack if continuously exposed; it is therefore important to dechlorinate feed streams.

Thin Film Composites

Work at the North Star Research and Development Institute, Minneapolis, in the mid-1960s (Francis 1966; Rozelle et al. 1967) funded by the U.S. Department of Interior, Office of Saline Water, led to the development of thin film composite membranes. Fluid Systems Division of Universal Oil Products (Riley et al. 1976) commercially introduced its TFC® (thin film composite) membranes in the mid-1970s, and FilmTec Inc. introduced its FT-30 composite membranes in the early 1980s (Cadotte et al. 1980). In these structures, an ultrathin membrane barrier layer (e.g., 0.2 μm) is formed on the surface of a microporous polysulfone that has been cast onto a porous fabric supporting layer. The barrier layer on the polysulfone is created by "in situ" interfacial polymerization techniques with polyamides or polyureas.

The advantages of the thin film composites depend on their chemistry, but their general characteristics are greater chemical stability and their ability to deliver high flux and high salt rejection at moderate pressures and to provide resistance to biological attack. They can operate continuously at temperatures in the range of 0°C to 40°C and pH in the range 2 to 12. Like the aramids, these materials have low resistance to chlorine and other oxidants.

MEMBRANE DEVICES AND CONFIGURATIONS

Membrane is a term commonly used to describe the complete membrane assembly that includes the membrane, membrane supports, flow distribution channels, provisions for brine and permeate outlets, and pressure vessels. Other terms commonly used are *permeators* for the complete assembly of hollow fiber membrane *bundles* enclosed in their pressure vessels; and *cartridges* or *elements* for spiral wound membrane assemblies that are inserted into pressure vessels or *shells*.

A typical RO system consists of either spiral wound or hollow fiber membrane elements housed in fiberglass-reinforced epoxy pressure vessels having burst strengths approximately four to six times normal operating pressures and containing the appropriate seals to prevent leakage. These devices are connected in parallel and/or in series to achieve the desired quantity and quality of product water. Feed water is introduced into the membrane device (see Figure 1.1) by a high-pressure pump at pressures in the range of 100 to 400 psig for brackish water and from 800 to 1,200 psig for seawater. The desalted product (permeate) is removed from the opposite side of the membrane at low pressure, and a flow-regulating valve on the brine side is used to create back pressure and control system recovery. The pressure drop on the brine side is minimal, which allows

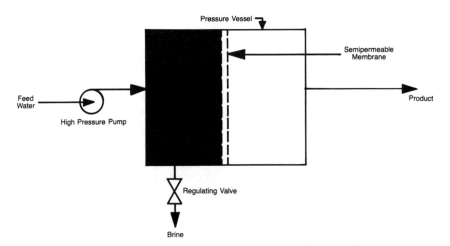

FIGURE 1.1.   Simplified RO flow diagram.   (From *Du Pont Permasep® Engineering Manual* 1987. Reprinted by permission.)

the high-pressure brine (reject) to be fed to successive RO units (stages) to increase productivity and system recovery. The brine volume is reduced in each stage, so the number of elements in successive stages is reduced (tapered) to maintain brine flow velocities specified by device manufacturers.

RO membranes are commercially available in spiral wound and hollow fiber configurations and, to a lesser extent, in tubular and plate and frame configurations. The following are desirable characteristics in a membrane device:

- Safe operation at high pressures
- No internal or external leaks
- Easy to flush or clean
- Minimal pressure drops on permeate and brine sides
- Made from inert, corrosion-resistant materials
- Long-term reliable operation

Spiral Wound

The spiral wound cartridge is widely used for brackish water and seawater desalting applications. It was originally developed in the mid-1960s at Gulf General Atomics through funding by the U.S. Office of Interior, Office of Saline Water (Sourirajan 1970c; Westmoreland 1968). Since then, numerous improvements have been made to the original design to allow higher operating pressures, improved flow channel designs and spacers, the use of new adhesives, improved seals, and better antitelescoping devices. The original spiral devices used cellulose acetate membranes, but today they also include cellulose triacetate and polyamide/polysulfone composite materials.

Structure
Spiral devices are made from flat film membranes that are wound around a perforated polyvinylchloride or polypropylene center permeate tube (see Figure 1.2). Two or more *leaves* (permeate *envelopes*) are attached to and wound around the center tube. Each leaf is made up of two membrane sheets supported and separated by a polyester tricot fabric, or another thin plastic net material, and is sealed at the edges by special epoxy or polyurethane adhesives, applied as the membrane assembly is wound around the center tube. A plastic net spacer is also sandwiched between adjacent leaves to provide flow channels for the feed on the outside of the permeate envelope. The permeate passes through the membrane into the envelope, where it spirals inward and enters the center tube through the perforations, and is removed through the permeate port. Toray (Japan) markets a spiral device in which the feed is introduced at the pressure vessel wall and travels in a spiral pattern to the center (parallel to the permeate), where it is removed from the edge of the membrane envelope.

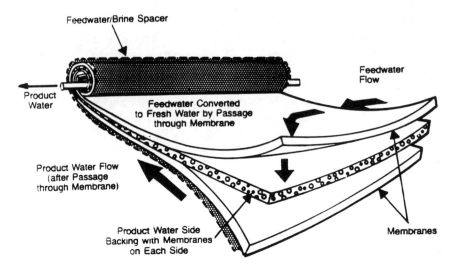

FIGURE 1.2.   Spiral wound reverse osmosis element.   (From *Du Pont Permasep®* *Engineering Manual* 1982. Reprinted by permission.)

After the spiral is wound the outside is wrapped with tape, fiberglass, or shrink wrap for mechanical strength. When the adhesive has cured the ends are cut square to specified lengths, (12 to 60 inches) and an antitelescoping device containing a U-cup brine seal is attached to the feed end. Diameters range from 1 to 12 inches. The ends of the permeate tubes are machined to accommodate O-rings and fit into a sleeve on the inside of the pressure vessel end plate or into a coupling between elements. There are various permeate tube connector designs, depending on the manufacturer.

Operation
Individual elements operate at approximately 8 to 10 percent recovery. Four to seven elements are connected in series in a single pressure vessel up to 22 feet long to achieve up to 50 percent recovery. Here, the slightly concentrated feed exiting one element feeds the next element. The permeate tubes for the individual elements are connected, and the discharge permeate is a blend of all elements. The desired system capacity and recovery are achieved by connecting pressure vessels in parallel and by reject staging.

Advantages
- Good resistance to fouling because of relatively open feed channels
- Easy to clean
- Easy field replacement

- Available in a wide variety of membrane materials
- Available from several manufacturers

Disadvantages
- Moderate membrane surface area to volume ratio
- Some tendency for concentration polarization to occur
- Difficult to troubleshoot individual elements in multiple-element tubes
- Difficult to achieve high recoveries in small systems—requires use of elements with several diameters

Hollow Fiber

Hollow fiber membrane configurations were pioneered by Du Pont in the late 1960s, followed by Dow Chemical Company and Toyobo (Japan). The construction of each manufacturer's device is similar, but they differ in their fiber dimensions, fiber support methods, and membrane materials. The Permasep® permeators, described as hollow fine fiber devices, are made from asymmetric aramid fibers 42 μm ID × 85 μm OD for B-9 brackish water permeators, and 42 μm ID × 95 μm OD for B-10 seawater permeators. The Dow permeators, no longer available, were made from cellulose triacetate fibers approximately 250 μm OD. Toyobo Hollosep® permeators are also made from cellulose triacetate, with fiber dimensions 70 μm ID × 165 μm OD.

Structure
Hollow fiber bundles are formed by orienting the fibers parallel to a perforated center feed tube (see Figure 1.3). The fibers are "potted" with special epoxy resins to create a tube sheet on one end and a "nub" on the opposite end. The tube sheet is machined to expose the ends of the fibers, and an O-ring groove is formed on the tube sheet circumference to prevent brine and permeate streams

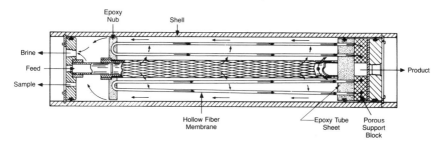

FIGURE 1.3.   Hollow fiber membrane device.   (From *Du Pont Permasep® Engineering Manual* 1987. Reprinted by permission.)

from mixing. These units are made in lengths from 18 to 60 inches and diameters from 4 to 10 inches. The permeators are assembled by inserting the bundles into pressure vessels. Unlike the spiral devices, the pressure vessel usually contains one fiber bundle, although recent designs use two bundles in one pressure vessel to minimize external connections.

Operation
Hollow fiber devices operate by introducing the pressurized feed into the center tube, where it is distributed along the entire length of the permeator and flows radially outward around the outside of the fibers. The product water permeates through the fiber walls into the bore and exits through the tube sheet into a permeate port. The brine flows between the outside of the fiber bundle and the inside wall of the pressure vessel to a brine port. The permeate exits at low pressure, but the brine undergoes a minimal pressure drop and is available to feed successive stages for additional recovery and greater system capacity. Individual hollow fiber permeators, for brackish applications, operate at approximately 50 percent recovery.

Advantages
- High membrane surface area to volume ratio
- High recovery in individual permeators
- Easy to troubleshoot
- Easy to change bundles in the field

Disadvantages
- Sensitive to fouling by colloidal and suspended materials
- Limited number of manufacturers and membrane materials

Tubular

Tubular configurations were some of the earliest practical RO devices first introduced by Glenn Havens of Havens Industries in 1961 (Sieveka 1966) and workers at the University of California, Los Angeles (McCutchan 1977). The tubular configuration was quickly displaced by spiral wound and hollow fiber devices, which offered greater membrane surface area to volume ratios and lower prices. Current applications are limited to feed waters with high levels of suspended solids, such as waste streams and food products (e.g., dewatering cheese whey or skim milk).

Structure
Tubular membrane devices are made by casting a membrane (e.g., cellulosic) onto porous supporting tubes having diameters ranging from .125 to 1.0 inch. These tubes are manufactured from fiberglass, ceramics, carbon, porous plastics,

and stainless steel and must be strong enough to contain the feed pressures. The tubes are pressed or molded into tube sheets on each end and are surrounded by a low-pressure jacket to collect the permeate. Each jacket may contain several tubes connected in series by external U-connections in order to achieve desired recoveries, as the recovery per tube length is very low.

Operation

High-pressure feed streams enter the tube bore, and the permeate passes through the membrane and supporting structure into the jacket, where it is removed through permeate ports. The feed, continually concentrated as it flows through the tube, exits the tube sheet on the opposite end where its flow is reversed by the U-connection and is passed through additional tubes to achieve the desired overall recovery.

Advantages
- Large well-defined flow passages
- Can achieve high flow velocities
- Low tendency to foul
- Easy to clean
- Membranes can be removed and reformed
- Can operate at very high pressures

Disadvantages
- Very low membrane surface area to volume ratio
- Expensive
- Minimal choice of membrane materials

Plate and Frame

Plate and frame devices use flat sheet membranes and constitute another early form of practical membrane devices and are modeled after plate and frame filter presses. The spiral wound device, in which flat membranes are wound around a mandrel, is an outgrowth of this concept. Like tubular devices, their applications in RO are limited because they are not as cost-effective as spiral wound and hollow fiber devices. The major use of plate and frame devices is to concentrate liquids containing high suspended solids or highly viscous liquids in the food, beverage, and pharmaceutical industries and in the treatment of waste waters. These configurations are, however, widely used in electrolytic membrane applications such as electrodialysis.

Structure

The simplest *stack* design consists of several sets of alternating frames, which support the membrane on the permeate side and separate the membranes on the feed side. The entire assembly is pressed between two end plates and held together with tie rods. The frames, in addition to supporting the membranes, have flow channels that collect the permeate and direct it to a permeate manifold. The feed frames are connected in parallel by a feed manifold and contain a spacer material to separate adjacent membrane sheets and provide flow channels.

There are design variations developed by De Danske Sukkerfabrikker (DDS), in which cylindrical pressure vessels contain circular envelopes, where the membranes sandwich a spacer net between them and are sealed around the circumference and near the center to form an annular membrane envelope. Permeate collection tubes extend into the envelope interior and are connected externally to a permeate collection manifold. Each envelope is stacked around a center support to achieve the desired membrane area for a pressure vessel of a given size. These devices can use any membrane material available in sheet form.

Operation

High-pressure feed is directed between the frames or spacers that separate the membranes, and the permeate is removed through permeate ports in the permeate frame, or through tubes in the permeate envelope in cylindrical devices. The brine exits the pressure vessel with minimal pressure drop so that it can be sent to additional stages to achieve higher recovery and capacity.

Advantages
- Open flow channels
- Low tendency to foul
- Easy disassembly for cleaning and membrane replacement
- Can use several membrane types

Disadvantages
- Low membrane surface area to volume ratio
- Potential for leaks between leaves
- Expensive

MEMBRANE DEVICE PERFORMANCE

Standard Rating Conditions

Salt rejection and product flow are dependent on the membrane material, membrane thickness, feed water concentration and the type of ions present, and operating conditions. All membrane manufacturers specify device capacity and salt rejection based on performance measurements with sodium chloride solutions under standard operating conditions. Such conditions vary, depending on the device configuration (e.g., spiral or hollow fiber) and their intended use (e.g., with brackish water or with seawater). The standards accepted by the industry for specifying salt rejection and capacity are described in the following paragraphs.

Spiral Wound

Spiral elements for high-pressure brackish water applications are rated using a 2,000 mg/l NaCl solution at 400 to 420 psig, 25°C, and 10 to 15 percent recovery. For low-pressure brackish applications, they are rated using a 2,000 mg/l NaCl solution at 225 psig, 25°C, and 10 to 15 percent recovery. For seawater applications ratings are made with the use of solutions having 30,000 to 35,000 mg/l NaCl and 800 to 1,000 psig. Typical salt rejections are greater than 97 percent for brackish water devices, and greater than 98 percent for seawater devices.

Hollow Fiber

Hollow fiber permeators for high-pressure brackish water applications are rated using a 1,500 mg/l NaCl feed solution at 400 psig, 25°C, and 75 percent recovery. For seawater applications, permeators are rated at 30,000 to 35,000 mg/l NaCl feed solution at 800 to 1,000 psig, 25°C, and 30 to 35 percent recovery. Typical salt rejections are greater than 90 percent for brackish permeators, and greater than 99 percent for seawater permeators.

Estimating Performance

Flow

For a particular membrane device the terms $K_w A / \tau$ (equation 1) can be estimated from the standard rating conditions, because the water flow rate, applied pressure terms, and osmotic pressure (equation 2) are known. With this constant established, the water transport can be estimated at any new set of operating pressures and solution osmotic pressures. Additional corrections are required for temperature effects and flow loss resulting from compaction.

Salt Passage

As in the equation for flow, the term $K_w A / \tau$ (equation 3) can be estimated for a particular device from the standard rating conditions, because the concentration difference and flow of salt across the membrane can be estimated (equations 1 and 5). With this constant established, membrane salt passage can be estimated at any new concentration. Additional corrections are required to reflect the passage of ions other than Na^+ and Cl^-.

RO SYSTEM DESIGN CONSIDERATIONS

The RO process should be viewed as a total system comprising the components and considerations summarized in Figure 1.4. The considerations are not all-inclusive, but are intended as a guide when planning or designing an RO system. Each component and each interface should have the proper controls and interlocks to ensure long-term performance and reliability. Each component and each system has economic/performance trade-offs that may be considered to satisfy the needs of particular users.

The following paragraphs present an overview of each component in reverse order.

End Use

The first consideration is the specific use of the product water, which determines the quality and quantity required to meet the user's needs. For potable applications, water quality is usually required to meet U.S. Public Health Standards or World Health Organization Standards (WHO 1984). For ultrapure electronics applications, water having a resistivity of 18 mega-ohms is required. Product specifications should not be more rigid than necessary, however, because a larger quantity or a higher quality than necessary can have a significant adverse impact on the cost of the water produced.

Posttreatment

RO permeate usually requires some posttreatment prior to its use. At a minimum, degasification is required to remove carbon dioxide generated during feed water acidification for scale control, and pH adjustment is required to prevent corrosion in downstream systems.

Posttreatment requirements depend on the application and must be determined on a case-by-case basis. For many industrial applications, posttreatment consists of demineralization with ion exchange and disinfection by ultraviolet radiation. For municipal applications, in addition to pH adjustment and degasification, disinfection with chlorine is required.

Membranes

Membranes are the heart of the system, and performance is affected by several factors not necessarily related to the membrane or its configuration, e.g., pretreatment and system operation and maintenance. However, careful consideration should be given to the selection of the membrane material and configuration

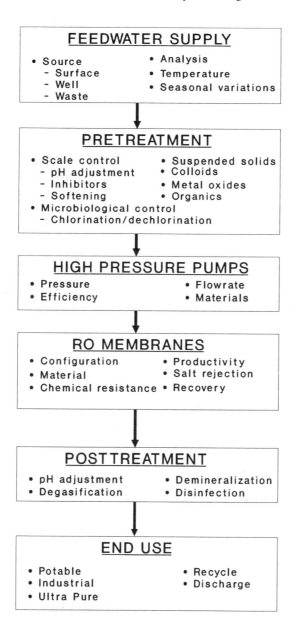

FIGURE 1.4.   Reverse osmosis system components.

based on end-use requirements and feed water quality. For the more common applications, membrane performance histories are frequently documented in trade journals and discussed at various trade meetings.

Information as to the optimum number of membrane elements, operating pressures, recovery, and expected product quality for a specific application can be obtained from membrane manufacturers and system designers, or can be calculated using computer software programs provided by many membrane manufacturers or independent software developers.

Operation and Maintenance

Operation and maintenance are key to successful system performance. Performance data should be collected and analyzed regularly in order to detect potential problems early. If problems occur, proper troubleshooting techniques should be used and the membrane manufacturer and/or systems designer consulted about proper corrective action. Uncontrolled scaling, fouling, or plugging requires frequent membrane cleaning to maintain performance. The irreversible accumulation of these materials, in the membrane device, leads to flow maldistribution and concentration polarization, which causes a gradual deterioration of membrane flux and salt rejection and sometimes leads to degradation of the membrane material (e.g., cellulosic membranes). This results in costly replacement of membrane elements.

Several techniques have been developed to restore poor membrane performance resulting from scaling or fouling and they are most effective if the need for cleaning is identified early. Detergents are used to remove particulate, colloidal, biological, and organic materials from membrane devices. This is usually done by circulating a cleaning solution at low pressure through the membrane device in the normal direction of flow until the foulants are removed. Backwashing membranes is seldom recommended.

Carbonate scale can be removed by circulating dilute hydrochloric or citric acid solutions. Sulfate scales are more difficult to remove. Citric acid or chelating agents such as sodium EDTA are often recommended. Metal oxides are removed with ammoniated citric acid, sodium hydrosulfite, sodium hexametaphosphate, or mixtures of citric acid and chelating agents.

Biological growths usually adhere tightly to a membrane surface and must be killed prior to removal. Because of the sensitivity of today's high-performance membranes to oxidizing agents, care must be taken not to destroy the membrane while killing biological growths. Thus, the use of chlorine or other strong oxidants should be avoided. Shock treatments with sodium bisulfite or peracetic acid, followed by detergent washing, provide an effective way to remove these materials. Periodic shock treatments with a biostat, e.g., sodium bisulfite or peracetic acid, during operation can minimize biofouling.

High-Pressure Pumps

High-pressure pumps provide the pressure necessary for the membranes to produce the required product flow and quality. Usual pump types are single-stage, high-speed centrifugal; positive displacement (plunger); or multistage centrifugal. Pump selection depends on the flow rate and pressure and the trade-off between initial cost and operating cost, determined by maintenance costs and operating efficiency. The single-stage centrifugal pump usually has the lowest efficiency, and the positive displacement pump the highest. For small systems high-speed centrifugal pumps are used, and for large systems multistage centrifugal pumps are preferred. The pumps are usually constructed of high-grade stainless steel.

Energy recovery systems should be considered if brine exits the RO system at 300 psig or more and if system recovery is less than 80 percent. This decision becomes dependent on the economic payback for the energy recovery equipment based on energy-cost savings. These systems are usually economically attractive in seawater RO systems of 250,000 gallons per day or greater.

Pretreatment

Pretreatment is the key to successful long-term RO performance, and its importance in system design should not be underestimated. The purpose of pretreatment is to guard against feed water upsets, remove suspended and colloidal material, prevent membrane scaling resulting from precipitation of sparingly soluble salts, and to prevent biological growth.

Much has been written about pretreatment throughout the years, and the techniques have become state-of-the-art for RO practitioners (Crossley 1983; Pohland 1987). Membrane manufacturers, systems designers and manufacturers, and/or consulting engineering organizations should be consulted about pretreatment requirements for specific feed waters and membrane types and configurations. Table 1.1 summarizes the pretreatment methods for RO systems.

Causes of Scaling and Fouling

Membrane surfaces can be fouled with colloidal materials, organics, metal oxides or hydroxides, and biological growths, and scaled by the precipitation of sparingly soluble salts from the concentrated brine. Colloidal materials are usually very fine clay particles; organics can be hydrocarbon oils and greases, humic acids, and tannins; metal oxides and hydroxides are generally formed from iron, manganese, and aluminum. The presence of suspended solids in the water, such as mud and silt, tend to cause gross plugging of the device rather than fouling of the membrane surface. Mineral scales usually consist of calcium

TABLE 1.1   Summary of Pretreatment Methods

Problem	Primary Pretreatment	Purpose	Counter Indications	Secondary Pretreatment	Purpose
(1) Ca/Mg bicarbonate scale	(a) Base exchange softening	Removes: Replaces Ca/Mg by Na which has a soluble bicarbonate	High TDS causes slip max. 800 mg/liter; too expensive at greater than 9,000 m³/day	Sequestrant	Delays any tendency for slip to cause precipitation
	(b) Lime softening	Removes: Precipitates Ca and bicarbonate as calcium carbonate and Mg as magnesium hydroxide	Not suitable for less than 5,000 m³/day	Sequestrant or acid dose	Prevents postprecipitation
	(c) Acid dose	Removes: Replaces the bicarbonate with the more soluble chloride or sulphate	Difficulty and cost of obtaining acid	Sequestrant	Backup if acid dosing fails
	(d) Add sequestrant (not recommended)	Stabilizes: Delays the formation of and modifies the nature of the precipitate	The other methods are more reliable; incorrect dosing, stagnant pockets, or incorrect shutdown procedures can cause scaling		
(2) Ca sulphate scale	(a) Base exchange	See (1a)			
	(b) Add sequestrant	See (1c)			
(3) Silica scale	(a) Raise temperature	Stabilizes: Increases solubility	Cost of heat		
	(b) Lime softening	Removes: Brings down some of the silica with the $CaCO_3$ and $Mg(OH)_2$	Not suitable for less than 5,000 m³/day		

(4) Iron precipitation	(a) Oxidize (aerate) and filter	Remove: Precipitates the iron as ferric hydroxide which is easily removed by filter	Presence of other oxidizable material, such as H_2S	Acid dose	Prevents further precipitation
	(b) Exclude oxidizing agents, e.g., air or Cl_2	Stabilizes: Keeps the iron in the ferrous state in which it is soluble	Not good for intermittent use or where operating and maintenance staff are of poor quality	Acid dose	Prevents postprecipitation of coagulant
	(c) Acid dose	Stabilizes: Keeps the iron in solution	pH needs to go to approximately 5, so acid is expensive		
(5) Colloids	(a) Coagulate and filter	Removes: Causes the colloids to form larger particles which can be filtered out		Acid dose (for some coagulants)	Prevents postprecipitation of coagulant
	(b) Base exchange softening	Stabilizes: Discourages coagulation as the solution is concentrated and the colloids go right through the RO plant into the concentrate	Not suitable for high TDS waters where slip through the softener causes more than 5 mg/liter total hardness in the softened water		
(6) Bacteria	Sterilize (Cl_2) and filter	Removes: Bacteria are killed and then removed by sand filtration		(a) Dose sodium metabisulfite (b) Carbon filter	Removes Cl_2 that would otherwise damage permeators

TABLE 1.1 (continued)

Problem	Primary Pretreatment	Purpose	Counter Indications	Secondary Pretreatment	Purpose
(7) Hydrogen sulfide	(a) De-gas and add Cl_2	Removes: Most of the H_2S that comes off as gas and the remainder oxidized to sulfate	Hard waters require acid dosing before de-gasser to prevent scaling of the de-gasser packing	See (6a) and (6b)	
	(b) Exclude oxidizing agents, e.g., air, Cl_2	Stabilizes: Stays in solution	Presence of bacteria or poor operating/maintenance staff	Permeate has to be de-gassed; reject disposed of away from habitation	The gas is a health hazard
(8) Chlorine or other strong oxidizing agent	(a) Dose sodium metabisulphate	Removes: Chemically destroys the Cl_2	Disastrous if dosing fails	Backup with second dose	Security; also aids sterility of the membranes
	(b) Carbon filter	Removes: By absorbing the oxidizing agent	Filter often becomes a breeding ground for bacteria; regeneration or replacement needed periodically		

From Crossley 1983. Reprinted by permission.

carbonate, calcium sulfate, and barium sulfate. Detailed descriptions of these phenomena and their effects on membrane performance can be found elsewhere (Matsuura and Sourirajan 1985; Amjad 1988; Paul and Rahman 1990; Amjad et al. 1991).

Pretreatment Techniques
The requirements for removal of colloidal or suspended solids are determined by measuring the Silt Density Index (SDI) of the feed (ASTM 1992). SDI is calculated from the ratio of the time required for 500 ml of feed water to pass through a 0.45-micron filter initially, and after 15 minutes of continuous feed-water flow through the filter at constant pressure. Membrane manufacturers generally specify the SDI required for their membranes. Techniques are available for achieving the required SDI through the use of media filters alone or in combination with coagulants. At a minimum, 5- to 10-micron cartridge filters are used prior to high-pressure pumps to remove the large particles and protect the close tolerance pumps and RO elements against feed-water upsets.

Calcium carbonate scale prevention is required on most all RO systems. Feed-water acidification is most common, which converts carbonates to carbon dioxide to achieve a negative Langelier Saturation Index in the RO brine in brackish water systems, or a negative Stiff and Davis Stability Index for seawater (Strantz 1982). In some cases, the need for acidification can be reduced or eliminated by softening to reduce the calcium hardness or by the addition of organic polymer inhibitors to retard precipitation. Scaling caused by other sparingly soluble salts such as calcium or barium sulfate must also be considered. The scaling potential for these materials is usually estimated from their solubility constants in the concentrated brine and can be retarded by adding inhibitors or by reducing system recovery. Guidelines for handling these various materials are available from membrane manufacturers.

Treatment for biological activity may be necessary, depending on the feed-water source. It is usually required for fresh surface waters and seawater. Chlorine injection is commonly used, but the optimum water pH, dosage, and contact time must be determined. The level of biological activity after chlorination should be monitored regularly to ensure that it is low and to adjust dosage rates for daily or seasonal variations. Many membranes do not tolerate chlorine; therefore, the feed water must be dechlorinated before it enters the RO membranes. Dechlorination is performed as close to the membranes as possible by injecting a sodium bisulfate solution. Shock treatments may also be given to membranes and upstream equipment to minimize biological growth.

Feed Water

Feed water is the source material for RO product water and has a direct influence on its quality. Well waters are generally free from suspended and colloidal materials, require minimal filtration, and are not subject to seasonal variations. Surface waters (fresh and seawater) are characterized by their level of suspended solids, microbiological activity, and seasonal variations. Before designing an RO system it is important to obtain a good feed-water analysis. At a minimum, the most common elements for which analysis is required, expressed in ppm or mg/l, are the following:

Cations	Symbol	Anions	Symbol
Calcium	Ca^{++}	Carbonate	CO_3^{--}
Magnesium	Mg^{++}	Bicarbonate	HCO_3^-
Sodium	Na^+	Chloride	Cl^-
Potassium	K^+	Sulfate	SO_4^{--}

In addition, it is important to determine the total dissolved solids (TDS), pH, silica (SiO_2), SDI, total organics (TOC) and temperature.

APPLICATIONS OF REVERSE OSMOSIS

The applications for RO have not changed substantially from those pioneered in the 1970s, but they now benefit from new membrane materials that operate over broader pH and temperature ranges, improved pretreatment, and better device designs. The major applications remain: desalting of brackish water and seawater for potable uses, ranging from ice manufacture to municipal water systems; pretreatment for ion exchange systems producing high-purity boiler feed water and ultrapure water for electronics manufacture and pharmaceutical manufacture; kidney dialysis and, to a limited degree, waste water reclamation and recycling.

Reverse Osmosis Versus Other Desalting Processes

RO is the newest of the commercially available water desalination processes, which are summarized in Table 1.2 (Hicks et al. 1972). From the technical and economic viewpoints, RO is the most versatile process as it can be used over a wide range of feed-water salinities, whereas the other technologies are applicable at either higher or lower salinities. The impact of RO as a worldwide desalting process is summarized in Table 1.3 (Wangnick 1989).

Ion Exchange
At low salinities, ion exchange (IX) is more economically attractive than RO to produce high-purity water. As salinity increases IX becomes less desirable because the need for regeneration chemicals increases and bed sizes must be larger to extend the time between regenerations. Waste disposal regulations and costs for regeneration chemicals must be considered before selecting IX. It has

TABLE 1.2 Summary of Guidelines for Desalting Process Selection

	Unit Size (mgd) vs. Time Period			Feed Water Characteristics For Most Favorable Economic Applications		Normal Product Water TDS (ppm)
	1972	1990	2000–2020	TDS (ppm)	Temperature	
Multistage flash distillation (MSF)	to 3.0	6.0	10.0	30,000–50,000	Cold	5–50
Multi-effect distillation (MED)	to 0.6	2.0	80.0	30,000–50,000	Cold	5–50
Mechanical vapor compression distillation (MVC)	to 0.5	0.5	0.5	10,000–50,000	Warm	5–50
Reverse osmosis (RO)	to 0.2	2.5	6.0	1,000–50,000	Warm	100–500
Electrodialysis reversal (EDR)	to 0.3	2.6	2.6	1,000–5,000	Warm	350–500
Ion exchange (IX)	to 1.0	1.0	1.0	<2,000	Warm	0–550

From Hicks et al. 1972 (updated by the authors). Reprinted by permission.

TABLE 1.3   Summary of Desalting Plants by Process (Systems Greater than 25,000 gallons per day)

Main Process	No. of Units	Capacity (MGD)
Reverse osmosis (RO)	5,154	1,343
Multistage flash distillation (MSF)	1,090	2,112
Electrodialysis (ED)	1,159	233
Vapor compression distillation	704	156
Multieffect distillation (MED)	606	192
Other	97	12
Hybrid (MSF and RO)	8	15
Freezing	1	.05

From Wangnick 1992. Reprinted by permission.

become standard practice to use RO as a "roughing" demineralizer prior to IX for high-purity water applications. RO removes 90 to 95 percent of the dissolved solids in feed water, which reduces regeneration frequency, bed size, and regeneration effluent quantity. The combination of RO and IX frequently provide lower capital and operating costs than the individual technologies.

Electrodialysis
Electrodialysis (ED) is a membrane process in which the driving force for separation is an electric field across alternating anion and cation exchange membranes. Cations and anions are selectively removed from the feed as they pass through their respective ion-selective membranes to form a concentrated brine. The applicable TDS range for ED is generally less than 5,000 mg/l because higher concentrations increase the electrical current required for ion removal. Improved ion-exchange membranes are capable of operating at higher current densities. This increases the salinity at which ED can operate economically. The choice between ED and RO becomes one of economics, desired water quality, and operating considerations.

Distillation
Thermal distillation techniques such as multistage flash distillation (MSF), multi-effect distillation (MED), vapor compression (VC), and various combinations of these technologies (e.g., VC-MED-MSF) are generally applicable to highly brackish waters or seawater. MSF has been the dominant technology used for large seawater desalting plants in the Middle East and elsewhere in the world. This technology predates RO and became well established in the 1970s while seawater RO was in its infancy. Today, seawater RO has become a viable economic alternative to thermal distillation because of its lower energy consump-

tion (see Table 1.4), improved membranes and system design, and long-term operating experience. There are also opportunities to combine RO with MSF or MED in hybrid systems to take advantage of both technologies. Thus, there are numerous considerations that must be weighed in designing new seawater desalting plants and for replacing or upgrading existing plants.

Seawater Reverse Osmosis

Seawater RO, which began with small systems in late 1973, is now used in plants with capacities up to 15 mgd. Because of the world's increasing need for fresh water and the aging of thermal distillation plants in the Middle East, this trend should continue.

In seawater RO systems, high pressures are necessary to overcome the high osmotic pressure of seawater (375 to 500 psi) and ensure high levels of productivity. Membranes for this application should be able to operate at pressures between 800 and 1200 psig and must have salt rejections of 99 percent or more in order to produce product water meeting World Health Organization (WHO) Standards. Overall system recovery is limited to about 50 to 60 percent, depending on feed-water concentration, allowable maximum brine concentration, and the maximum feed pressure the membrane device can tolerate. Sometimes the product water from the seawater membranes is repressurized and passed through a brackish water system (second pass) to reduce the product TDS to more acceptable values.

TABLE 1.4   Energy Consumption for Distillation and Reverse Osmosis for Seawater Desalting. (Basis, Dual Purpose Power and Water Plant)

Process	Total Power Consumption kWh/1,000 gal Product	Maximum Operating Temperature, °C
MSF distillation, performance ratio 12		95
Pumping power	12	
Decrease in power plant net cycle output		
For steam extraction for distillation	24	
MSF total	36	
MED distillation, performance ratio 15		75
Pumping power	8	
Decrease in power plant net cycle output		
For steam extraction for distillation	18	
	26	
Vapor compression distillation	35	100
RO with energy recovery	18	45

Pretreatment
The techniques for pretreatment of seawater are basically the same as those for brackish water. However, careful attention should be paid to seawater pretreatment because seawater is frequently obtained from surface sources, which are subject to seasonal variations in salinity, temperature, and biological activity, and to upsets caused by storms. An alternative is to construct sea wells in or near the sea, which provide a feed water free of suspended matter and are less susceptible to upsets and seasonal variations. Pretreatment systems should be carefully designed, monitored, and operated to avoid problems. The control of biological activity warrants attention, but scale control is less problematic.

Energy Recovery
Energy recovery systems should be considered in the design of seawater RO systems because of their large volumes of high-pressure brine. In such systems the high-pressure brine exiting the final RO stage is fed to an energy recovery unit. These units may be hydroturbines directly coupled to the shaft of the high-pressure feed pump or impulse turbines (or Pelton wheels), which are usually connected to induction motor/generators that generate electricity to be returned to the power grid (Sackinger 1980). The use of energy recovery units can reduce energy consumption in seawater systems up to 30 percent, depending on system recovery and the type of energy recovery unit used.

Combination of RO and Distillation
Until the 1980s, thermal distillation was the technology of choice for large seawater desalting plants. The rapid growth of this technology took place in the 1970s and early 1980s in the Middle East as the result of oil exporting activity. The predominant technology used was multistage flash (MSF) combined with electrical generating plants. These dual-purpose plants use oil or natural gas to produce high-pressure steam to drive electricity-generating turbines whose low-pressure exhaust steam supplies heat to the MSF plants. Operation of such plants is not very flexible owing to the need to produce water and electricity simultaneously, whereas the seasonal demands for water and electricity do not always coincide (Brandt and Battey 1983). Seawater RO offers flexibility in the Middle East by diversifying away from predominantly MSF technology, using electricity from the grid which has been more efficiently generated than in a dual purpose plant, and being able to follow daily and seasonal demands for water (Brandt 1985; Al-Sofi 1991).

In the Middle East, the combination of seawater RO plants with existing MSF plants offers an attractive opportunity to expand water production by blending the higher TDS RO product water (350 mg/l or greater) with MSF distillate (25 mg/l or less) to meet WHO standards. In Saudi Arabia, the decision to use seawater RO plants or dual-purpose plants is based on the selling price of export electricity generated by dual-purpose plants. If the price is greater than the direct cost to produce it, dual-purpose plants are justified; if the selling price is less than the direct production costs, then this electricity could be more economically used for seawater RO plants (Al-Sofi 1991).

In summary, the choice of RO versus thermal processes for seawater desalting is an economic decision contingent on site-specific factors. The selection must be based on ability to meet the economic requirements and other needs of the user. However, RO is generally acknowledged to have the following advantages over thermal processes:

- It requires less than 50 percent of energy required by thermal plants.
- It operates at ambient temperature, requiring no phase change.
- It uses noncorrosive polymeric materials rather than metallic materials for all but pumps and piping.
- It requires a fraction of the space required for thermal systems.
- It is built in modular components for quick delivery, easy expansion, and operating flexibility.
- It enables quick start-up and shutdown.
- It requires approximately one-third of the feed water required for thermal processes, which reduces intake and pumping costs.
- It is easy to operate with minimal operator training.

Other Applications

In addition to potable applications, several others have been summarized (Belfort 1984). In the food and beverage industry, RO has been used to concentrate cheese whey, fruit and vegetable juices, maple sap, coffee, and sugar solutions. It has also been used to concentrate, recycle, and recover valuable products from waste streams such as agricultural drainage waters, cooling tower blowdown, electropainting rinse waters, textile wastes, electroplating wastes, pulp and paper streams, photographic wastes, acid mine drainage, municipal waste water reclamation, and ground water remediation. It is used in the pharmaceutical industry for fermentation broth clarification and concentration.

These applications are all unique, because each stream has its own special characteristics (e.g., analyses, time variations, impurities, upsets). "Standard" systems are usually unacceptable, and custom systems are required for each application based on extensive laboratory and pilot testing to obtain basic data. In these systems, the desired product may be both the concentrate and the permeate. Membrane longevity may be 6 months or less but is justified, versus other processes, by the value of the materials recovered or disposal costs avoided.

RECENT DEVELOPMENTS

Membrane Materials

The patent literature is full of new membrane materials. The commercial viability of these materials will depend on their improvement over existing materials. Current membranes already offer high salt rejection and high flux rates. Any improvements are likely to be marginal, and in the case of spiral devices there will be a greater tendency toward concentration polarization as flux rates increase.

Perhaps the most significant recent advance in commercial membranes has been the introduction, by several manufacturers, of nanofiltration or softening membranes. These are characterized by their high water flux, low salt rejection, and the ability to operate at pressures of approximately 100 psig. With use of these membranes, electricity is reduced to approximately 25 percent of older brackish water membranes. Salt rejection rates of nanofiltration membranes are between those of RO and ultrafiltration (Wilf et al. 1991). Their ability to reject divalent ions (e.g., calcium and sulfate) and organics (e.g., halocarbons) makes them desirable for potable purposes. Here, low salinity is not required, but eliminating the potential for trihalomethane formation upon chlorination is important.

Pretreatment

New scale inhibitors that affect crystal formation of sulfate and carbonate scales and allow system operation at higher brine concentrations without acidifying the feed streams have been developed and are undergoing testing (Amjad 1988; Ayoub et al. 1991; Evans and Finan 1991). Improved methods of biological disinfection have also been demonstrated in the laboratory to minimize biological fouling in seawater plants (Applegate et al. 1989), but need to be tested for their effectiveness in commercial plants.

Waste Treatment

New applications for RO tend to be focused on the treatment of waste streams for recovery, concentration, or recycle. The applications are fragmented, driven by environmental regulations, and are generally kept confidential by the developers and end users for competitive purposes. Generically, the system concepts have not changed since the 1970s, but the broad variety of membranes on the market today and the ability to tailor membrane properties to specific separations is opening up new applications. The systems are being applied upstream in manufacturing processes rather than at the end of the pipeline. Applications of RO to wastes is an empirical science in which testing of each stream is required to develop the proper pretreatment, membrane type, and configuration. Frequently, a hybrid of several technologies is required to achieve the separation goal (Ray et al. 1991). Several "boutique" membrane companies are in business to take a complete systems approach to solve these problems. Because of the increasing concern about toxic wastes, water reuse, and other environmental issues, this area is likely to be the greatest challenge and an opportunity of high value for RO in the next decade.

Food Processing

Recent activity has centered on tailoring membranes (Strantz 1990) and systems design (Walker and Ferguson 1990; Walker 1990) to achieve concentrations of citrus juice up to 60° Brix (60 percent sugar) at 1,000 psig versus 30° Brix (30 percent sugar) previously attained by RO (Cross 1989). This has been accomplished by the combination of more open membrane structures and various system staging configurations. In addition to achieving higher juice concentrations, this process is able to maintain the juice's flavor components.

Dealcoholization of alcoholic beverages has become important during the last 5 years as the result of a health-conscious public and strong enforcement of drunk driving regulations. The J. Lohr Winery (California) has conducted numerous studies on the dealcoholization of red and white wines using RO membranes and has commercialized nonalcoholic wines. Similar efforts are under way in Europe. The details of these applications are kept confidential for commercial reasons, but the key is to remove alcohol without removing the bouquet and flavor from the wine. RO is preferred to thermal techniques because its low-temperature operation results in less destruction of the sensory components.

REVERSE OSMOSIS MEMBRANE MANUFACTURERS—A WORLDWIDE PERSPECTIVE

The following is a list of the major companies that manufacture membranes and membrane devices, and their respective locations.

Spiral Wound Manufacturers

Desalination Systems, Inc.	Escondido, California
FilmTec Corporation (Subsidiary of Dow Chemical Company)	Midland, Michigan
Fluid Systems Corporation (Subsidiary of Allied Signal Inc.)	San Diego, California
Hydranautics, Inc. (Subsidiary of Nitto Denko, Japan)	San Diego, California
Nitto Denko Corp.	Tokyo, Japan
Osmonics, Inc.	Minnetonka, Minnesota
Separem, Spa.	Biella, Italy

Toray Industries Inc.	Otsu-shi, Shiga, Japan
TriSep	Goleta, California

Hollow Fiber Manufacturers

E. I. Du Pont De Nemours & Co. Inc.	Wilmington, Delaware
Toyobo Co. Ltd.	Ohtsu, Shiga, Japan

Tubular Manufacturers

PCI Membrane Systems Ltd.	Whitchurch, Hampshire, United Kingdom
Sumitomo Chemical Co.	Osaka, Japan

Plate and Frame Manufacturers

Dow Plate and Frame (Formerly DDS) (Subsidiary of Dow Chemical Company)	Midland, Michigan
ROCHEM Separation Systems	Torrance, California

SUMMARY

Current State of the Industry

Today, RO is a mature industry. Desalination capacity has increased dramatically in the past 20 years. In 1960, worldwide desalination plant capacity was about 50 million gallons per day (mgd). By 1970 total capacity had reached about 200 mgd, and by 1990 almost 3,500 mgd. Membrane processes, not yet commercial in 1960, now represent 36 percent of the total capacity (RO, 31 percent and electrodialysis, 5 percent).

In the last 5 years, RO has been used in about 85 percent of the new desalination plants constructed. This trend toward greater market share for membrane processes, particularly for RO, is expected to continue. This has come about as the result of some remarkable improvements in technology over the past 20 years, such as the following:

- Broadened applications for RO through introduction of improved thin film composite membranes.
- Increased membrane life through improved knowledge of pretreatment requirements and improved pretreatment processes.
- Greater reliability resulting from better design and assembly of membrane devices, including pressure vessels, end plates, and membrane connection components.
- Membrane improvements permitting reduction in operating pressures, often to only 25 percent of earlier designs.
- New membranes for seawater, permitting increases in operating pressures to 1,200 psi and increases in recovery to 50 percent.

- The recent application of energy recovery systems for seawater RO units. With current equipment, the energy requirements for seawater RO have been reduced from approximately 30 kWh/1,000 gallons to less than 18 kWh/1,000 gallons.

Improved membrane technology has brought about corresponding reductions in the total cost of producing fresh water from seawater. In 1979 the U.S. Office of Water Research and Technology published a "cost update." In 1991 actual direct capital costs for seawater RO plants were 50 percent lower as compared with the 1979 cost, updated by inflation.

Forecast

Seawater and Brackish Water Desalination
The potential for growth in this field is substantial and will be sustained by the need for replacement of existing units after a service life of 10 to 20 years, and to satisfy the new demands as population increases. Through the end of the twentieth century, the annual number of new units should easily equal 15 percent, compounded, of the total installed RO units. For 1992 this indicates a total market of about 900 units with a capacity of 25,000 gallons per day or greater worldwide.

Industrial Applications
Demand will increase for RO units for production of process water and, when coupled with polishing demineralizers, to produce high-purity water for power plants and ultrapure water.

Other Applications
The concentration of waste streams for recovery, recycling, and disposal will continue to be driven by environmental regulations. New membrane materials, the ability to tailor membranes to different applications, and improved knowledge and experience with pretreatment techniques should make this a growing field for RO in the 1990s and beyond.

References
Al-Sofi, M.A.K. 1991. *Water and power co-production cost*. Technical Proceedings, vol. 1. International Desalination Association World Conference on Desalination and Water Reuse, 25–29 August 1991, at Washington, D.C.
Amjad, Z. 1988. Mechanistic aspects of reverse osmosis mineral scale formation and inhibition. *Ultrapure Water* 5(6):23–28.
Amjad, Z., J. Hooley, and J. Pugh. 1991. Reverse osmosis failures: Causes, cases, and prevention. *Ultrapure Water* 8(9):14–21.

Applegate, L.E., C.W. Erkenbrecher, Jr., and H. Winters. 1989. New chloramine process to control aftergrowth and biofouling in Permasep B-10 RO surface seawater plant. *Desalination* 74:51–67.

ASTM. 1992. Silt density index (SDI) of water, D-4189-82 (rev 1987). Annual Book of ASTM Standards, 11.01:299; Philadelphia: American Society for Testing and Materials.

Ayoub, M.A., J.D. Stowell, and K.I. Al-Mansour. 1991. Evaluation of polymer based antiscalant (Flocon 100) for RO plants. Technical Proceedings, vol. 2. International Desalination Association World Conference on Desalination and Water Reuse, 25–29 August 1991, at Washington, D.C.

Belfort, G. 1984. *Desalting experience in hyperfiltration (reverse osmosis) in the United States. Synthetic Membrane Processes, Fundamentals and Water Applications,* edited by Georges Belfort. Orlando, Fla: Academic Press, p. 229.

Brandt, D.C., and R.F. Battey, 1983. Dual purpose desalting, RO versus MSF . . . an economic comparison. *NWSIA Journal* 10(1):35–43.

Brandt, D.C. 1985. Seawater reverse osmosis—an economic alternative to distillation. *Desalination* 52:177–86.

Cadotte, J.E., R.J. Peterson, R.E. Larson, and E.E. Erickson. 1980. A new thin-film composite membrane for seawater desalting applications. *Desalination* 32:25.

Chemical Engineering. 1971. Why hollow-fiber reverse osmosis won the top CE prize for Du Pont. *Chemical Engineering* 78(27):54–59.

Cross, S. 1989. Membrane concentration of orange juice. *Proceedings of the Florida State Horticultural Society* 102:146–52.

Crossley, I.A. 1983. "Desalination by reverse osmosis." In *Desalination technology developments and practice,* edited by Andrew Porteous. Essex, England: Applied Science Publishers, Ltd, pp. 205–248.

Evans, C.K., and M.A. Finan. 1991. *Comparisons of the use of different antiscalant treatments in a pilot reverse osmosis plant at Ras Abu JarJur, Bahrain.* Technical Proceedings, vol. 2. International Desalination Association World Conference on Desalination and Water Reuse, 25–29 August 1991, at Washington, D.C.

Francis, P.S. 1966. *Fabrication and evaluation of new ultrathin reverse osmosis membranes.* National Technical Information Service, Report No. PB-177083, Springfield, Va.

Hicks, R.D., C.B. Masin, L.A. Haugseth, T.A. Logan, R.F. Wilson, C. van Hoek, J.J. Strobel, F. O'Shaughnessy, and E.F. Miller. 1972. *Desalting handbook for planners.* Denver, Colo.: U.S. Department of Interior, Bureau of Reclamation and Office of Saline Water.

Hwang, S.-T., and K. Kammermeyer. 1984. *Membranes in separations.* vol 8, *Techniques of chemistry.* Malabar, Fla.: Robert E. Krieger, p. 25.

Loeb, S. 1981. "The Loeb-Sourirajan membrane and how it came about." Chap. 1 in *Synthetic membranes,* edited by A.F. Turbak, vol. 1, Desalination, ACS Symposium Series 153. Washington, D.C.: American Chemical Society.

Loeb, S., and S. Sourirajan. 1963. Sea water demineralization by means of an osmotic membrane. *Advances in Chemistry Series* 38:117.

Lonsdale, H.K. 1982. The growth of membrane technology. *Journal of Membrane Science* 10:81.

Lonsdale, H.K. 1986. "Reverse osmosis." In *Synthetic membranes: Science, engineering and applications*, edited by P.M. Bungay, H.K. Lonsdale, M.N. de Pinho. Holland: D. Reidel, pp. 307–342.

Mason, E.A. 1991. From pig bladders and cracked jars to polysulfones: An historical perspective on membrane transport. *Journal of Membrane Science* 60:125–45.

Matsuura, T., and S. Sourirajan. 1985. "Process technologies." In *Theory and Practice of Reverse Osmosis*, edited by R. Bakish. Topsfield, MA: International Desalination Association, pp. 234–236.

McCutchan, J.W. 1977. "Tubular reverse osmosis plant design." Chap. 27 in *Reverse osmosis and synthetic membranes*, edited by S. Sourirajan. Ottawa: National Research Council of Canada.

Merten, U., ed. 1966. *Desalination by reverse osmosis*. Cambridge, Mass.: MIT Press.

Paul, D.H., and A.H. Rahman. 1990. Membrane fouling—the final frontier? *Ultrapure Water* 7(3):25–36.

Permasep Engineering Manual. 1982. E.I. Du Pont de Nemours, Co. Inc., Wilmington, Del. 1982 ff.

Pohland, H.W. 1987. Reverse osmosis. Chap. 8 in *Handbook of water purification*. 2d ed., edited by W. Lorch. New York: Halstead Press.

Ray, R., R.W. Wytcherley, D. Newbold, S. McCray, D. Friesen, and D. Brose. 1991. Synergistic, membrane-based hybrid separation systems. *Journal of Membrane Science* 62:347–69.

Reid, C.E., and E.J. Breton. 1959. Water and ion flow across cellulosic membranes. *Journal of Applied Polymer Science* 1:133–143.

Richter, J.W., and H.H. Hoehn. 1971. *Permselective, aromatic, nitrogen-containing polymeric membranes*. U.S. Patent 3,567,632.

Riley, R.L., R.L. Fox, C.R. Lyons, C.E. Milstead, M.W. Seroy, and M. Tagani. 1976. Spiral-wound poly(ether/amide) thin-film composite membrane systems. *Desalination* 19:113.

Rozelle, L.T., J.E. Cadotte, R.D. Corneliussen, and E.E. Erickson. 1967. *Development of new reverse osmosis membranes for desalination*. Report No. PB-206329. Springfield, Va: National Technical Information Service.

Sackinger, C.T. 1980. The energy requirements of seawater desalination: reverse osmosis vs. multi-stage flash. *Middle East Water and Sewage*, April/May.

Sieveka, E.H. 1966. "Reverse osmosis plants." Chap. 7 in *Desalination by reverse osmosis*, edited by U. Merten. Cambridge, Mass.: MIT Press.

Sourirajan, S. 1970a. *Reverse osmosis*. London: Logos Press.

Sourirajan, S. 1970b. Appendix II. *Reverse osmosis*, London: Logos Press, pp. 561–565.

Sourirajan, S. 1970c. Some applications and engineering developments in the reverse osmosis separation process. Chap. 8 in *Reverse osmosis*. London: Logos Press.

Strantz, J.W. 1982. *Predicting $CaCO_3$ scaling in seawater RO systems*. Technical Proceedings, vol. 1, 10th Annual Conference and Trade Fair of the Water Supply Improvement Association, 25–29 July 1982, at Honolulu, Hawaii.

Strantz, J.W. 1990. *Treatment for reverse osmosis membranes*, U.S. Patent 4,938,872.

Walker, J.B., and R.R. Ferguson. 1990. *Process to make juice products with improved flavor*. U.S. Patent 4,933,197.

Walker, J.B. 1990. *Reverse osmosis concentration of juice products with improved flavor*. U.S. Patent 4,959,237.

Wangnick, K. 1992. *IDA worldwide desalting plants inventory*, Report 12, December 31, 1991. Topsfield, Mass.: International Desalination Association.

Westmoreland, J.C. 1968. *Spirally wrapped reverse osmosis membrane cell*. U.S. Patent 3,367,504.

Wilf, M., P. Chebi, P. Lange, P. Laverty, J.C. Whitmar, and G.A. Morgon, Jr. 1991. *Design and performance of a large reverse osmosis softening plant*. Technical Proceedings, vol. 1, International Desalination Association World Conference on Desalination and Water Reuse, 25–29 August 1991, at Washington, D.C.

World Health Organization. 1984. *Guidelines for drinking water quality*. vol. 1, *Recommendations*. Geneva: World Health Organization.

2

Future Trends in Reverse Osmosis Membrane Research and Technology

Takeshi Matsuura

Faculty of Engineering
University of Ottawa

INTRODUCTION

Almost 30 years have passed since Loeb and Sourirajan developed the first cellulose acetate membrane for seawater desalination (Loeb and Sourirajan 1963). The progress made in the field during the past 30 years is indeed spectacular, even when the scope is limited to reverse osmosis (RO) alone. Since the inception of the composite membrane by Cadotte (see for example, Rozelle et al. 1977) a variety of polymeric materials have been used for the RO membrane. The principle that underlies the entire RO membrane development remains, however, the same. As clearly stated in Sourirajan's first book (Sourirajan 1970), the interfacial fluid, the composition of which is different from the bulk fluid, is forced through channels, however small they may be, penetrating across the membrane. The essence of the membrane design is, therefore, to control the size of the fluid channel and to control the membrane material-solution interaction by which the composition of the interfacial fluid is governed. This concept, vividly described in Sourirajan's preferential sorption–capillary flow mechanism, is focused in this chapter. The author believes that the future of RO depends on the further realization of this concept in industrial practice.

RECENT DEVELOPMENT OF MEMBRANES AND MEMBRANE MATERIALS

In earlier stages of RO development, there were only a limited number of commercial products, and these were mostly based on cellulose acetate materials and were of asymmetric structure. A large number of membrane and membrane

---

Current Address: Department of Chemical Engineering, University of Ottawa.

module products were announced recently. This was possible primarily due to the introduction of the composite structure to the membrane, which opened up the choice of the top skin layer to a variety of organic polymeric materials. Although some of the membrane materials have not been disclosed by the membrane manufacturers, many of them are polyamides, as shown in Table 2.1, and are based on polyethylene-imine, piperazine, and fully aromatic structures. Table 2.1 indicates also that they are either positively or negatively charged. Because of the thinness of the skin layer and because of the ionic character which enables a high salt rejection owing to the co-ion repulsion, these composite membranes exhibit fluxes higher than the conventional asymmetric RO membranes at a given level of reference sodium chloride rejection. Comparison is made in Figure 2.1 of a number of commercial spiral membrane elements with respect to the rejection-productivity relationship. Figure 2.1 clearly indicates that the productivity of composite membranes more than doubled that of the reference cellulose acetate membrane. Improved chlorine resistance and pH resistance were also achieved for the composite membranes. In a word, new membrane development was focused on (1) high salt rejection, (2) low-pressure operation, and (3) high membrane durability.

TABLE 2.1   Classification of Composite Reverse Osmosis Membranes

Class	Name	Manufacturer	Charge
Polyamides			
Heterocyclic/aromatic	NTR-729HF	Nitto Denko	Negative
	NTR-739HF	Nitto Denko	—
	UTC-60	Toray	—
	UTC-20HF	Toray	—
	NF-40	Film Tec	—
	NF-40HF	Film Tec	—
Aromatic	NTR-759HR	Nitto Denko	Negative
	CPA2	Hydronautics	—
	UTC-70	Toray	Negative
	BW-30	Film Tec	—
	NF-50	Film Tec	—
	NF-70	Film Tec	—
Aromatic/alicyclic	A-15	DuPont	Negative
Polyurea			
Aliphatic/aromatic	NTR-719HF	Nitto Denko	Positive
	IFC1	Hydronautics	—
	TFC Series	UOP	—
	PA300	UOP	Positive
	UTC-40HR	Toray	—
Polyethers			
Sulfonated aromatic	NTR-7400	Nitto Denko	Negative

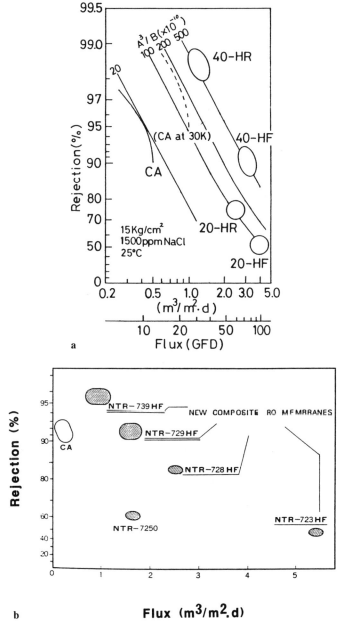

FIGURE 2.1.   Performance data of some composite membranes. (a) High separation membranes, feed NaCl concentration, 1,500 ppm; pressure, 1.47 MPag (213 psig).   (From Kurihara 1985. Reproduced by permission of Elsevier Science Publishers); (b) High flux membranes, feed NaCl concentration, 1,500 ppm; pressure, 1 MPag (143 psig).   (From Kawada et al. 1987. Reproduced by permission of Elsevier Science Publishers.)

The earlier development of RO membranes was targeted at the seawater desalination process. More than 99 percent of sodium chloride rejection was required for the membranes. As the application of RO membranes was extended to other sections of industries, it was found that lower salt rejections are required in many of such applications. In particular, many valuable products from biotechnological processes are oligomers of monosaccharides or amino acids, and the concentration of such product compounds can be performed effectively by membranes whose pore sizes are larger than those of seawater desalination membranes. Simultaneous passage of the salts that are added as nutrients can also be effected. Therefore, within the pharmaceutical and biotechnological industries there was a high demand for the development of membranes with pore sizes between those of RO membranes ($<10 \times 10^{-10}$ m) and ultrafiltration membranes ($>20 \times 10^{-10}$ m). Although these membranes were called nanofiltration membranes, because the pore sizes were considered to be in the range of 1 nanometer ($= 10 \times 10^{-10}$ m), by the FilmTec Company, many of the composite membranes developed by other companies belong to this category. Table 2.2 compares the rejection data of some inorganic and organic solutes by three nanofiltration membranes (NF70, XP45, and XP20) manufactured by FilmTec, together with the rejection data of RO membrane FT30 (Cadotte et al. 1988). (*Rejection* and *separation* are interchangeably used in this chapter. They are equal to [solute concentration in the feed – solute concentration in the permeate]/ solute concentration in the feed.) Obviously, the rejection of the electrolytes by the nanofiltration membranes is much lower than that of FT30 RO membranes. The anionic nature of NF70 is obvious, because the rejection of NaCl salt with a monovalent cation is higher than that of $MgCl_2$ with divalent cation. In addition,

TABLE 2.2   Rejections of Electrolyte Solutes by
Nanofiltration Membranes

Solutes	FT-30	Rejection, % NF-70	XP-45	XP-20
NaCl	99.5	75	50	20
$MgCl_2$	>99.5	70	83	—
$MgSO_4$	>99.5	97.5	97.5	85
$NaNO_3$	90	50	<20	0
Ethylene glycol	70	13	24	11
Glycerol	96	25	44	15
Glucose	99	93	95	60
Sucrose	100	100	100	89

Pressure for inorganic solutes, applied to give a flux of 10 μm/s ($=20$ gfd); pressure for organic solutes, unknown.

the rejection of NaCl is affected strongly by the feed pH, whereas the rejection of $CaCl_2$ is not affected. This also indicates the anionic nature of the membrane. Every membrane exhibits a high rejection for sucrose solute. Therefore, the almost complete rejection of the oligomers of monosaccharides can be expected for these membranes. Table 2.3 shows the net driving pressure (= operating pressure − osmotic pressure difference on both sides of the membrane) required to produce 10 μm/s (= 20 gallon/ft^2 day) of the permeate flux. It is clear that less than 1.034 MPag (150 psig) is enough to create the given flux for nanofiltration membranes.

Another interesting report in the area of the membrane development has been made on the spiral RO element for use at elevated temperatures. Universal Oil Products (UOP) claims that the HT-RO element can be used at a temperature close to the boiling point of water (Chu, Campbell, and Light 1988). According to the authors, the advantages of such membranes include the following:

1. The energy content of a hot fluid can be conserved.
2. RO units can be used for hot solutions, for which temperatures must be maintained because of the solubility limit of some solutes.
3. RO units can be used for processing of liquid food at a temperature suitable for pasteurization.

The membrane is based on nitrogen-containing polymers supported on a fabric-reinforced, porous ultrafiltration membrane. Modules were specially constructed to withstand operating temperatures and pressures up to 90°C and 6.895 MPag (1,000 psig), respectively. Some experimental data are shown in Table 2.4 for the separation of sodium chloride and magnesium sulfate solutes at 1.724 MPag (250 psig). Feed concentrations were kept at 1,000 ppm for sodium chloride and 200 ppm for magnesium sulfate, respectively. Operating tempera-

TABLE 2.3   Net Driving Pressures Required to Obtain 10 μm/ (=20 gft) for Various Membranes

Membrane	Pressure × 10^{-6}, Pa
FT-30	1.4
NF-70	0.4
XP-45	0.7
XP-20	1.0
CA	1.7–3.5

TABLE 2.4   Some Performance Data for HT-RO Elements[1]

Temperature	Permeation Rate gfd (Lmh)	Rejection, % NaCl	MgSO₄
25	24 (41)	99.2	>99
50	26 (44)	—	—
60	27 (46)	97.0	>99
65	29 (49)	97.3	>99
70	28 (48)	96.7	>99
80	25 (42)	96.3	>99
88	25 (42)	96.5	>99
25[2]	<10 (16)	90.0	>99

[1]Pressure, 1.724 MPag (250 psig); 1 Lmh = 2.778 × 10⁻⁷ m³/m² ·s.
[2]After exposure to high temperatures.

ture was changed from 25°C to 88°C. Table 2.4 shows that the permeation rate did not change with an increase in temperature, whereas the sodium chloride separation decreased from 99.2 to 96.5 percent. Magnesium sulfate separation was maintained at more than 99 percent at all levels of temperature. These membrane units were applied for hot wash water recovery, dewatering of sugar solution, and thin sugar beet concentration.

As mentioned earlier, there has recently been greater emphasis on the development of nanofiltration membranes because of the possible applications of membrane separation technology for biotechnology. On the other hand, there is no indication in the literature that another aspect of RO membrane application has been explored intensively, which is the development of membranes with very small pore sizes to enable the rejection of small organic molecules. Major progress in this area occurred much earlier in connection with the separation and concentration of ethyl alcohol product in the fermentation broth. Typical experimental data are summarized in Table 2.5 for Toray's PEC-1,000 membrane (Kurihara et al. 1981). The latter membrane not only showed extremely high ethyl alcohol rejection, but was also very effective for the separation of many organic solutes in the dilute aqueous solution.

Although the development of inorganic membranes for RO has long been awaited, particularly for the purpose of separating hydrocarbon mixtures, the goal has not been fully achieved. The smallest pore radius generated so far is 20 × 10⁻¹⁰ m (2 nm), which is still too large for RO purposes. An attempt has been made to decrease the pore size as much as possible. Interestingly, the method of decreasing the pore size is very similar to that involved in the preparation of

TABLE 2.5   Separation of Organic Solutes by PEC-1000 Membrane*

Solute	Rejection %	Feed Concentration wt.%	Solute	Rejection %	Feed Concentration wt.%
Alcohols			*Ketones*		
Methanol	41	5	Acetone	97	4
Ethanol	92	5	Methyl ethyl ketone	98	4
2-Propanol	99.5	5			
n-Butanol	99	5	*Esters*		
Benzyl alcohol	82	4	Ethyl acetate	99.2	4
Ethylene glycol	94	5	n-Butyl acetate	99.6	1
Propylene glycol	99.7	5			
Glycerine	99.9	5	*Ethers*		
Phenol	99.0	1	Tetrahydrofuran	99.8	5
			1,4-Dioxane	99.9	5
Carboxylic acids					
Formic acid	34	5	*Amines*		
Acetic acid	91	5	Ethylenediamine	99.5	5
Propionic acid	97	5	Aniline	95	1
Oxalic acid	99.1	0.5			
			Amides		
Aldehydes			Urea	85	1
Formaldehyde	69	5	N,N-dimethylformamide	98	5
Acetaldehyde	89	3	N,N-dimethylacetamide	99.6	5
			ε-Caprolactam	99.9	5
			Sulfoxide		
			Dimethylsulfoxide	99.6	5
			Sugar		
			Lactose	>99.9	5

*At 5.516 MPag (800 psig) and 25°C.

polymeric membranes. Membranes are prepared using precursor sols which contain extremely small sol particles (Zeltner and Anderson 1989), because the interstitial void space is the nascent pore of the resultant membrane. This requires methods of synthesis to prepare "Q particles," including low-temperature hydrolysis under pH conditions that lead to highly charged primary particles and dialysis of the resulting sol to reduce the ionic strength of the final sol. Zeltner and Anderson prepared titania sols for which the hydrodynamic diameters of the primary particles were about 30×10^{-10} m (3 nm). Xerogels prepared from this sol had pore radii less than 10×10^{-10} m (1 nm), and the membranes fired at 250°C had pore radii less than 15×10^{-10} m (1.5 nm). Separation data for these membranes have not yet been presented. The RO data presented by Leenaars, Keizer, and Burggraaf (1985) is still the only record available in the literature. According to their data, the RO separation of raffinose is 28 percent.

DEVELOPMENT OF MATERIALS FOR HIGH CHLORINE RESISTANT MEMBRANES

Although membranes prepared from aromatic polyamide polymer have excellent RO characteristics, the polymer is strongly susceptible to degradation by chlorination. The chemical interaction of aromatic polyamide polymer with halogen compounds was therefore studied thoroughly by Glater and Zachariah (1985). They chose benzanilide as a model compound because it can represent the repeat unit of poly-m-phenylene-isophthalamide, which is the polymeric material constituting the Du Pont B-9 separator. From the infrared (IR) spectra of both unexposed and bromine-exposed benzanilide they have concluded that halogen substitution occurred at the aromatic ring. Whether this is direct halogenation of the aromatic ring or a result of the Orton rearrangement (attack of the amide nitrogen followed by rearrangement involving the aromatic ring substitution) is not clear. As for the ring substitution, it is a mixture of meta, para, and 1, 2, 4 substitutions. It seems natural inasmuch as one ring is activated by the –NHCOϕ group, which is ortho-para director, whereas the other ring is activated by the –CONHϕ group, which is meta director. Considering the substitution pattern of the model compound as well as the change in IR spectra induced by halogen exposure of B-9 aromatic polyamide polymer, they have proposed the halogen substitution pattern of the latter polymer as indicated in Figure 2.2. They have also proposed that the intermolecular hydrogen bonding of

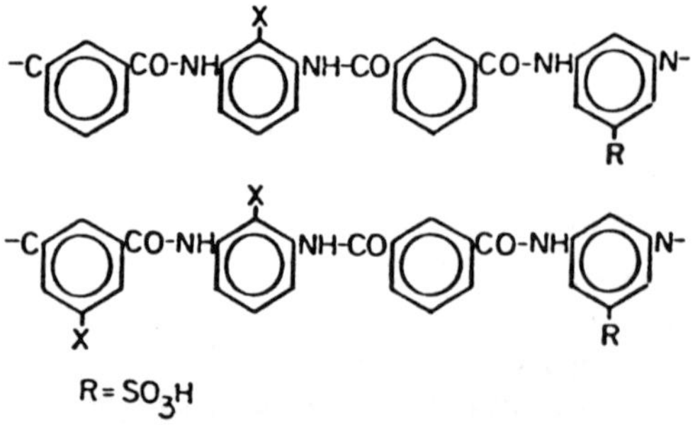

FIGURE 2.2.   Proposed halogen substitution patterns for B-9 polymer.   (From Glater and Zachariah 1985. Reprinted with permission from ACS Symposium Series. Copyright 1985 American Chemical Society.)

the polymer was weakened by halogenation, which in turn led to the deformation of the polymer chain. Disintegration of the polymer took place as a result of the chain deformation.

The previously mentioned study suggests the following two steps for chlorination of aromatic polyamide polymer.

1. Replacement of hydrogen in NHCO group by Cl.
2. Rearrangement in the position of Cl. Cl moves into the aromatic ring. The ortho position of the benzene nucleus of an aromatic diamine is substituted by Cl.

Based on this mechanism, Konagaya et al. designed structures of the aromatic polyamide polymer that were resistant to chlorine attack (1989). There are three basic strategies for the design of the molecule.

1. Substitute hydrogen in NHCO with CH_3.
2. Substitute hydrogen in the ortho position in the benzene nucleus of aromatic diamine with CH_3.
3. Insert a functional group with an electron-withdrawing effect to prevent the ortho substitution.

They have synthesized various polymers with the structure illustrated in Figure 2.3. The results are summarized in Table 2.6. Comparison of polymer 2 and polymer 5 indicates that the amount of chlorine absorbed became less than half when H in the R_1 position was substituted by CH_3. Comparison of polymer 2 and polymer 4 indicates that the amount of chlorine absorption was also halved when all four ortho positions, in relation to the amide group, were substituted by CH_3. Furthermore, the insertion of $-O-$, $-CH_2-$, and $-SO_2-$ groups at the X position decreased chlorine absorption in the above order. SO_2 insertion is especially effective when $-CONR_1-$ groups are at 3,3' positions in the aromatic rings.

The results shown in Table 2.6 indicate that poly (isophthaloyl-4,4'-diaminodiphenylsulfone) is one of the most chlorine-resistant polymers. Therefore, membranes were produced from the latter polymer using an N,N-dimethylacetamide (DMAc) solvent and a lithium chloride nonsolvent additive.

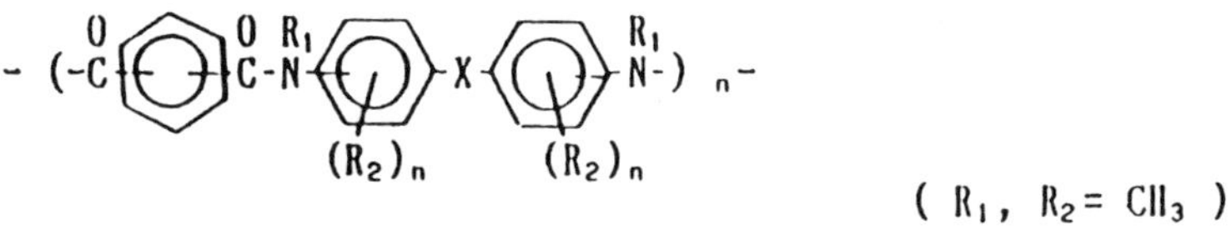

$$(R_1, R_2 = CH_3)$$

FIGURE 2.3.   **Structure of aromatic polyamide polymer.**   (From Konagaya et al. 1989. Reproduced by permission of the authors.)

TABLE 2.6   Chlorine Resistance of Aromatic Polyamide

		Positions of Amide Bonds and Substituents and Species of X, R_1, R_2 *CON(R_1)-*	*Diamine Component* *-X-*	*R_1*	*R_2*	*Chlorine Resistance Amount of Cl^+ Absorbed (mol Cl^+/ mol unit)*
1	*2*					
1	IPC	4,4'-	-O-	H	H	1.35
2		4,4'-	-CH$_2$-	H	H	1.17
3				H	3,3'-(CH$_3$)$_2$	1.36
4				H	3,3',5,5'-(CH$_3$)$_4$	0.5
5				CH$_3$	H	0.5
6		4,4'-	-NHCO-	H	H	1.74
7		4,4'-	-SO$_2$-	H	H	0.3
8		3,3'-		H	H	0.1
9	TPC	4,4'-	-SO$_2$-	H	H	0.5
10		3,3'-		H	H	0.1

[1]Membrane number.
[2]Diacid component, IPC: isophthaloyl dichloride, TPC: terephthaloyl dichloride.

Unfortunately, the RO performance of the membrane was not quite satisfactory. Sodium chloride rejection of 67.5 percent and the flux of 7.176×10^{-7} m^3/m^2 ·s (62 L/m^2 ·day) at 5,393 MPa (55 kg/cm^2) were obtained. In order to improve the RO performance, a copolymer of 4,4'-diaminodiphenylsulfone (80 mol%) and piperazine (20 mol%) was synthesized. The comonomer piperazine is known to be hydrophilic and to react with an aromatic diacid chloride forming a secondary amide, =NCO-, which is insensitive to chlorine. The membrane produced from this copolymer exhibited a sodium chloride rejection of 99.3 percent and a flux of 7.523×10^{-7} m^3/m^2·s (65 L/m^2·day). Long-term tests were performed with a membrane from this type of copolymer, which was called PAM-1, and comparison was made with a membrane prepared from NOMEX-polymer. The results are illustrated in Figure 2.4. It is obvious that the PAM-1 membrane is superior to the NOMEX membrane in the chlorine environment.

ORIGIN OF THE MEMBRANE PORE

The functionality of synthetic membranes is broadly classified by their pore sizes. According to Kesting (1989), the approximate diameters are as follows: microfiltration (MF) (200 to 100,000 $\times$ 10^{-10} m); ultrafiltration (UF) (10 to 200 $\times$ 10^{-10} m); RO (3 to 10 $\times$ 10^{-10} m); gas separation (GS) (2 to 5 $\times$ 10^{-10} m). On

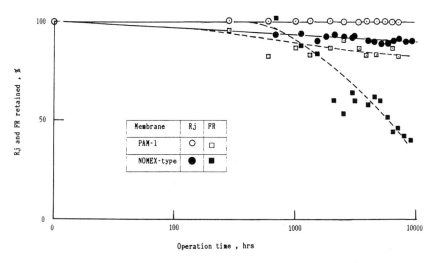

FIGURE 2.4.   Long-term test of RO membranes for chlorine resistance Cl$^+$ concentration, 0.1–0.4 ppm; feed NaCl concentration, 3,500 ppm; pressure, 5.39 MPa (55 kg/cm^2).   (From Konagaya et al. 1989. Reproduced by permission of the authors.)

the basis of observations by transmission, scanning, and electron micrographs, Kesting has further suggested that the pores are surrounded by intramolecular chain segments (GS); intermolecular chain segments in interstices between nodules (RO); nodule aggregates (UF); and aggregates of nodule aggregates (MF). Whether the space surrounded by intramolecular chain segments contributes to RO pores is still open to discussion (Sourirajan and Matsuura 1985). Bimodal pore size distributions are possible owing to the simultaneous presence of two different pore types.

In the phase inversion technique of membrane preparation, a polymer solution must be prepared. The polymer solution is in the state of sol, and is then cast into a film. During the subsequent gelation step, a change in composition of the polymer solution occurs either by solvent evaporation or by exchange of solvent and nonsolvent in the gelation bath, and the polymer solution passes through the liquid-liquid phase separation before the transition from sol to gel occurs. The structure of the polymer in the sol immediately before, and that of the polymer in the gel immediately after the sol-gel transition are very similar. Therefore, the structure of the polymer solution (sol) is regarded as the nascent membrane structure. In a casting solution, the polymer concentration of which is usually below 25 percent, polymer molecules are believed to exist as individual separate entities. As the concentration of the polymer increases, either by solvent evaporation or by solvent-nonsolvent exchange, polymer entanglement starts to occur and several tens of polymer molecules merge into a nodule. (A nodule is also called a polymer aggregate. It is not clear how many polymer molecules are in one nodule.) Further loss of the solvent may even lead to the gathering of several polymer nodules to form a nodule aggregate. The progress from individual polymer molecules, to polymer nodules, and to nodule aggregates may be interrupted by the phase separation and the subsequent gel formation at any stage, depending on the circumstances surrounding the polymer solution. Of course, solvent loss occurs most rapidly at the polymer solution/air (in case of solvent evaporation) or the polymer solution/nonsolvent (in case of solvent-nonsolvent exchange) interface. Therefore, the process proceeds most rapidly at the interface. The polymer concentration also becomes highest at the interface. As a result, the polymer solution is covered by a single layer of closely packed nodules or nodule aggregates at the interface. Thus, the skin layer is formed. The loss of solvent takes place more slowly below the skin layer than at the skin layer. More solvent is available at the time of the polymer solidification, and there is also enough time for nodule aggregates to increase in size. Therefore, the nodule aggregates are larger and packed more loosely. Thus is formed the asymmetric structure of the membrane.

Polymer nodules and nodule aggregates were observed by electron micrograph. Panar, Hoehn, and Hebert (1973) observed spherical objects of 400 to 800 × 10^{-10} m in diameter both at the surface of the polyamidehydrazide solution and at the skin layer of the asymmetric polyamidehydrazide membrane (see Figure 2.5). They thought that these spherical objects were the same as micelles of much smaller size (about 188 × 10^{-10} m), found earlier by Schultz and Asunmaa (1970) in the skin layer of cellulose acetate membranes. Observing similar objects as those noted by Schultz and Asunmaa, Kesting has concluded that the objects seen by himself and Schultz and Asunmaa under the electron microscope are polymer nodules, whereas those observed by Panar and colleagues are nodule aggregates.

Kesting considered that the interstitial domain among individual polymer nodules was of the lower density regions and formed the pore of RO membranes

polymer nodule

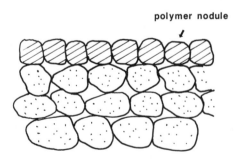

FIGURE 2.5.   Schematic representation of the polymer nodules at the membrane surface.

(1989). According to Kesting, the dominant pores in the skins of RO membranes are the spaces between chain segments in the low density domain between nodules. On the other hand, the pores in the skin of the UF membranes are formed from the interstitial domain between nodule aggregates. Because nodule aggregates are larger than individual nodules, UF membrane pores are larger than RO membrane pores. Sometimes these two pores may be present simultaneously, exhibiting a bimodal pore size distribution.

Nguyen, Matsuura, and Sourirajan (1990) have investigated the performance of membranes prepared from poly-m-phenylene-iso(x)-co-tere(100 − x)-phthalamide copolymers (abbreviated hereafter as PA). Membranes were cast from casting solutions including polymer, DMAc solvent, and lithium chloride nonsolvent additives. Different membranes were prepared by changing the value of x, which represents the isophthaloyl content of the polymer, and the LiCl/polymer weight ratio in the casting solution. The performance results of nine membranes are shown in Table 2.7 for the separation of seven organic solutes of different sizes. It is interesting to note that the separation of organic solutes whose Stokes' law radii are equal to or less than 2.8×10^{-10} m shows a maximum near membrane number 5 whose isophthaloyl content is 70 percent. On the other hand, a maximum separation is achieved by membrane number 9, whose isophthaloyl content is 30 percent, with respect to the solutes whose Stokes' law radii are more than 2.8×10^{-10} m. These data can be understood only when a bimodal pore size distribution is assumed. In Table 2.8 the average pore sizes are listed for each membrane. Note that all membranes possess two types of pores, one that is in the range of 4 to 7×10^{-10} m and the other in the range of 20 to 26×10^{-10} m. Although the first type of pore has a minimum near membrane number 5, the second type decreases in radius as the isophthaloyl content progressively decreases. The h_2 value shown in the last column of the table is the ratio of the number of the second pore to that of the first pore; h_2 decreases with a decrease in the isophthaloyl content. The separation data shown in Table 2.7 can be explained by the pore size distribution given in Table 2.8.

Small solute molecules (Stokes' radii $< 2.80 \times 10^{-10}$ m) pass through the second type of pores (radius $\bar{R}_{b,2}$) freely, and therefore separations by these pores are zero. These molecules can, however, be separated by the first type of pores (radius $\bar{R}_{b,1}$). The separation of small solute molecules is, therefore, governed by the first type of pores. This is why small organic molecules are separated most effectively when isophthaloyl content is 70 percent. Remember that $\bar{R}_{b,1}$ showed a minimum at this isophthaloyl content. With respect to large solute molecules, on the other hand, separation should be governed by the second type of pore, inasmuch as separations of these molecules by small pores are 100 percent. It would then be only natural if large organic molecules are separated progressively more effectively with a decrease in isophthaloyl content, since $\bar{R}_{b,2}$ decreases monotonically with a decrease in the isophthaloyl content.

TABLE 2.7   **Separation of Organic Solutes by Aromatic Polyamide Membranes**

Film No.	Membrane*	Ethanol	Trimethylene Oxide	1,3-Dioxolane	p-Dioxane
			Stokes' Radius, $\times 10^{10}$ m		
		1.94	2.30	2.41	2.80
			Solute Separation, %		
1	PAI(100)-0.5	—	55.6	54.0	71.3
2	PAI(100)-0.7	—	57.7	56.9	81.0
3	PAI(100)-1.0	—	58.8	62.1	80.5
4	PAI(70)-0.5	41.0	59.1	65.0	89.7
5	PAI(70)-0.7	43.8	58.4	65.9	89.7
6	PAI(50)-1.0	39.2	53.8	—	78.1
7	PAI(50)-1.0	40.0	55.1	66.2	90.9
8	PAI(30)-1.0	27.0	—	56.4	77.6
9	PAI(30)-1.5	29.7	—	53.4	83.1

Film No.	Membrane	12-Crown-4	15-Crown-5	18-Crown-6
		Stokes' Radius, $\times 10^{10}$ m		
		4.00	4.65	5.30
		Solute Separation, %		
1	PAI(100)-0.5	83.8	87.6	87.6
2	PAI(100)-0.7	87.8	87.9	88.6
3	PAI(100)-1.0	86.8	86.9	87.5
4	PAI(70)-0.5	95.2	96.9	—
5	PAI(70)-0.7	97.6	96.9	—
6	PAI(50)-1.0	98.7	98.6	98.9
7	PAI(50)-1.0	98.3	99.0	99.4
8	PAI(30)-1.0	98.3	99.2	99.4
9	PAI(30)-1.5	98.7	99.1	99.7

*PAI(100)-0.5 means that the isophthaloyl content of the copolymer is 100% and the weight ratio of LiCl additive in the casting solution to the polymer is 0.5.

MEMBRANE TRANSPORT

As stated earlier, both pore size and interaction force govern the RO separation of the solute molecule from the solvent. Therefore, these two aspects must be incorporated in the membrane transport. Such an attempt was made in the preferential sorption-capillary flow (PSCF) mechanism (Sourirajan 1970) in a qualitative way, and later the surface force-pore flow (SFPF) mechanism, which gives a more quantitative expression to the PSCF mechanism, was developed (Sourirajan and Matsuura 1985). A further modification of the model was made

TABLE 2.8   Average Pore Size and Pore Size Distribution of the Aromatic Polyamide Membranes

Film No.	$\bar{R}_{b,1} \times 10^{10}$, m	$\bar{R}_{b,2} \times 10^{10}$, m	h_2
1	5.7	26	0.007
2	5.4	25	0.005
3	4.7	25	0.004
4	4.7	24	0.001
5	4.8	23	0.001
6	5.8	22	0.001
7	5.8	—	0.001
8	7.1	20	0.001
9	7.0	20	0.001

by Mehdizadeh and Dickson (1989). Because a rigorous mathematical presentation requires a large space, only a pictorial sketch of the mechanism is attempted
(Matsuura and Sourirajan 1989).

When there is no pressure difference between both ends of the pore, the solute
starts to flow from the high-concentration side (usually the feed-solution side of
the pore) by diffusion, and the solute flow stops when the solute concentrations
on both sides of the pore become equal. Therefore, the solute separation is zero
at the stationary state. This should be true for any solutes, regardless of the type
of sorption occurring at the membrane-solution interface. When the pressure on
the feed-solution side of the pore is increased, the solvent starts to flow through
the membrane. Then, the effect of the concentration profiles in the pore, depicted
in Figure 2.6a and Figure 2.7a, on the solute separation becomes more explicit.
Consider first the preferential sorption of water. Electrolyte solutes such as
sodium chloride provide typical examples. The concentration profile is illustrated in Figure 2.6a. When the distance from the pore wall to the center of the
solute is denoted as d, the center of the solute cannot enter into the range $0 < d <$
D owing to the steric repulsion, and therefore the solute concentration is zero in
this range. **D** can be equated to the radius of the hydration sphere of an ion. From
the distance **D**, the solute concentration increases as the distance from the pore
wall, d, increases and finally reaches the bulk concentration c_b, which is the
same concentration as that of the feed solution. Therefore, there is deficiency of
the solute in the pore as compared with the concentration of the feed solution.
When the solution in the pore is driven by pressure and leaves the pore outlet as
permeate, the permeate concentration is less than that of the feed, and solute
separation occurs. The effect of the concentration profile in the pore becomes
more explicit as pressure increases, and, therefore, the separation increases with
an increase of the pressure. It is known that the permeate concentration becomes

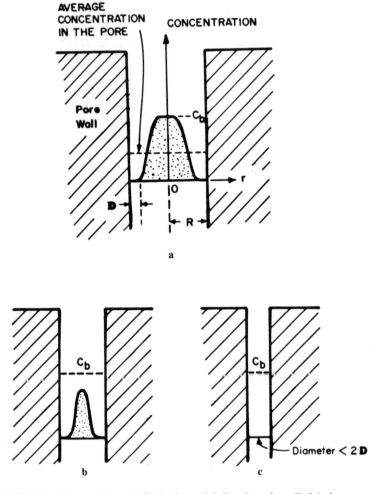

FIGURE 2.6.   Concentration profile in the radial direction of a cylindrical pore present on a membrane surface—preferential sorption of water.   (From Matsuura and Sourirajan 1989. Reprinted by permission.)

equal to the average concentration in the pore when the pressure is infinity, and this sets the limit to solute separation. In the case of solute preferential sorption, the concentration profile in the pore is as illustrated in Figure 2.7a. The system p-Cl phenol-water-cellulose acetate is a typical example. In a range of the distance from the pore wall, d, from zero to D, the solute concentration is zero because of the steric hindrance. At the distance D, the solute concentration is very high because of the strong adsorption of the solute at the pore wall. The

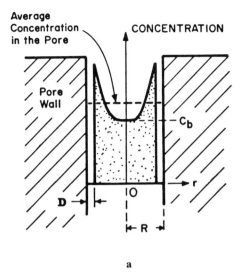

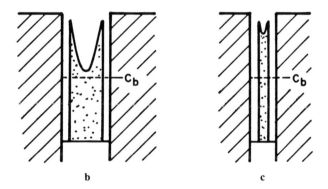

FIGURE 2.7.   Concentration profile in the radial direction of a cylindrical pore present on a membrane surface—preferential sorption of solute.   (From Matsuura and Sourira-jan 1989. Reprinted by permission.)

concentration starts to decrease as the distance increases and finally becomes c_b at the center of the pore. The solute is in excess in the pore. When the solution in the pore leaves the pore under pressure from the pore outlet, the solute is enriched in the permeate. The solute separation becomes negative. When the pressure is increased, the solute enrichment in the permeate is intensified and the permeate concentration ultimately becomes equal to the average concentration in the pore. This sets the limit to the solute enrichment. The change in the concentration profile with a decrease in the pore radius is illustrated in Figure 2.6b and c for the case of preferential sorption of water. The average concentration in the pore decreases with a decrease in the pore radius, and, as a consequence, the solute separation increases progressively. The concentration profiles for the smaller pores are illustrated in Figure 2.7b and c for the case of preferential sorption of solute. The average concentration of the pore b is higher than that of the pore a because the high solute concentration region in the vicinity of the pore wall controls the average concentration. The average concentration of the pore c is lower than that of the pore b because the zero solute concentration region (the distance from the pore wall, from zero to D) governs the average concentration in the pore. When the pore size is even smaller than D, the solute cannot enter into the pore because of the steric hindrance. The average concentration in the pore becomes zero. Therefore, the change in the solute separation expected when the pore radius is decreased is as illustrated in Figure 2.8b. The separation is nearly equal to zero when the pore size is very large. The solute enrichment in the permeate (the negative solute separation) is enhanced with a decrease in the pore radius. After passing a minimum, the solute separation starts to increase and becomes positive. The solute separation increases further with a decrease in the pore radius. The patterns illustrated in Figure 2.8a and b were indeed observed experimentally.

PORE SIZE CONTROL OF REVERSE OSMOSIS MEMBRANE

As noted in the earlier discussion on the formation of a membrane pore, it is possible to control the size of the pore by controlling the size of the polymer

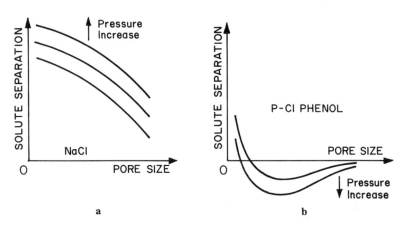

FIGURE 2.8.   Effect of the operating pressure and the pore size on the solute separation. (a) Preferential sorption of water; (b) Preferential sorption of solute.   (From Matsuura and Sourirajan 1989. Reprinted by permission.)

nodule, because the interstitial void created between polymer nodules becomes the membrane pore. Therefore the size of the supermolecular aggregate, defined as the size of a group of polymers that move together in a solution as if they were one single entity, of poly-m-phenylene-iso(70)-co-tere(30)-phthalamide polymer (MW 31,300) in DMAc solvent was determined by viscosity measurement (Nguyen, Matsuura, and Sourirajan 1987b). Calcium chloride was added to the polymer solution as a nonsolvent additive. The compositions of the polymer solution studied are indicated in Figure 2.9 as twenty-five points on a triangular composition diagram. Moving along a solid line on the diagram, the polymer wt.% is increased at a fixed calcium chloride/polymer weight ratio. It was found that the aggregate radius, $< \bar{S} >$, ranges from 50 to 73 $\times$ 10^{-10} m. The radius increases with an increase in calcium chloride/polymer ratio and levels off when the ratio is more than 0.3. The aggregate radius decreases with an increase in the polymer wt. % for a fixed calcium chloride/polymer ratio. Whether the aggregate under consideration is the same as the polymer nodule that Panar, Hoehn, and Hebert (1973) observed under the electron microscope is not certain. Because

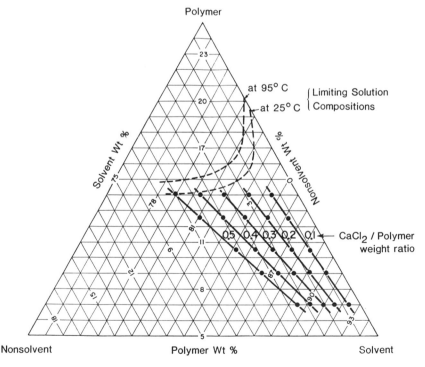

FIGURE 2.9.   **Triangular diagram of casting solution compositions for aromatic polyamide membrane.**   **(From Nguyen, Matsuura, and Sourirajan 1987a. Reproduced by permission of Gordon and Breach Science Publishers S.A.)**

the radius of the polymer aggregate is almost of an order of magnitude smaller than the nodule size observed by Panar, Hoehn, and Hebert, it is considered that the aggregate radius determined is the size of a polymer aggregate into which several polyamide polymers were assembled by the binding power of calcium chloride additive. Membranes were cast from the polymer solutions so prepared and characterized by the radius of the first pore, $\bar{R}_{b,1}$, and the second pore, $\bar{R}_{b,2}$. As illustrated in Figure 2.10, strong linear correlations were found between $ln\bar{R}_{b,2}$ and $ln <\bar{S}>$. This result shows that the control of the pore size is possible by controlling the size of the polymer aggregate in the casting solution.

The size of the nodule at the surface skin layer was observed by Kazama, Kaneta, and Sakashita in their Cardo-type polyamide membrane (1989). While investigating hollow fiber membranes, they have observed both inner and outer skin layers. In particular, the topmost part of the inner skin layer consisted of two

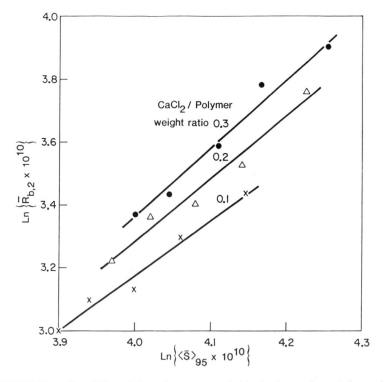

FIGURE 2.10.   Correlation of the polymer aggregate size in the casting solution and the radius of the second (larger) pores on the surface of aromatic polyamide membranes. (From Nguyen, Matsuura, and Sourirajan 1987b. Reproduced by permission of Gordon and Breach Science Publishers S.A.)

FIGURE 2.11.   Transmission electron micrographic image of inner skin layer of Cardo membrane.   (From Kazama, Kaneta, and Sakashita 1989. Reproduced by permission of the authors.)

layers of closely packed polymer nodules (see Figure 2.11). The thickness of each layer was 80×10^{-10} m (8 nm). Each polymer nodule was football-shaped. The nodule width of 80×10^{-10} m (8 nm) was observed when the membrane was sectioned vertical to the hollow fiber spinning direction, whereas the width was about 150×10^{-10} m (15 nm) when the membrane was sectioned parallel to the spinning direction. These sizes could be changed by changing the membrane casting conditions. These authors used these membranes for O_2/N_2 separation after drying. The distribution of the nodule size and the oxygen/nitrogen selectivity were further correlated. As Figure 2.12 shows, the selectivity was high when the nodule size was small. They have argued that the space between the nodules can be closed more easily when the nodule size is smaller, leading to a higher selectivity. This is the first report in which the relation between nodule size and selectivity was reported.

As for control of the distance between intramolecular chain segments in the polymer nodule, it can be achieved by changing the size of cations and anions that constitute the electrolyte additives in the casting solution (Nguyen, Matsuura, and Sourirajan 1987a).

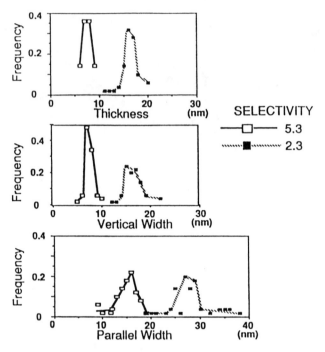

FIGURE 2.12.   Distribution of nodule size and O_2/N_2 selectivity.   (From Kazama, Kaneta, and Sakashita 1989. Reproduced by permission of the authors.)

MEMBRANE FOULING

It is generally recognized that the growth of the membrane-based industry has been hampered by membrane fouling. Since such fouling is often caused by the adsorption of solutes not only on the membrane surface but also inside the membrane pore, identification of the adsorbates and the adsorption sites has been the subject of research concerning membrane fouling (Hanemaaijer et al. 1989; Bhattacharyya and Madadi 1988). The interaction force working between the adsorbate and the membrane material should therefore be the central issue of the fouling. Another approach is to describe membrane fouling mathematically, and the gel-model is used most often for this purpose (Blatt et al. 1970). The model starts from a mass balance equation established in the laminar boundary layer:

$$-D dc/dz + uc = 0 \tag{1}$$

Equation 1 means that the amount of the solute approaching the membrane surface by convective flow is counterbalanced by the amount of solute moving to the reverse direction by diffusion. The integration of equation 1 yields

$$u = (D/\delta)ln(c_2/c_1) \tag{2}$$

where c_2 and c_1 are the concentration at the membrane surface and the feed concentration, respectively. δ is the thickness of the boundary layer, D is the diffusion coefficient of the solute, and D/δ is often defined as k and called *mass transfer coefficient*. The wall concentration, c_2, increases rapidly with an increase in permeation velocity, u, and reaches ultimately gel concentration, c_g, where the solution is no longer fluid. At this moment u reaches a limiting value, u_{inf}, therefore,

$$u_{inf} = kln(c_g/c_1) \tag{3}$$

Equation 3 predicts a linear plot for u_{inf} versus lnc_1 with a slope equal to $-k$; extrapolation to $u_{inf} = 0$ yields the lnc_g value. Although equation 3 is convenient for correlation of experimental data, it is not always correct. Because c_g is a property of the solution, c_g obtained experimentally has to be independent from the membrane and the operating conditions. However, different authors find significantly different c_g values for an identical solution (Wijmans, Nakao, and Smolders 1984). Unfortunately, no term for the interaction between the solute and the membrane surface is included in equation 3. It is therefore unrealistic to expect that equation 3 can describe membrane fouling, because the interaction is known to play an important role in the solute adsorption. None of the models alternative to the gel-model include the interaction term (Wijmans, Nakao, and Smolders 1984). An attempt was made recently to include the interaction term in the mass balance equation (Zhang 1990). Assuming that forces working on the solute are the diffusive force and the membrane-solute interaction force, the following equation was derived.

$$-Ddc/dz + \alpha c/(\delta - z) + uc = 0 \tag{4}$$

Note that the effect of the interaction is included as the second term of the mass balance equation. An assumption was made that the interaction force is inversely proportional to the distance between the membrane and the solute ($\delta - z$). The constant α includes the proportionality constant for the interaction force and the friction coefficient between the solute and the solvent. Integration of equation 4 yields

$$c_{inf} = c_g(\mathbf{D}/\delta)^{(\alpha/D)} \tag{5}$$

where c_{inf} is the feed concentration at which the permeation velocity becomes zero. $\mathbf{D}$ is a distance from the pore wall that is characteristic of the solute. Equation 5 predicts that c_{inf} is not equal to c_g but decreases with an increase in

α, which means an increase in the interaction force, and with an increase in the boundary layer thickness, δ, which is caused by less turbulence of the feed solution near the membrane surface. Figure 2.13 illustrates the positions of c_{inf} for different membrane materials. c_{inf} is much smaller for a polyethersulfone (PES) membrane than cellulose acetate (CA) membrane in the RO concentration of tea juice (Zhang 1990). It is known that a PES membrane is contaminated by the tea juice components more strongly than a CA membrane, and therefore the solute-membrane interaction is stronger for a PES membrane. The above data indicate clearly that c_{inf} should depend on the membrane material.

MODULE DESIGN

Since the inception of RO technology in the 1960s, four module configurations have been devised and utilized: plate-and-frame, tubular, spiral wound, and hollow fiber. Interestingly, no new module configurations have been devised since then. Because concentration polarization is one of the major concerns when the module is used for a specific application, the solute mass transfer has been studied intensively. These results are summarized in Cheryan (1986). Although there are only a few documents in the literature, the industry has been making continuous efforts to improve the module design. One report was presented by

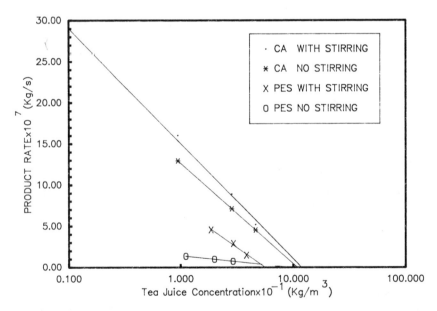

FIGURE 2.13.   Position of c_{inf} for different polymeric materials.   (From Zhang 1990. Reproduced by permission of the author.)

Parise, Parekh, and Smith (1985) at the Symposium on Reverse Osmosis and Ultrafiltration held at Philadelphia in 1984. A spiral wound cartridge was designed specifically to minimize bacterial growth in the module and to make sanitation of the module as easy as possible. Figure 2.14 illustrates the details of the new device. A wheel-shaped cap is placed, usually at the feed side of the spiral wound cartridge. A chevron seal is also attached to the edge of the wheel-shaped cap to prevent the bypass of the feed solution through the channel between the cartridge surface and the cartridge housing. This device, however, creates a dead space behind the seal. When the module is sanitized, the solution of the sanitizing agent does not flow into the dead space. Therefore, Parise and colleagues made four slots on the edge of the wheel-shaped cap. The slots allow a controlled portion of the feed flow (3 to 6 percent) to pass beneath the chevron seal to the space between the cartridge and the cartridge housing. A flow channel was also incorporated between the outer wrap and the cartridge surface. This bypass stream was very effective to flush the solution in the cartridge. According to Parise and colleagues, less than 1 minute was required to flush the solution in the annular space between the cartridge and the cartridge housing as evidenced by a visual test using methylene blue dye at the normal flow rate of 4 to 10 gpm.

Another report on the improvement of the module was presented by Paulson, Phelps, and Gach (1989) at the Symposium on the Advances in Reverse Osmosis and Ultrafiltration held in Toronto in 1988. To facilitate the fluid turbulence on

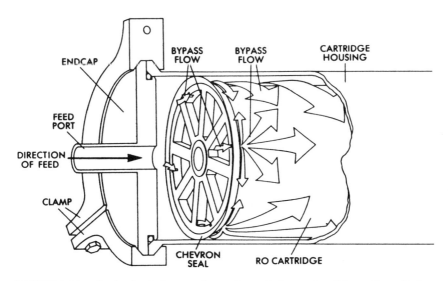

FIGURE 2.14.   Device for feed stream bypass in a spiral wound module.   (From Parise, Parekh, and Smith 1985. Reprinted with permission from ACS Symposium Series. Copyright 1985 American Chemical Society.)

the membrane surface, the spacers are sandwiched between membranes. The channel spacers are, almost without exception, coaxially extruded, large-fiber, diamond-mesh netting of polypropyrene. The tortuous flow caused by the presence of the spacer increases the turbulence of the solution. There are some shortcomings in this device. When the solution is highly concentrated with particulate matters, the polymer mesh is plugged and the channel is blocked. The pressure drop becomes very high when a viscous solution is treated. Paulson and colleagues made two new devices. One is called the *full-fit construction*. Without the brine seal and outer wrap, a spiral wound cartridge is mounted into a cartridge housing. The presence of the dead space between the membrane cartridge and the cartridge housing is thus minimized. Furthermore, the cartridge is allowed to expand slightly in the radial direction. When the feed solution flows through the channel space between two membranes, the distance between the membranes expands slightly, leaving some space between the polymer mesh and the membrane. The flow regimes for the standard spacer construction and the full-fit construction are depicted in Figure 2.15. Because of the larger space opened to the feed solution flow, the plugging of the solution channel is minimized. Another spacer device called a *tubular spiral* is depicted in Figure 2.16. The feed solution flows through a triangular space without much obstacle to the flow. This device is constructed with triangular tubes built inside the spiral wound module. Because of the open channel volume/membrane area ratio, which is greater than those of the standard mesh and the full-fit designs, this allows greatly increased cross-flow rates, and the pressure drop is low. Figure

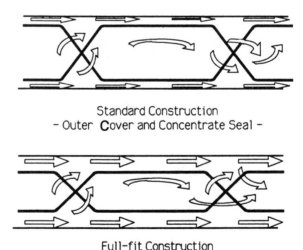

Standard Construction
– Outer Cover and Concentrate Seal –

Full-fit Construction

FIGURE 2.15.   Fluid flow dynamics.   (From Paulson, Phelps, and Gach 1989. Reproduced by permission of the authors.)

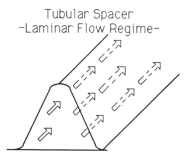

FIGURE 2.16. **Flow in a tubular spacer.** (From Paulson, Phelps, and Gach 1989. Reproduced by permission of the authors.)

2.17 illustrates some experimental results on the pressure drop. It is clear that the tubular spiral module shows the least pressure drop, whereas the standard spiral wound module exhibits the highest pressure drop.

The effect of turbulence on the membrane surface was studied more systematically by placing CA membrane on a surface of a corrugated plate (van der Waal and Racz 1989). Corrugated plates were produced by cutting cylindrical PVC bars of 3 mm diameter in an axial direction. These half-cylindrical corrugations were glued onto a PVC plate at a given distance perpendicular to the flow direction. A flow pattern on the corrugated surface is depicted in Figure 2.18.

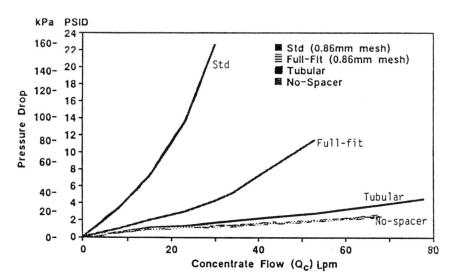

FIGURE 2.17. **Pressure drop versus concentrate flow rate for different spacers.** (From Paulson, Phelps, and Gach 1989. Reproduced by permission of the authors.)

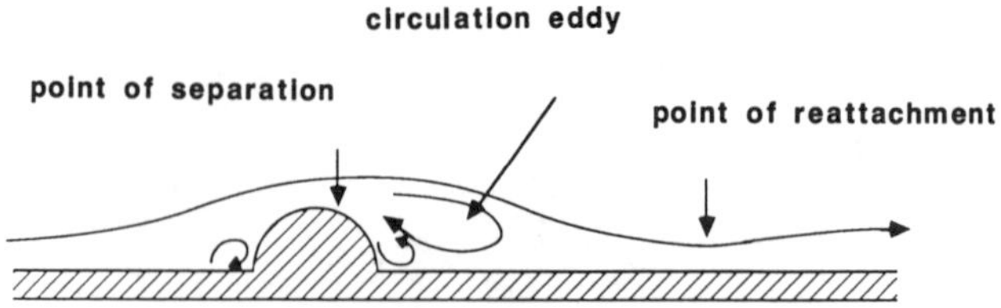

FIGURE 2.18.   Schematic representation of stream lines for flow over corrugated plates.   (From van der Waal and Racz 1989. Reproduced by permission of Elsevier Science Publishers.)

The main stream line moves upward at the point of separation owing to the presence of the corrugation. The stream line slowly comes down and reattaches the plate surface. A circulation eddy is formed about 10 to 15 mm behind the corrugation. An optimum distance between two corrugations is obvious from the flow pattern. When a circulation eddy fills the distance between two corrugations, the turbulence on the surface of the membrane placed on the top of the plate surface is expected to be the most vigorous.

MEMBRANE APPLICATIONS

Membranes for Ultrapure Water Production

In the process of ultrapure water production, RO membranes are used for the primary treatment of potable tap water preceding use of ion exchange columns. The salt concentration of tap water depends on the location of the household, but is in the range of 100 to 700 ppm. Increasingly, RO units are placed after the ion exchange columns in order to remove the organic and particulate contaminants originating from the ion exchange resin beds. Moreover, there is a tendency to replace the RO/ion exchange combination by the RO/RO combination, which is called a two-stage RO system, for ultrapure water production. For such an application, the membrane performance data for very low salt concentrations are necessary. Most of the membranes have been so far characterized by the rejection of the reference sodium chloride solute at the feed concentration of either 1,500 or 2,000 ppm, which is a suitable range for brackish water treatment. It is expected that salt rejection deviates at very low salt concentration from the salt rejection rated in the aforementioned range of feed salt concentration, particularly because of the ionic nature of many composite membranes. The rejection of individual ions of electrolyte solutes was studied by two groups: Toray, using the SU-700 module, and Nitto Denko, using the NTR-759HR module (Toray 1987; Kamiyama et al. 1989). Both modules consist of spiral wound elements. Both groups obtained very similar results for the rejection of individual ions and for the effects of the salt concentration and pH on the rejection. Typical examples are illustrated in Figure 2.19 and Figure 2.20. In Figure 2.19 the concentration of sodium chloride was changed from 1 to 10 ppm. The rejection of the chloride ion is much higher than that of the sodium ion, reflecting the anionic character of the membrane. The broken line shows the salt rejection when the feed sodium chloride concentration is 1,500 ppm. The rejection of the chloride ion changes from 99.4 percent to 97 percent when the feed sodium chloride concentration is decreased from 10 to 1 ppm. The ionic character of the membrane is also reflected in the rejection data at various feed solution pH concentrations. Figure 2.20 illustrates the results for a feed sodium chloride concentration of 1 ppm. The rejections of sodium and chloride ions cross over at pH 4.2, which is the same as the pK_a of the carboxylic group of benzoic acid.

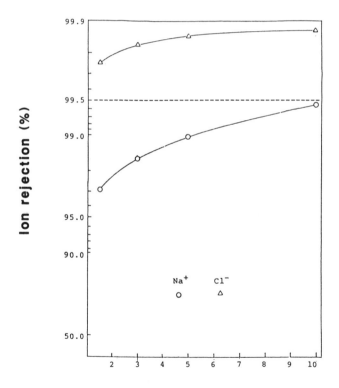

NaCl concentration in feed (ppm)

FIGURE 2.19.   Sodium chloride separation data of a charged composite membrane. Membrane, NTR-759HR spiral element; pressure, 1.5 MPag (215 psig); pH, 6.5.   (From Kamiyama et al. 1989. Reproduced by permission of the authors.)

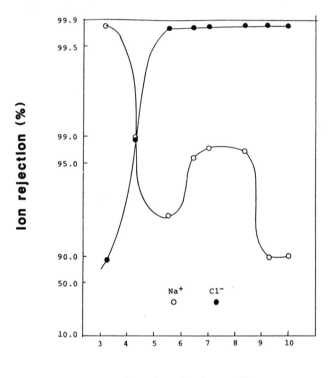

Feed solution pH

FIGURE 2.20.   Effect of pH on the separation of ions. Membrane, NTR-759HR spiral element; feed NaCl concentration, 1 ppm; pressure, 1.5 MPag (215 psig).   (From Kamiyama et al. 1989. Reproduced by permission of the authors.)

Reverse Osmosis Separation of Nonaqueous Solutions

RO separation of nonaqueous solutions is not new in the literature (Farnand 1983); however, its industrial applications are almost unexplored, primarily because of the unavailability of membranes suitable for the treatment of non-aqueous solutions. As mentioned earlier, the development of inorganic RO membranes would have a major impact in this field. There are, however, several attempts at industrial applications using organic polymeric membranes. Lubrication oil contains waxes that solidify in cold weather and cause difficulties in starting automobile engines. Therefore, the dewaxing of lubrication oil is necessary. This process was conventionally performed by mixing solvent with the lubricant oil and cooling the mixture down to −20°C, at which temperature crystallization of wax occurs. However, the solvent and lubricant oil must be separated for solvent recovery by flashing and stripping, and the cost of this process is high, owing to its high energy consumption. Replacement by a cheaper separation process is desirable, and RO might be an attractive alternative. Bitter, Haan, and Rijkens used silicone rubber membranes for this purpose and obtained promising results (1989). When a fluorinated siloxane polymer is used for the membrane material, a flux of 1.736×10^{-5} m^3/m^2 ·s (1.5 m^3/m^2 ·day) at 4 MPa (40 bar) and a selectivity of more than 50 was obtained for a solvent mixture of methyl ethyl ketone, toluene, and a residual lubeoil fraction (volume ratio $1:1:1$). The membrane was swollen 100 vol.% in the solvent/lubeoil mixture. This is in contrast to the result obtained for a membrane produced from polydimethylsiloxane polymer. The membrane flux was 4.630×10^{-5} m^3/m^2 ·s (4 m^3/m^2 ·day) at 4 MPa (40 bar), and the selectivity was less than 7. The swelling was more than 300 vol.%. The lower swelling of the fluorinated siloxane polymer is due to the higher degree of cross-linking and the lower swelling by toluene. The fluorinated siloxane polymer was swollen 30 vol.% in toluene, whereas the polydimethylsiloxane polymer was swollen 800 vol.%.

The RO process is limited by the osmotic pressure of the feed solution. It is interesting to note that the solvent concentration in the feed may go down to 30 wt.%, where the flux at 4 MPa (40 bar) becomes zero. This is because of the osmotic pressure of nonaqueous organic solutions, which is much lower than that of aqueous solution owing to the larger molar volume of the organic solvent.

A variety of polysulfone membranes were evaluated for the processing of crude oil feedstocks. The products of the process are a permeate, which is the equivalent of a gas oil, and a concentrate enriched in asphaltenes (Hazlett, Kutowy, and Tweddle 1989). Membranes were either laboratory produced from polysulfone materials or commercial membranes of polyvinylidene fluoride, polysulfone, and sulfonated polysulfone materials. The molecular weight cutoffs ranged from 6,000 to 50,000 when they were used in an aqueous environment. A high temperature operation, from 50°C to 80°C, was required because of the high viscosity of the feed crude oil. Although all membranes used in this study were ultrafiltration membranes, the performance data look more like those of RO membranes. A typical example is given in Figure 2.21 as molecular weight distributions of the feed crude oil and the permeate. The average molecular weight of the permeate sample was below 1,000 dalton, which is far lower than the nominal molecular weight cut-off of the aqueous solutions. Moreover, the molecular weight distribution of the permeate sample was strikingly similar, regardless of the molecular weight cut-off of the ultrafiltration membranes. The similarity of the permeate composition is also indicated in Table 2.9, where the percent removal of nickel and vanadium is listed for various membranes. It is obvious that the concentrations of these metals in the permeate are almost the same for every membrane. From these observations authors have concluded that membrane separation resulted from the formation of a dynamic gel-polysulfone

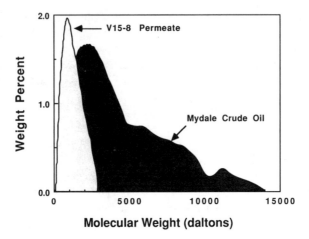

Molecular Weight (daltons)

FIGURE 2.21.   **Molecular weight distributions of feed crude oil and permeate from V 15-8 membrane. Temperature, 55°C; pressure, 1.5 MPa; feed flow rate, 5.7 L/min. (From Hazlett, Kutowy, and Tweddle 1989. Reproduced by permission of the American Institute of Chemical Engineers.)**

composite membrane. The major difference between various membranes was the permeation rate. A linear relation between operating pressure and the flux was found in the case of GR81 membrane, the pore size of which was the smallest among the membranes tested. The circulation flow rate of the feed hardly affected the flux. The flux versus operating pressure curves for V15-0 mem-

TABLE 2.9   Nickel and Vanadium Removal for Mydale Crude Permeates*

Sample	Ni, ppm	Rejection, %	V, ppm	Rejection, %
Feed	29		53	
V15-8	3	90	3	94
V16-16	2	93	3	94
U16-16	2	93	3	94
V16-16	2	93	3	94
R16-16	2	93	3	94
FS60	3	90	4	92
GR51	2	93	3	94
GR61	2	93	3	94
GR81	2	93	3	94
GS61	2	93	4	92

*Temperature = 55°C, pressure = 1.5 MPa, feed flow rate = 5.7 L/min, product recovery = 40%.

Adapted from Hazlett, Kutowy, and Tweddle 1989. Reprinted by permission.

brane, the pore size of which was one of the largest, showed a typical concentration polarization effect; i.e., the flux leveled off as the operating pressure was increased. The flux value of the plateau also increased with an increase in the circulation rate of the feed crude oil. These experimental results indicate that the inherent resistance of membranes controls the membrane permeation rate at lower operating pressures (<1.0 MPa). The layer that is formed on the surface of the membrane as a result of the concentration polarization, on the other hand, controls the permeation rate when the pore size of the membrane is very large and the operating pressure is high.

Dynamic Membranes

The discovery of dynamically formed membranes is generally attributed to Marcinkowsky of the Oak Ridge National Laboratories (Marcinkowsky et al. 1966). Although the development of dynamic membranes was not as spectacular as that of RO membranes of polymeric materials, applications of the process to the treatment of industrial effluents have been continued in North Carolina, in South Africa, and in China. Several papers were presented at the Symposium on Advances in Reverse Osmosis and Ultrafiltration in Toronto in June 1988. To form dynamic membranes, a dilute solution (10^{-4} molar) of one or more specific additives is passed over the surface of a porous support. The most typical membranes are hydrous zirconium (IV) oxide layer and hydrous zirconium oxide (IV)/polyacrylic acid dual layer. Zirconium (IV) species are known to polymerize in aqueous solutions. In particular, the degree of polymerization increases with a decrease in the acidity of the solution. The zirconium polynuclear hydrous species has an ability to react with oxygen-containing organic compounds. As such, organic compounds, those including at least two OH functional groups, or at least one CO functional group in the form of aldehyde or carboxylic acids, or one CONH group as polypeptides, are required. Therefore, polyvinyl alcohol (Wang, Xu, and Wang 1986), polyacrylic acid (Johnson, Minturn, and Wadia 1972), and polyacrylate are often used for the second additive. Although in earlier days dynamic membranes were formed on the inner surface of a carbon or ceramic tube, stainless steel tubes are used in most of the recent works. The dynamic membranes are considered to have the following advantages (Buckley et al. 1989).

1. High temperature stability
2. Long service life of the support tube
3. Ability to replace the dynamic membranes in situ
4. Availability of a range of dynamic membranes for tailoring to a particular application
5. High flux rates

Johnson and colleagues (1989) reported on the RO performance data of two systems for water treatment of nuclear plants, each system made up of 34 modules. Hydrous zirconium (IV)/polyacrylate membranes were formed on the inside of porous stainless tubes with an inner diameter of ⅝ inch. Reproducibility of the dynamically formed membranes has always been questionable since its inception. Moreover, reproducibility of the membrane has special significance, inasmuch as membrane replacement is customarily performed on site. Johnson and colleagues have tested the reproducibility of dynamic membranes of pilot plant scale. The test was performed by the separation of $NaNO_3$ (2,000 ppm) and Na_2SO_4 solutions under 6.205 to 6.895 MPag (900 to 1,000 psig) at 60°C. Figure 2.22 shows that the separation of $NaNO_3$ changes quite significantly from module to module. Na_2SO_4, on the other hand, exhibited separations in a narrow range. The flux data were also found to be in a narrow range.

Van Reenen and Sanderson (1989) synthesized acrylic acid/vinyl acetate copolymers (I) with 2.5 to 56.5 percent vinyl acetate content. The polymers were further hydrolyzed fully to give corresponding acrylic acid/vinyl alcohol copolymers (II). Dynamic membranes were then prepared on the basis of hydrous zirconium oxide (IV). Sodium nitrate (0.0235 M) separation was determined at the operating pressure of 6 MP. The best data obtained were the sodium nitrate separation of >95 percent and 5,700 Lmd for the hydrous zirconium oxide (IV)/(I) membrane when vinyl acetate content was 10 percent. The state-of-the-art membrane based on hydrous zirconium oxide (IV)/Acrysol A3 polymer (polyacrylic acid) exhibited sodium nitrate rejection of 89 to 90 percent and a flux of 2.083 to 2.431 $\times$ 10^{-5} m^3/m^2 ·s (1,800 to 2,100 Lmd). The improvement of the performance data was significant.

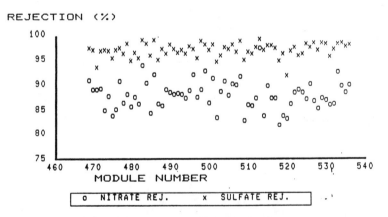

FIGURE 2.22. Reproducibility of salt rejections for different dynamic membrane modules prepared under the same conditions. Feed sodium nitrate concentration, 0.024 M; feed sodium sulfate concentration, 0.006 M; operating pressure, 6.21–6.89 MPag (900–1,000 psig). (From Johnson et al. 1989. Reproduced by permission of the authors.)

Buckley and colleagues (1989) investigated the treatment of wool-scouring effluent containing 33 g/L total solids, 21 g/L emulsified grease, and 7 g/L suint salts, using zirconium oxide membranes formed on 16 mm ID sintered stainless steel tubes. A total membrane area of 25 m^2 was installed in a module used for a pilot plant. The inlet pressure was 5 MPa. The rejection of grease was 100 percent, whereas the point rejection values based on the conductivity, the total solid content, and the total carbon content were 50 to 60 percent, 85 to 90 percent, and 95 percent, respectively. The permeate flux was stabilized by chemical cleaning in which hydrogen peroxide was used at pH 8.0 to 8.5 for 30 to 60 minutes at every 24 to 48 hours, and could be expressed as

$$J = 50 - 0.18TS \qquad (6)$$

Where

J = permeate flux (L/m^2 ·h)
TS = total solid content (g/L)

On the basis of the pilot plant data a demonstration plant of 3.2 m^3/h capacity was constructed.

CONCLUSION

In light of the recent developments in RO science and technology, the following advances in the field are expected in the near future.

1. Polymeric membranes to remove small chlorinated organic molecules will be developed.
2. Inorganic RO membranes to separate hydrocarbon mixtures will be developed.
3. Membranes of inorganic and organic hybrid materials will be developed, based on the dynamic membrane.
4. More sophisticated physical methods will be used to learn the structure of the membrane and the structure of the liquid in the membrane.
5. The pore size of the membrane will be controlled more rigorously by controlling the size of the polymer nodule and the polymer density in the nodule.
6. The concept of the polymer nodule will also be applied for the design of composite membranes.
7. A more rigorous transport theory will be developed, based on the pore size of the membrane and the polymer-solution interaction.
8. Membranes capable of fractionating mixed solutes will be developed by controlling the membrane pore size and the membrane-solute interaction.

9. Membrane fouling will be controlled by membrane design and membrane module design.
10. Hybrid separation systems of RO and other separation processes will increasingly penetrate the chemical industry and related industries.
11. Reaction and membrane separation will be combined in many chemical and biological syntheses.

References

Bhattacharyya, D., and M.R. Madadi. 1988. Separation of phenolic compounds by low pressure composite membranes: mathematical model and experimental results. In *New membrane materials and processes for separation*. AIChE Symposium Series 261, vol. 84, edited by K.K. Sirkar and D.R. Lloyd. New York: American Institute of Chemical Engineering, pp. 139–157.

Bitter, J.G.A., J.P. Haan, and H.C. Rijkens. 1989. Solvent recovery using membranes in the lubeoil dewaxing process. In *Membrane separations in chemical engineering*, AIChE Symposium Series 272, Vol. 85, edited by A.E. Fouda, J.D. Hazlett, T. Matsuura, and J. Johnson. New York: American Institute of Chemical Engineering, pp. 98–100.

Blatt, W.F., A. Dravid, A.S. Michaels, and L.M. Nelsen. 1970. Solute polarization and cake formation in membrane ultrafiltration: causes, consequences, and control techniques. In *Membrane science and technology,* edited by J.E. Flinn. New York: Plenum.

Buckley, C.A., F.G. Neytzell-de Wilde, R.B. Townsend, M.P.R. Cawdron, C. Steenkamp, M.P.J. Simpson, and T.L. Boschoff. 1989. Dynamic membrane treatment of wool-scouring effluents. In *Advances in reverse osmosis and ultrafiltration,* edited by T. Matsuura and S. Sourirajan. Ottawa: National Research Council of Canada, pp. 317–333.

Cadotte, J., R. Forester, M. Kim, R. Petersen, and T. Stocker. 1988. Nanofiltration membranes broaden the use of membrane separation technology. *Desalination* 70:77–88.

Cheryan, M. 1986. *Ultrafiltration handbook*. Lancaster, Pa: Technomic.

Chu, H.C., J.S. Campbell, and W.G. Light. 1988. High-temperature reverse osmosis membrane element. *Desalination* 70:65–76.

Farnand, B.A. 1983. A study of reverse osmosis separation involving nonaqueous solutions. Ph.D. diss., University of Ottawa.

Glater, J., and M.R. Zachariah. 1985. A mechanistic study of halogen interaction with polyamide reverse-osmosis membranes. In *Reverse osmosis and ultrafiltration*. ACS Symposium Series 281, edited by S. Sourirajan and T. Matsuura. Washington, D.C.: American Chemical Society, pp. 345–358.

Hanemaaijer, J.H., T. Robbertsen, Th. van den Boomgaard, and J.W. Gunnink. 1989. Fouling of ultrafiltration membranes: the role of protein adsorption and salt precipitation. *Journal of Membrane Science* 40:199–217.

Hazlett, J.D., O. Kutowy, and T.A. Tweddle. 1989. Processing of crude oils with polymeric ultrafiltration membranes. In *Membrane separations in chemical engineering,* AIChE Symposium Series 272, vol. 85, edited by A.E. Fouda, J.D. Hazlett, T.

Matsuura, and J. Johnson. New York: American Institute of Chemical Engineering, pp. 101–107.

Johnson, S.J., Jr., R.E. Minturn, and P.H. Wadia. 1972. Hyperfiltration. XXI. Dynamically formed hydrous Zr(IV) oxide-polyacrylate membranes. *Journal of Electroanalytical Chemistry* 37:267.

Johnson, S.J., Jr., H.G. Spencer, D.B. McClellan, N.D. Ellis, D.A. Jernigan, D.N. Mahony, and J. Jones. 1989. Reproducibility of frame-in-place membranes on practical modules. In *Advances in reverse osmosis and ultrafiltration*, edited by T. Matsuura and S. Sourirajan. Ottawa: National Research Council of Canada, pp. 357–363.

Kamiyama, Y., R. Lesan, T. Shintani, and J. Tomaschke. 1989. A comparison of different classes of spiral wound membrane elements at low concentration feeds. In *Proceedings of Ultrapure Water, EXPO'89—West Conference on High Purity Water, November 13–15, 1989, San Jose, California*, Littleton, Colorado: Tall Oakes Publisher, pp. 79–96.

Kawada, I., K. Inoue, Y. Kazuse, H. Ito, T. Shintani, and Y. Kamiyama. 1987. New thin-film composite low pressure reverse osmosis membranes and spiral wound modules. *Desalination* 64:387–401.

Kazama, S., T. Kaneta, and M. Sakashita. 1989. Gas separation membranes of Cardo type polyamides. Paper read at International Symposium on Gas Separation Technology, 10–15 September, at Antwerp, Belgium.

Kesting, R.E. 1989. The nature of pores in integrally skinned phase inversion membranes. In *Advances in reverse osmosis and ultrafiltration*, edited by T. Matsuura and S. Sourirajan. Ottawa: National Research Council of Canada, pp. 1–13.

Konagaya, S., H. Kuzumoto, K. Nita, M. Tokai, and O. Watanabe. 1989. New reverse osmosis (RO) membrane materials with high chlorine resistance. Paper read at International Conference on Membrane Separation Processes, 24–26 May, at Brighton, England.

Kurihara, M., N. Harumiya, N. Kanamaru, T. Tonomura, and M. Nakasatori. 1981. Development of the PEC-1000 composite membrane for single-stage seawater desalination and the concentration of dilute aqueous solutions containing valuable materials. *Desalination* 38:449–460.

Kurihara, M., T. Uemura, Y. Nakagawa, and T. Tonomura. 1985. The thin-film composite low pressure reverse osmosis membranes. *Desalination* 54:75–88.

Leenaars, A.F.M., K. Keizer, and A.J. Burggraaf. 1985. Structure, permeability, and separation characteristics of porous alumina membranes. In *Reverse osmosis and ultrafiltration*. ACS Symposium Series 281, edited by S. Sourirajan and T. Matsuura. Washington, D.C.: American Chemical Society, pp. 57–68.

Loeb, S., and S. Sourirajan. 1963. Seawater demineralization by means of an osmotic membrane. *Advances in Chemistry Series* 38:117.

Marcinkowsky, A. E., K.A. Kraus, H.O. Phillips, J.S. Johnson, Jr., and A.J. Shor. 1966. Hyperfiltration studies. IV. Salt rejection by dynamically formed hydrous oxide membranes. *Journal of the American Chemical Society* 88:5744.

Matsuura, T., and S. Sourirajan. 1989. Preferential sorption—capillary flow mechanism and surface force-pore flow model-applicability to different membrane separation processes. In *Advances in reverse osmosis and ultrafiltration*, edited by T. Matsuura and S. Sourirajan. Ottawa: National Research Council of Canada, pp. 139–175.

Mehdizadeh, H., and J.M. Dickson. 1989. The role of membrane potential functions in determining reverse osmosis transport phenomena. In *Proceedings of Second International Conference on Separation Science and Technology, Hamilton, Canada, October 1–4, 1989*, edited by M.H.I. Baird and S. Vijayan. Ottawa: Canadian Society for Chemical Engineering, pp. 9–17.

Nguyen, T.D., T. Matsuura, and S. Sourirajan. 1987a. Effect of non-solvent additives on the pore size and the pore size distribution of resulting aromatic polyamide membranes. *Chemical Engineering Communications* 54:17–36.

Nguyen, T.D., T. Matsuura, and S. Sourirajan. 1987b. Effect of the casting solution composition on pore size and pore size distribution of aromatic polyamide RO membranes. *Chemical Engineering Communications* 57:351–369.

Nguyen, T.D., T. Matsuura, and S. Sourirajan. 1990. Effect of iso- and tere-phthaloyl content on the pore size and the pore size distribution of aromatic polyamide RO membranes. *Chemical Engineering Communications* 88:91–104.

Panar, M., H. Hoehn, and R. Hebert. 1973. The nature of asymmetry in reverse osmosis membranes. *Macromolecules* 6:777.

Parise, P.L., B.S. Parekh, and R. Smith. 1985. Development of sanitary reverse osmosis systems for the pharmaceutical industry. In *Reverse osmosis and ultrafiltration*. ACS Symposium Series 281, edited by S. Sourirajan and T. Matsuura. Washington D.C.: American Chemical Society, pp. 297–312.

Paulson, D.J., B.W. Phelps, and G.J. Gach. 1989. Design innovations for processing high-fouling solutions with spiral-wound membrane elements. In *Advances in reverse osmosis and ultrafiltration*, edited by T. Matsuura and S. Sourirajan. Ottawa: National Research Council of Canada, pp. 499–515.

Rozelle, L.T., J.E. Cadotte, K.E. Kobian, and C.V. Kopp, Jr. 1977. Nonpolysaccharide membranes for reverse osmosis: NS-100 membranes. In *Reverse osmosis and synthetic membranes*, edited by S. Sourirajan. Ottawa: National Research Council of Canada, pp. 249–261.

Schultz, R., and S. Asunmaa. 1970. Ordered water and ultrastructure of the cellular plasma membrane. *Recent Progress in Surface Science* 3:291–332.

Sourirajan, S. 1970. *Reverse osmosis*. New York: Academic.

Sourirajan, S., and T. Matsuura. 1985. *Reverse osmosis and ultrafiltration/process principles*. Ottawa: National Research Council of Canada.

Toray. 1987. Toray "ROmembra" UTC membrane reverse osmosis module (SU-Series). *Toray Technical Bulletin*, June 1987.

van Reenen, A.J., and R.D. Sanderson. 1989. Copolymers for dynamically formed membranes with enhanced rejection and flux properties. In *Advances in reverse osmosis and ultrafiltration*, edited by T. Matsuura and S. Sourirajan. Ottawa: National Research Council of Canada, pp. 587–598.

van der Waal, M.J., and I.G. Racz. 1989. Mass transfer in corrugated-plate membrane modules. I. Hyperfiltration experiments. *Journal of Membrane Science* 40:243–260.

Wang, Y., X. Xu, and J. Wang. 1986. Controlling pore diameter in dynamically-formed membranes. In *Proceedings of the International Membrane Conference on the 25th Anniversary of Membrane Research in Canada, September 24–26, 1986, Ottawa*, edited by M. Malaiyandi, O. Kutowy, and F. Talbot, Ottawa: National Research Council of Canada, pp. 129–144.

Wijmans, J.G., S. Nakao, and C.A. Smolders. 1984. Flux limitation in ultrafiltration: osmotic pressure model and gel layer model. *Journal of Membrane Science* 20:115–124.

Zeltner, W.A., and M.A. Anderson. 1989. Chemical control over ceramic membrane processing: promises, problems and prospects. Paper read at the First International Conference on Inorganic Membranes, 3–6 June, at Montpellier, France.

Zhang, S. Q., A. E. Fouda, T. Matsuura, and K. Chan. 1991. Some experimental results and design calculations for reverse osmosis concentration of green tea juice. *Desalination* 80:211.

3

The Economics of Desalination Processes

Gregory A. Pittner

Arrowhead Industrial Water, Inc.
A BFGoodrich Company

INTRODUCTION

Most chemical processes can be carried out by more than one means, and desalination is no exception. Several techniques are commonly used, differing in the relative amounts of equipment, energy, labor, and maintenance each requires. The most economically attractive process is determined by the nature of the particular application. For instance, a short duration desalination requirement would most appropriately be satisfied by a process that has low capital requirements but higher operating cost for energy, labor, and maintenance.

A proper economic comparison of alternatives requires two basic ingredients. The first is intimate knowledge of each process, so that all cost ingredients can be ascertained with sufficient accuracy. The second requirement, which often receives less attention than it should, is the use of a meaningful set of rules and methods for measuring the relative costs. Not all methods of differentiating capital projects are equally suitable.

Many difficulties arise in evaluating costs which can lead to erroneous results. To determine the economics of all available competing processes, each must be designed in sufficient detail to firmly establish costs. Sometimes shortcuts are taken, which can cause capital requirements to be underestimated or operating difficulties to be overlooked. Depending on the extent of the inaccuracy, it is quite possible for the outcome of a comparison to be in error. Typical areas of difficulty frequently encountered include the forecast of corrosion rates and leakages in distillation equipment. In reverse osmosis (RO), an incorrect estimate of membrane-fouling rates can cause cleaning and membrane replacement costs to exceed the original estimate by 200 percent or more. The economic consequences of these inaccuracies can be devastating.

A second source of difficulty relates to the state of the art of the technology. Economic comparisons should always be conducted without regard to the likelihood that changes in the state of the art will occur in the future. For instance, the technology of distillation is changing very slowly, as compared with RO, in which membrane improvements have been frequent. Although improvements may have been rapid in the past, one should not assume that this rate of improvement will continue in the future. Except in very unusual circumstances, the probability of changes in the state of the art should not be reflected in a relative cost analysis.

In the event that a financial evaluation leads to a relatively close rating of two or more alternatives, it is necessary to do a sensitivity analysis to determine the degree to which an inaccuracy in the estimate would change the outcome. For instance, if a 20 percent change in the expected RO feed-water pressure requirements would upset the comparison between RO and electrodialysis, a firmer understanding of the projected RO membrane operating characteristics may be in order.

Finally, whichever method of financial comparison is used, it should be suitable for the audience for which it is being prepared. Many corporations have standardized methods for financial analysis for capital projects, allowing all projects to be compared on an equal basis.

METHODS OF FINANCIAL ANALYSIS

The advantages and disadvantages of four common methods of differentiating capital projects are reviewed in this chapter: Payback Period, Accounting Rate of Return, Net Present Value, and Internal Rate of Return.

Payback Period

Of the four methods of differentiation, Payback Period is probably the simplest to calculate. It is defined (Brigham and Gapenski 1985) as the time required for the cash flows of a project to sum to zero. In other words, the time required for net income to equal the sum of capital and operating costs is equal to the payback period.

Although generally applied to situations in which an investment leads to reduced costs or increased profits, the payback period is easily used to compare alternatives that provide the same benefit—in this case, desalination of a fixed quantity of water. Assume Projects A and B, with Project A having a capital cost of $2,000,000 and operating costs of $10,000 per month, and Project B with a capital cost of $1,500,000 and operating costs of $24,000 per month. The differential capital requirement is $500,000, representing a negative cash flow when using Project A as the basis. The difference in operating costs is $14,000 per month, considered a positive cash flow, resulting in a payback period of 35.7 months. If this period meets the required standards for project financing, Project A, with the higher capital costs, would be chosen over Project B. If this payback period is in excess of the maximum allowed, Project B (or neither project) should be undertaken.

The concept of Payback Period can be applied quickly to simple situations. In complex situations, this method is likely to be inadequate or to provide an unsuitable answer. One notable shortcoming can be eliminated by the calculation of the Discounted Payback Period (Brigham and Gapenski 1985) in which all net cash flows are discounted for the time value of money using an appropriate interest rate, before calculating the period over which the cash flows equal the original investment.

Accounting Rate of Return

The Accounting Rate of Return (ARR) (Brigham and Gapenski 1985) is also easy to apply; however, it is not sufficiently accurate for many comparisons. The ARR is equal to the ratio of the average annual income of the project divided by the average investment, with the result expressed as a percentage. Average annual income is defined as average cash flow, less annual depreciation expense. This is divided by the amount of the average investment, defined as the original cost of the project, less 50 percent of the expected total depreciation.

Referring to the previous example of Projects A and B, Table 3.1 has been constructed to show the cash flows of each, together with a third, Project (A-B), which is an imaginary project made up of the differential between Projects A and B. The average annual income for Project (A-B) is equal to the average cash inflow, $168,000 per year, less the annual depreciation, which would be $100,000 per year assuming no salvage value at the end of the 5-year service life. This would provide an average annual net income of $68,000. The average investment is equal to the capital investment of $500,000, less 50 percent of the

TABLE 3.1   Application of Accounting Rate of Return Method

	Project A	Project B	Project (A-B)
Capital Investment	($2,000,000)	($1,500,000)	($500,000)
Operating Costs			
Year 1	(120,000)	(288,000)	168,000
Year 2	(120,000)	(288,000)	168,000
Year 3	(120,000)	(288,000)	168,000
Year 4	(120,000)	(288,000)	168,000
Year 5	(120,000)	(288,000)	168,000

depreciation, equaling $250,000. The accounting rate of return is therefore $68,000 divided by $250,000, which equals 27.2 percent.

A serious flaw in this method is similar to that in the Payback Period. ARR does not account for the time value of money, valuing income from future years to the same extent as income received in earlier years.

Net Present Value

The Net Present Value (NPV) (Brigham and Gapenski 1985) of a project recognizes the changes in value of money depending on the time at which that money is received. The NPV method evaluates the present value of the cash flow stream evaluated at an appropriate cost of capital for the project, determined over the entire life of the project. The sum of all discounted cash flows is added to obtain a total NPV. If the NPV is greater than zero, after accounting for capital requirements, the project should be accepted. If two projects have different NPVs, the project with the greater NPV should be selected.

Table 3.2 has been constructed to provide an example of the manner in which NPV would be applied to Project (A-B). The differential capital outlay of $500,000 occurs in year 0, before the first operation of the system. Operating costs are assumed to occur evenly during each year and, thus, for the purposes of discounting, are treated as if occurring at the end of the sixth month of the year. Discounting was performed at an interest rate of 20 percent, which would be a reasonable rate of return for a low-risk industrial venture. Since the resulting NPV is greater than zero, Project B would be preferred over Project A.

Internal Rate of Return

The method of comparison believed by some to be the most informative is the Internal Rate of Return (IRR) (Brigham and Gapenski 1985), which is defined as the discount rate at which the NPV of all cash flows equal zero. This method can be the most difficult to calculate but provides very useful information.

Returning to the previous example of Projects A and B, Table 3.3 has been

TABLE 3.2   Application of the Net Present Value Method to Project (A-B)

	Year 0	Year 1	Year 2	Year 3	Year 4	Year 5
Capital:	($500,000)	-0-	-0-	-0-	-0-	-0-
Operating Costs:	-0-	$168,000	$168,000	$168,000	$168,000	$168,000
Discounted Operating Costs:		$152,727	$152,727	$106,060	$ 88,382	$ 73,652
NPV:	$48,093					

TABLE 3.3   Application of the Internal Rate of Return Method to Project (A-B)

	Year 0	Year 1	Year 2	Year 3	Year 4	Year 5
Capital:	($500,000)					
Operating Costs:	-0-	$168,000	$168,000	$168,000	$168,000	$168,000
Discounted Operating Costs:						
at 20%		$152,727	$127,272	$106,060	$ 88,382	$ 73,652
at 22%		$151,351	$124,058	$101,687	$ 83,350	$ 68,320
at 24%		$150,000	$120,967	$ 97,554	$ 78,673	$ 63,446
at 26%		$148,672	$117,994	$ 93,646	$ 74,322	$ 58,986
NPV at 20%	$48,093					
NPV at 22%	$28,766					
NPV at 24%	$10,640					
NPV at 26%	($6,380)					

constructed to show the calculation of the IRR for Project (A-B). Because the IRR is equal to the interest rate at which the NPV is 0, the NPV has been calculated for several interest rates ranging from 20 to 26 percent, over which it changes from a positive $10,640 at 24 percent to a negative $6,380 at 26 percent. Based on simple interpolation, the IRR is found to be approximately 25.25 percent.

ENERGY CONSUMPTION IN DESALINATION

All desalination processes must satisfy the four laws of thermodynamics, which means simply that all methods consume energy. In addition, thermodynamics can be used to calculate the minimum energy required for separation, which is equal to the energy released upon dissolution of the salts being removed. Unfortunately for those paying for the power, the actual energy consumed in desalination typically exceeds the minimum energy by several hundred percent. The osmotic pressure of a typical supply of brackish water is equal to approximately 10 psi per 1,000 parts per million total dissolved solids (TDS) (as sodium chloride). One could theoretically desalinate water using RO (assuming a perfect RO membrane) with a supply pressure just slightly above the osmotic pressure, accomplishing separation at approximately the thermodynamic minimum energy. As a practical matter, however, pressure levels of 2,500 percent of the minimum are commonly employed to achieve separation in real situations. Similarly, distillation processes are conducted at temperatures that exceed the boiling point of the brine by 20° or more. This is done to conduct heat more rapidly, using a smaller heat-transfer surface area. In summary, the extent to which actual operating conditions differ from thermodynamic minimum conditions is a measure of the extent to which the actual energy exceeds the thermodynamic minimum energy. The potential for energy consumption can also be judged by the difference between normal operating conditions and thermodynamic minimum conditions. By this measure, RO shows considerable room for cost improvement, whereas ion exchange and distillation demonstrate less potential (Smith 1959).

OVERVIEW OF DESALINATION PROCESSES

Reverse Osmosis

RO involves separation of dissolved and particulate matter from water by selective passage of the water through a membrane, constructed so that dissolved ions, large organic molecules, and particles are excluded. Currently available membranes can eliminate as much as 99 percent of the mineral content of the water in one pass. Because the cost of RO treatment increases very little with increased TDS, it is economically attractive for the desalination of both brackish water supplies at TDS concentrations of 100 parts per million and seawater supplies at 45,000 parts per million.

A typical RO system consists of pretreatment, high-pressure feed pump(s), an array of membranes housed in pressure vessels, and a network of piping and valves to divide the feed water and collect the permeate product. A typical RO system is shown in Figure 3.1. A cartridge filter is always included, primarily to remove large particles that could block the relatively small spaces between the membrane layers, but also to reduce particulate fouling of the membrane surface. The membrane arrangement shown in Figure 3.1 is referred to as a 3-2-1 array, described as such because the brine from the first three vessels is combined and divided to feed the second two, and the brine from the second two vessels is subsequently combined to feed the last vessel. This particular array is often used in practice because a relatively high water recovery can be ob-

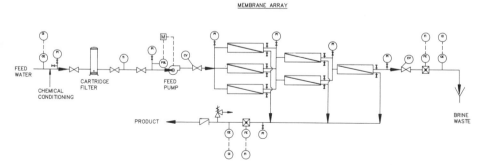

FIGURE 3.1.   Typical reverse osmosis system.

tained while maintaining brine flow rates within the minimum and maximum limits imposed by the membrane manufacturer.

Ion Exchange

Ion exchange (IX) has been widely used for many years; thus it has been established as one of the most important desalination processes. IX utilizes a synthetic plastic resin containing fixed chemical groups that interact with either the anions or the cations dissolved in the water. When water is treated by this process, undesirable ions from the liquid phase are exchanged for more desirable ions and held within the resin bed. Complete desalination is accomplished by exchanging sodium, calcium, magnesium, and other metal cations for hydrogen ions; and exchanging anions such as chloride, sulfate, or phosphate for hydroxide. The hydrogen and hydroxide then combine to form water. Because undesirable salts are removed from the water and fixed in the resin bed, exhaustion of the available resin sites eventually results. The resins can be reused for another cycle after regeneration, in which the exchange is reversed by contacting the cation resin with acid, and the anion resin with sodium hydroxide.

IX is not normally used to remove salts from water containing more than 1,000 parts per million of dissolved solids. Because the regeneration frequency is in direct proportion to the TDS of the feed water, the amount of product water produced per pound of regenerant chemical consumed decreases sharply with increasing TDS. Moreover, each regeneration consumes substantial amounts of water, so at higher TDS levels the amount of waste water can equal or exceed the quantity of product water.

Figure 3.2 illustrates a typical IX system. The unit depicted is a simple two-bed system consisting of a cation exchanger and anion exchanger. Each

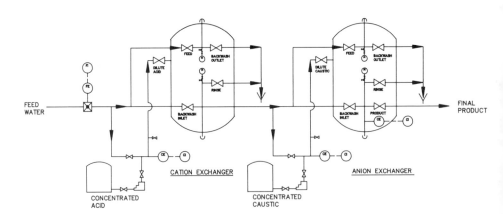

FIGURE 3.2.   Typical ion exchange system.

vessel has its own regenerant preparation equipment in which the concentrated chemical regenerants are diluted prior to contact with the exchange resin. Most of the piping, valving, and controls provided with an IX system is required for the regeneration process. The regeneration sequence involves, first, a backwash step to remove accumulated suspended solids and purge the bed of resin fines. The next step is chemical addition, followed by a rinse prior to return to service.

IX is a batch process, and the chemistry and physical design of the apparatus is complicated. Equipment purchase costs can vary by a factor of 3 or more owing to differences in the quality of the construction, optional equipment, and sophistication of the control and monitoring instrumentation. Cost estimating for an IX system requires a complete understanding of the construction standards to be employed in the system.

Electrodialysis

Electrodialysis (ED) involves the removal of ions from water by transport through an ion-permeable membrane, driven by the force of an electrical field. An ED unit is depicted in Figure 3.3, and is normally constructed with hundreds of compartments operated in parallel, sandwiched together into a large array. Within one-half of the compartments the water is demineralized, whereas the salt content of the water in the adjacent compartments is increased. Note that the desalination process occurs during the flow of feed water past, not through, the ED membrane. This eliminates many membrane-fouling mechanisms common

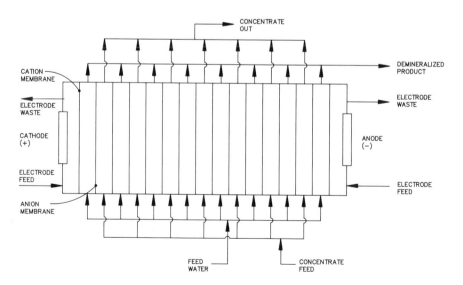

FIGURE 3.3.   Typical electrodialysis unit.

to RO, but there is no improvement in the particulate quality of the water. Ion transport consumes electrical energy in direct relation to the amount of salt transported from the feed water, governed by Ohm's law and Faraday's law. The efficiency is controlled by the effectiveness of the membranes in conducting electricity and blocking the passage of oppositely charged ions.

Because the degree of desalination normally achieved in one stage is about 50 percent, several stages in series must be utilized to obtain high degrees of separation. Designed for only partial TDS removal, ED is efficiently applied in situations that require 50 percent to 75 percent salt removal, including production of potable water from a brackish supply or pretreatment prior to IX (Rice 1991). ED may also be used for polishing the product of a seawater RO unit, from which even high rejection RO elements cannot always provide the degree of separation needed to achieve a product of potable quality.

Depending on the quality of the water supply, salts can form at the surface of the membranes, causing process inefficiencies. A refinement of ED, generally referred to as electrodialysis reversal (EDR), is thus frequently employed. In EDR the electrical potential of the terminals alternates approximately every 15 minutes (Meller 1984), changing each compartment from demineralization service to brine formation and eliminating salt precipitation. Each time the polarity of the terminals is alternated, the entire unit must be thoroughly flushed.

ED is not used for seawater desalination because the amount of electricity consumed is in direct proportion to the amount of salt removed. Evaporation and RO are less sensitive to the feed-water salt concentration and are, therefore, considerably more economical. ED and EDR act only on ionized materials. Thus, uncharged materials, including many organics, are not removed by this process.

Phase Change Processes

Water can be separated from dissolved contaminants by two processes involving a phase change; these are evaporation and crystallization. In evaporation, heat is added to cause the water to vaporize. Dissolved materials are less soluble in the vapor phase because the greater distance between the molecules does not allow for the hydrogen bonding that stabilizes ions in liquid water. The solubility of the dissolved salts in steam is so vastly less than in water that the separation factor of this process, defined as the concentration of the salts in the vapor phase divided by the concentration of the salts in the liquid phase, is $1:1000$ or greater.

Multiple Effect Evaporation

The details of the steps in which water is converted to vapor, and is then recondensed, differ from plant to plant. A common configuration is known as multiple-effect evaporation, where each "effect" of the system is equal to one distillation unit. The feed water to all but one unit is the brine from an adjacent

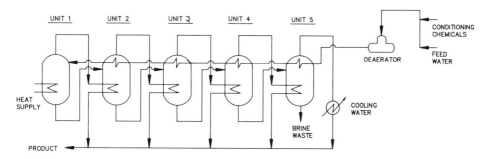

FIGURE 3.4.   Multiple-effect evaporator.

unit. The heat source is product vapor, which is received from the adjacent unit on the opposite side. Figure 3.4 provides a flow diagram of a multiple-effect evaporator. Each effect operates at a different temperature; the lowest temperature is in the fifth unit in the series. Using product vapor as the heat source for four of the five units, this type of system achieves high energy efficiency. Whereas 1 pound of steam would be required to produce a pound of product in a single-effect evaporator, the efficiency of a multiple-effect system is proportional to the number of effects. A five-effect system would produce slightly less than 5 pounds of product per pound of steam supplied.

Vapor Recompression

A second evaporation method, which may be cost-effective where electricity is readily available, is vapor recompression. In this process, water is evaporated and then compressed, causing it to condense at a higher temperature, thus providing a source of latent heat for further evaporation. The temperature difference between the vapor and the liquid is normally kept relatively small, reducing compression requirements but requiring a relatively large heat-exchange surface area. A diagram of this system is shown in Figure 3.5 (King

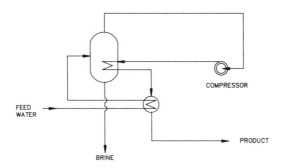

FIGURE 3.5.   Vapor recompression distillation system.

1971). This process is appropriate for producing relatively small amounts of purified water where thermal energy is not readily available.

Flash Evaporation

Both vapor recompression and multiple-effect distillation require heat transfer through an isolating medium, such as a metal tube. A third alternate is flash evaporation, in which the water is first heated under pressure and then the pressure is suddenly reduced. This results in the evaporation of a percentage of the water. A typical plant configuration is shown in Figure 3.6 (Aqua Chem a). Flash evaporation is generally accomplished in several stages, each stage involving relatively small temperature and pressure differences (King 1971).

Unfortunately, the percentage of the feed water that is evaporated in a multistage flash plant is relatively small, whereas the pretreatment and conditioning requirements are similar to those for a large quantity of water per pound of product. A flash evaporation plant requires approximately four times the number of stages required by multiple-effect evaporators to achieve the same steam efficiency; however, a single vessel can house numerous stages, greatly reducing the cost per stage.

An alternate method of economizing the basic multistage flash plant involves reuse of a portion of the brine waste as feed, thus reducing the amount of feed water that must be pretreated. The heat source for one multistage flash plant can be the product from a second plant, providing an overall system in which the concepts of multistage flash and multiple-effect evaporation are combined to achieve greater utilization of heat energy and reduced construction costs (Aqua Chem b)

Crystallization

Freezing can also be used as a method of separation, as the crystal formation process excludes foreign materials from the solid lattice. This generates a two-phase system of liquid brine and pure water ice. The two are then physically separated, and purified water is recovered by melting.

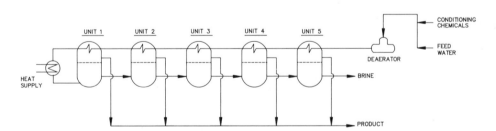

FIGURE 3.6.   Multistage flash evaporator.

Freezing is not employed as a desalination technique nearly as often as evaporation owing to the difficulties involved in solids handling and separating the ice crystals from the liquid. The maximum purity that can be obtained in one stage is controlled primarily by the extent to which the ice crystals can be separated from the brine. Cooling and heating are also complicated, inasmuch as heat transfer to and from solids occurs more slowly than with liquids.

Figure 3.7 shows a typical crystallizing desalination unit. This system has the added feature of a vapor recompression unit, which removes sufficient vapor from the refrigeration unit to both cool the feed water and provide latent heat for melting the final product (King 1971).

TYPICAL CAPITAL AND OPERATING COSTS

Reverse Osmosis

The total cost for any desalination process is the total of the operating costs, plus depreciation expense for the equipment and installation, and interest charges on the money required to build the system (Birkett 1988).

The equipment costs are the total of costs for the purchase of the RO unit itself, pretreatment, polishing, and servicing equipment. Some pretreatment is normally required upstream of the RO unit. This may include granular media filtration, activated carbon filtration, IX softening, clarification, or ultrafiltration. Polishing equipment may consist of ion exchange for the production of pure water, or a second RO unit or ED system may be used when a lower degree of polishing is needed. The IX polishing unit is most frequently a mixed bed, which

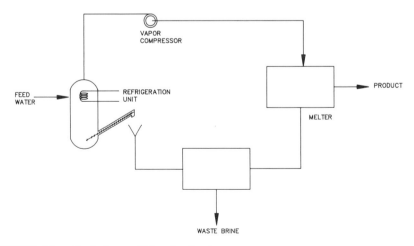

FIGURE 3.7.   Typical crystallizing desalination system.

is generally less expensive than a two-bed system and can provide a higher quality product.

A rule of thumb, in common use since the 1970s, is that the cost of a brackish RO unit can be approximated at $1.00 per gallon per day of capacity. A unit capable of producing 120,000 gallons per day would therefore be priced at $120,000. Remarkably, over the last 15 years, improvements in membrane performance and efficiency in design and construction of RO equipment have largely offset inflation so that this standard remains valid.

The cost for pretreatment and polishing equipment varies substantially from case to case. Most well water sources do not require extensive pretreatment, whereas seawater desalination systems require careful design of intake structures and filtration units to control the suspended solids concentration (Polidoroff 1987). Assuming a water supply that requires moderate pretreatment, the cost can be estimated at 30 percent of the cost of the RO unit (Epstein 1978).

Polishing costs are also highly variable, but if a regenerable mixed-bed ion exchange system is utilized, its cost can range from 30 to 50 percent of an RO system. The alternative of acquiring service-contract ion exchange polishing is becoming increasingly popular owing to economics in capital and operational expenses.

Installation costs for an RO unit and associated equipment typically equal 30 percent of the total equipment cost. The total site expense, including real estate, foundations, buildings, and support facilities, can generally be estimated at 150 percent of the equipment cost.

Some cost elements are variable, and thus change in direct proportion to the amount of product produced. Others are fixed, required at the same rate whether the plant is operating or not. The operating cost for a complete RO system can be broken down as described in the following paragraphs.

Chemical Conditioning

Several chemicals may be used in an RO plant. The pretreatment system may require chemical coagulants such as alum or ferric chloride, flocculating poly-electrolytes, or chlorine. If the system includes a softener, salt will be required. A granular media filter should not, under most circumstances, require frequent replacement of the filter media. However, an activated carbon filter will require media replacement at an interval that is subject to variation with the quality of the influent water. If the activated carbon unit is being used primarily for the removal of dissolved organics, the exchange can be as frequent as once every month. Carbon filters are often used for removal of chlorine prior to thin film composite RO units. In such cases, the activated carbon reacts with the active chlorine, forming chloride and converting the carbon to carbon dioxide. As a result, the volume of carbon in the vessel should be checked at least once a year to ensure that sufficient carbon remains for adequate chlorine control. In some cases, the carbon may become ineffective through exhaustion with dissolved organics. Thus, regular analysis for free and combined chlorine in the carbon filter product is required (Coulter 1980).

Immediately upstream of an RO unit, chemical conditioning agents such as sulfuric acid may be required to lower the pH in order to avoid calcium carbonate precipitation and to minimize the rate of hydrolysis in cellulose acetate membranes; a reducing agent such as sodium bisulfite may be needed to eliminate trace oxidants; an antiscalant may be required to control precipitation of sparingly soluble salts, including strontium sulfate, barium sulfate, and calcium carbonate; and, in some cases, a dispersant is beneficial to reduce the fouling of the membrane with suspended solids (O'Brien and LaTerra 1981). The cost of chemicals is subject to fluctuation with market conditions, packaging requirements, and shipping expense. Thus, the costs for these agents should be determined for each location. All chemical costs are generally considered variable.

Electricity

Electric power is an important cost component, as it provides the energy for the operation of the main feed pump, which provides the water pressure that drives the separation process. The electrical requirement for a single-stage, thin film composite RO can be as low as 4 kWh per 1,000 gallons of product, assuming a pump efficiency of 78 percent and a motor efficiency of 93 percent. Secondary pumps required to operate pretreatment and polishing units add 1 to 2 kWh per 1,000 gallons of product (Hickman and Conlon 1984). A double-pass RO system, including all auxiliary equipment, can be operated with 8 kWh per 1,000 gallons of product (Hickman and Conlon 1984).

A seawater RO unit requires a higher operating pressure and is generally operated at a lower recovery, both acting to increase the electrical consumption to the range of 25 to 30 kWh per 1,000 gallons of product. Unless the cost of electricity is unusually low, brine energy recovery devices (which can reduce electric consumption by up to 45 percent) may be desirable. Electricity costs can vary widely from site to site. In 1991 the range of $.07 to $.12 per kilowatt hour covered 90 percent of applications (Polidoroff 1988).

Membrane Cleaning and Replacement

Membrane replacement typically accounts for 6 to 10 percent of the total cost of producing water by RO. The membrane itself should be considered a wear item, which chemically deteriorates from attack by the chemicals in the influent water and the chemicals used for cleaning (Vera 1981).

The cost of the first charge of membranes furnished with the original equipment is normally included in the initial purchase cost, so must be separated out and depreciated over a shorter lifetime, usually 3 to 5 years. The cost of membranes for a brackish system (based on 8″ × 40″ thin film composite modules producing 5,000 gallons per day and assuming $1,200 per module and a 3-year life) would be $.31 per 1,000 gallons of product. The cost of membranes for a seawater system, assuming the same operating conditions, would be $.44 per 1,000 gallons of product.

Membrane cleaning is normally required every 6 months. However, systems equipped with a very effective pretreatment process, or operated on high-quality water supplies, have been known to operate for 3 or more years between cleanings. Unfortunately, the other extreme is also encountered, with some systems requiring cleaning as frequently as once a month. On average, when cleaning an RO system containing more than 50 modules, $50 per membrane for chemicals and one-half hour of labor should be assumed for each $8'' \times 40''$ module.

Operating Labor

Operating labor is required to take readings of pressure, conductivity, and flow and to adjust valves to ensure that membranes are consistently operated within design parameters. Depending on the size of the RO system, operating labor can vary from 5 hours a week for a unit of less than 10,000 gallons per day capacity, to 3 hours a day for units of 400,000 gallons per day. In the event that the influent water quality is poor or highly variable, the amount of operating labor can easily be twice the norm.

Repair and Maintenance

In general, 2 percent of the total equipment costs can be assumed for repair and maintenance of an RO system with pretreatment and polishing units. This covers repair and calibration of instruments, rebuilding pumps, and replacement of O-ring seals on the interconnector between individual RO elements. Because of the extensive corrosion of piping in seawater systems, a cost of 4 percent should be used to estimate these systems.

Polishing Costs

Estimation of IX polishing costs is a complicated process. The cost elements include acid and caustic for regeneration of the resins, resin rebed and attrition losses, operating and maintenance labor, and neutralizing chemicals if required to neutralize the regenerate waste prior to discharge. Total polishing costs range from $0.30 to $1.00 per kilograin removed.

Ion Exchange

An IX unit generally includes a pretreatment section, a main IX unit, and a neutralization system. Pretreatment in many cases is not required. However, carbon filtration is frequently used on potable water supplies for chlorine removal, and granular media filters are used for surface water supplies to remove small amounts of turbidity. In cases of higher turbidity, clarification systems with chemical feed, followed by filtration, are employed.

The IX system itself can consist of a single cation unit and single anion unit; the two together are referred to as one train. Additional trains may be included; some systems consist of eight or more. If the product water from a two-bed system is not sufficient, a mixed-bed unit may be employed for polishing. Often one mixed-bed unit is furnished per train. However, two mixed beds may be sufficient for three or four trains, depending on the flow rates of the application.

The size of each IX unit can be controlled by the regeneration frequency or by flow rate. The usual and customary flow rate for IX resin is 2 to 5 gallons per minute (gpm) per cubic foot, but higher flow rates can be used. Because the regeneration process requires approximately 4 hours for a single train, it is unusual to design a system that requires no more than two regeneration sequences per day.

To estimate the cost of an IX system with a reasonable degree of accuracy, it is necessary to determine the standard to which the units will be constructed. Less expensive systems utilize fiberglass vessels with PVC multiport valves, controlled by a simple electromechanical panel. Alternatively, the equipment can be constructed from ASME code-stamped pressure vessels, which are lined with $\frac{3}{16}$-inch natural rubber or a similar coating, equipped with face piping of either #16-inch stainless steel or polypropylene-lined steel, and include control systems that are programmable and capable of fully automatic operation with manual backup. The cost of these systems can vary by a factor of 3 or more. The less expensive systems operate for 5 to 10 years, during which time replacement of key components may be required. It is not unusual for higher quality systems to operate for 20 to 30 years without overhaul.

Because of the wide variation in the cost of IX equipment, any cost estimating standards must be applied with great care. In the absence of better information that would provide for a more accurate cost estimate, the cost of an IX system can be estimated at approximately seven times the IX resin cost. In 1991 the cost for strong acid cation resin was approximately $45 per cubic foot, and the cost of strong base Type 2 anion resin was approximately $160 per cubic foot. A system that required 100 cubic feet of each type of resin would therefore cost approximately $140,000, equal to seven times $20,500. This rule should be restricted to units containing between 100 and 500 cubic feet of resin to avoid substantial inaccuracy.

The installed cost of a system includes the costs for real estate, buildings, and site installation. Installation costs usually are approximately 30 percent of the purchase cost of the equipment, including both the IX system and any pretreatment or neutralization equipment. When all site costs are included, the total plant cost is approximately 250 percent of the equipment cost.

Operating costs of an IX system are discussed in detail in the following paragraphs.

Energy Costs

Electrical energy is required for pumping water through IX units. Although the pressure drop is normally small (in the range of 20 pounds per vessel), power constitutes 3 to 5 percent of the operating costs of the system.

Chemical Costs

Chemicals for regeneration make the greatest contribution to the operating cost of the system. Chemicals normally cost between 6 and 25 percent of the total, varying in direct proportion to feed-water quality.

Resin Attrition and Replacement

IX resin degrades chemically, causing it to lose active chemical groups and thus exchange capacity. The plastic beads on which the chemical groups are attached also depolymerize, resulting in swelling and increasing moisture content, which also decreases the capacity per cubic foot. Moreover, a portion of the resin is lost from the bed during each backwash, resulting in physical losses. Attrition losses each year generally represent 3 to 5 percent of the total quantity of resin. The entire bed must be replaced periodically. Cation units are normally expected to last at least 5 years, whereas anion beds are often replaced every 3 years.

Labor Costs

Labor required to monitor the operation of IX units can be estimated at 2 hours a day per train; slightly less time is required for smaller units in less critical applications. In most systems, additional labor is required to monitor regeneration cycles; 1 to 3 hours of operation labor is assumed per regeneration.

Maintenance

Maintenance costs should be assumed at 2 percent of the capital cost. These costs include replacement of wear components on valves, repair and calibration of instrumentation, and rebuilding of rotating equipment such as mixers, pumps, and chemical feeders.

Taxes and Insurance

In most areas, a rate of 2 percent of the original capital cost of the system is a reasonable estimate of the costs of taxes and insurance. However, because they can contribute 10 percent to the total annual cost, an effort should be made to identify these costs closely for each application.

Distillation/Evaporation

Evaporation is often used for seawater desalination where larger flow rates of water are involved. Table 3.6 contains cost information obtained from a system designed in 1987 for producing up to 566,000 gallons per day of water containing between 10 and 25 ppm of dissolved solids, from seawater containing 37,000 ppm as the feed supply. The plant was designed as a once-through, cross-tube type multistage flash evaporator. It consumed 3,400 gpm of seawater, including cooling water for condensers. Steam consumption was estimated at 36,100 pounds per hour, providing a product rate of 5.45 pounds per pound of steam. The total equipment cost was estimated to be $1,800,000, with an additional $400,000 required for installation. The total facility cost, including real estate, foundation, and seawater intake structure, was estimated to be a total of $4.2 million.

   In Table 3.4, a 20 percent interest rate was used to convert the capital cost to a cost per 1,000 gallons. The labor component assumed that four persons would be employed full time to operate the plant and provide repair labor. The cost for repairs and pump maintenance are for purchased parts only. The steam cost is evaluated at $1.75 per thousand pounds, and electricity cost is evaluated at $.07 per kilowatt hour. The taxes and insurance were assumed to equal 2 percent of the total installed cost.

Electrodialysis

The total cost of an ED system must include pretreatment, the ED unit itself with associated feed pumps, piping and valves, electrical controls, and any polishing

TABLE 3.4   394 gpm Multistage Flash Distillation Unit Capital and Operating Costs per 1,000 Gallons of Product

Chemicals	$0.218
Cartridge filters	$0.016
Pump maintenance	$0.282
Labor	$0.769
Repairs	$0.324
Steam	$2.680
Electricity	$0.840
Total operating costs	$5.129
Taxes and insurance	$0.435
Capital recovery	$4.465
TOTAL COST	$10.029

equipment. Although ED is a membrane process, the water does not flow through the membrane but rather past it, so there is less tendency for suspended solids to collect on the membrane surface. Thus, pretreatment requirements are not as stringent as those for RO systems. The primary operating cost of an ED unit is electrical power. A small amount of chemical conditioning of the water is sometimes required. Periodic membrane cleaning is also necessary (Codina and Curry 1987).

The majority of ED units provided worldwide are currently furnished by the Ionics Corporation in Massachusetts. Most of these units are designed after the principle of EDR, originally patented by Ionics during the 1960s. Table 3.5 shows the capital and operating requirements for a total of eight EDR designs based on four different well water analyses that appear in Table 3.6. At the bottom of Table 3.5, the total cost per thousand gallons has been calculated, based on a 20 percent interest rate, 10-year equipment life, $.07 per

TABLE 3.5   Summary of Electrodialysis Capital and Operating Costs

	Case 1		Case 2		Case 3		Case 4	
TDS removal	50%	80%	50%	80%	50%	80%	50%	80%
EDR cost (1000)	$430	$720	$430	$720	$450	$770	$470	$790
kwh/day[1]	1144	2549	1455	3436	1940	4645	2362	5726
Operation								
Labor/day (hrs)	1	1	1	1	1–5	1–5	2	2
Cleaning cost per month								
Labor	2	2	2	2	3	3	3	3
Chemicals	$1000	$1000	$1000	$1000	$1500	$1500	$2000	$2000
Annual repair and maintenance cost as % of								
EDR cost[2]	5	5	5	5	5	5	5	5
Annual membrane replacement allowance								
Labor (hrs)	1.0	1.0	1.0	1.0	1.5	1.5	2.0	2.0
Material (1000)	$20	$40	$20	$40	$20	$40	$25	$45
Conditioning								
Chemical cost per month	0	0	0	0	0	0	0	0
Total production								
GPD (1000)	858	806	861	806	861	806	861	806
Overall water recovery, %[3]	85	80	85	80	85	80	85	80
Total cost per 1000 gallons	$1.12	$2.02	$1.14	$2.10	$1.28	$2.35	$1.39	$2.55

[1]kWh/day is based on EDR power consumption and total production.

[2]Exclusive of membrane replacement.

[3]Total production and overall water recovery are based on EDR production and blended flow, if any.

TABLE 3.6   Water Analysis for Electrodialysis Cost
Estimation

	Case 1	Case 2	Case 3	Case 4
TDS	480	960	1,440	1,920
Ca^{++}	180	200	220	240
Mg^{++}	200	220	240	260
Na^+	100	540	980	1,420
HCO_3^-	210	220	230	240
SO_4^-	70	140	210	280
Cl^-	191	582	973	1,364
NO_3^-	9	18	27	36
SiO_2	30	30	30	30
TOC (ppm)	3	3	3	3
Turbidity (FTU)	1	1	1	1

kilowatt hour of electricity, labor cost of $30 per hour, and total facility construction cost equal to 250 percent of the equipment cost.

COMPARISON OF RO AND ION EXCHANGE FOR PRODUCTION OF HIGH PURITY WATER

There is a large United States market for the purification of well, surface, and municipal water supplies containing less than 500 parts per million TDS for the production of high purity water suitable for steam generation and a variety of process applications. The usual specifications are for conductivity of less than 0.1 micromhos, silica concentration of less than 20 ppb (parts per billion), sodium less than 50 ppb, and hardness levels less than 10 ppb.

For this particular application, the two most popular processes have been RO and IX. EDR has been sporadically employed at TDS levels within and slightly above 500 ppm. Since the commercialization of RO, improvements in membrane cost, membrane performance, and applications technology have occurred more rapidly than in the field of IX, resulting in changing the comparative costs for these two processes.

In 1987 the Dow Chemical Company prepared RO versus IX cost information, which was presented at the first annual Ultrapure Water Conference and Exposition (Whipple et al. 1987). The methods of analysis and presentation used by Dow have been updated and presented in this publication using current cost information. The original publication included a comparison of RO polished with ion exchange, IX only, and double-pass RO systems. The third alternative, double-pass RO, has been deleted from this presentation for simplicity. However, the double pass concept is viable in applications requiring superior control of particulates and organics.

Design Basis

The conditions for raw water turbidity, flow rate, and system construction quality assumed as a basis for this comparison are important factors in determining the outcome of the analysis. In cases in which the source water requires only moderate amounts of pretreatment, such as granular media filtration, RO has a distinct advantage. Water sources that have marginal quality, however, including many surface water supplies, are suitable for use with IX, but are unsuitable for direct feed to an RO unit. This comparison assumes a water quality suitable for RO with pretreatment consisting of clarification and filtration. The flow rate for the system has been chosen at 1 million gallons per day, equal to 694 gpm, which would be typical of a moderate size chemical processing plant or small refinery. The composition of the four alternate supplies, together with costs assumed for supplies such as energy, chemicals, and labor, are summarized in Table 3.7.

The level of quality assumed for both systems is typical of systems customarily purchased by chemical processing facilities. The IX equipment would utilize polypropylene-lined steel pipe with lined cast iron fittings and rubber-lined steel vessels. Vessel internals would be stainless steel, with little or no PVC. The RO

TABLE 3.7   Assumed Costs and Conditions for Comparison

Case 1:		TDS = 79.3 ppm
Ca^{++} 31.0		HCO_3^- 55.5 SiO_2 (as SiO_2) 5.0
Mg^{++} 32.5		SO_3^- 11.8 Temp 55°F
Na^+ 15.8		NO_3^- 1.5 pH 7.6
Case 2:		TDS = 160 ppm (Case 1 × 2)
Case 3:		TDS = 320 ppm (Case 1 × 4)
Case 4:		TDS = 480 ppm (Case 1 × 6)
Costs: Electricity		$0.07/kWh
Steam		$1.75/1000 lb
Caustic soda		$0.18/lb
Sulfuric acid		$0.038/lb 100% basis
Antiscalant		$1.45/lb
Lime		$0.02/lb
SAC resin		$70.00/cu ft
SBA resin		$170.00/cu ft
8″ × 40″ TFC membranes		$1,200 each
10″ long, 5 micron cartridge filter		$2.00 each
Burdened labor rate		$30.00/hour
Interest rate		0.0%
Variable conditions:		
Cation resin life		5 years
Anion resin life		3 years
Mixed-bed resin life		3 years
Membrane life		3 years
System life		10 years

system would be constructed with PVC piping upstream of the high-pressure feed pump, with stainless steel used for high-pressure lines. PVC would be used for permeate piping from the pressure vessels to the storage tank, and polypropylene lined piping would be used for polishing demineralization units. A factor of 2.5 times the equipment cost has been assumed for real estate, buildings, and installation.

The IX system flow diagram appears in Figure 3.8. For all cases, two trains with a single degasifier and final storage tank is the basis for the equipment cost estimate. Pretreatment consists of sand filters and carbon beds, and a neutralization system with lime feed is included.

By virtue of the water analysis, the anion and cation units of the ion exchange system are out of balance, and thus would not be regenerated simultaneously. Since the bicarbonate content of the water exceeds 50 percent of the total anionic load, exchangeable anions are approximately 50 percent of the exchangeable cations. In Case 1, the load on the anion unit is so small as to have the anion exchanger design driven primarily by flow rate, and not by exchange frequency. Only in Cases 3 and 4 does exchange frequency become important for this system. The strong acid cation unit size is dictated primarily by exchange frequency, especially in Cases 3 and 4. The time between unit regeneration is just over 6 hours for Case 4, which approaches the maximum regeneration frequency considered suitable for most applications. A complete summary of the design specifications for the IX system appears in Table 3.8.

The RO system has been assumed to require more extensive pretreatment than the IX system; specifically, a flocculation clarifier and gravity filter unit are included. The RO system also includes a neutralization unit; however, it is smaller than assumed for the IX system, because of the smaller exchange units. A complete summary of the design conditions of the RO/IX system appears in Table 3.9, and a system flow diagram is illustrated in Figure 3.9.

With the analysis performed at the base conditions (graphed in Figure 3.10),

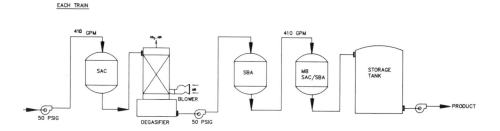

FIGURE 3.8.   Ion exchange purification system.

TABLE 3.8 Assumed Design for Ion Exchange System

	Case 1	Case 2	Case 3	Case 4
Working Cation				
Number	2	2	2	2
Diameter, ft	8.0	9.0	10.0	10.0
Resin, cu ft ea.	175	300	500	660
Acid dose	6.0	6.0	6.0	6.0
Regens per day	1.83	2.22	2.80	2.56
F-D Degasifier				
Number	1	1	1	1
Diameter, ft	8.0	8.0	8.0	8.0
Working Anion				
Number	2	2	2	2
Diameter, ft	8.0	8.0	8.0	9.0
Resin, cu ft ea.	175	175	200	254
Caustic dose	6.0	6.0	6.0	6.0
Regens per day	0.67	1.36	2.48	3.05
Polish Mixed Beds				
Number	2	2	2	2
Diameter, ft	7	7	7	7
Cat. Resin, cu ft ea.	81	81	81	81
An. Resin, cu ft ea.	54	54	54	54
Acid dose	8.0	8.0	8.0	8.0
Caustic dose	6.0	6.0	6.0	6.0
Regens per day	0.03	0.06	0.12	0.24

TABLE 3.9 Assumed Design for Reverse Osmosis/IX System

Membrane Array
    Type: FilmTec thin-film composite
               BW30-8040, spiral wound
    Number: 204
    Rejection after 3 years: 97% at 2000 ppm NaCl feed
    Recovery 75%, two stages
Feed Pump:
    Type: Centrifugal
    TDH: 285–300 Psig
Polish Ion Exchange System:
    Number: 2
    Diameter, ft: 7
    Cation resin, cu ft ea.:   127
    Acid dose: 8
    Anion resin, cu ft ea.:   191
    Caustic dose: 6
    Regens per day: .072–.632

Labor costs for both systems were evaluated using the same standard. It was assumed that 2 hours per day per ion exchange train would be required for monitoring, an additional 3 hours were estimated for each regeneration sequence, and 5 hours per day were assumed for monitoring the RO system. These labor requirements were in addition to maintenance, which was assumed at 2% of the capital cost per year.

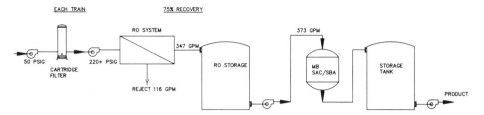

FIGURE 3.9.   Reverse osmosis/ion exchange purification.

the crossover point between these two processes has remained largely unchanged since the results were originally published in 1987. In all but very low TDS supplies, it is now more cost-efficient to utilize RO with polishing IX rather than to rely on IX for the total treatment. In Tables 3.10 and 3.11, the cost breakdown for the two processes is given to allow closer examination of the relative cost contributions of each input. The change in both capital and operating costs with RO is relatively small over this TDS range, whereas the cost for exchanger chemicals rises in proportion with TDS, and labor increases with the frequency of regeneration.

In Figures 3.11 through 3.14, the effects of membrane life, caustic price, electricity cost, and interest rate have been evaluated. It is noteworthy that the effect of interest rate is the most pronounced of the four. Requirements for an acceptable return on investment are generally necessary for approval to construct or expand most water treatment facilities. Figure 3.14 has been derived using the base condition assumed in Figure 3.10, and at interest levels of 0, 10, 20, and 25 percent. An interest rate between 20 and 25 percent would be typical of most corporations.

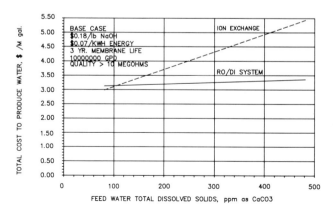

FIGURE 3.10.   Comparison of IX and RO/IX desalination cost—base cost basis.

TABLE 3.10   Total Cost to Purify Water by Ion Exchange

	$ per 1000 Gallons (% of Total)							
	Case 1		Case 2		Case 3		Case 4	
Electricity	.094	(3.2)	.098	(2.7)	.106	(2.4)	.114	(2.1)
Sulfuric acid	.074	(2.5)	.153	(4.3)	.320	(7.2)	.537	(9.9)
Caustic soda	.128	(4.3)	.260	(7.3)	.538	(12.1)	.839	(15.5)
Lime	.016	(0.5)	.034	(1.0)	.074	(1.7)	.137	(2.5)
Subtotal	.312	(10.5)	.545	(15.3)	1.038	(23.3)	1.627	(30.1)
Resin replacement	.110	(3.7)	.121	(3.4)	.148	(3.3)	.182	(3.4)
Labor	.408	(13.7)	.505	(14.1)	.658	(14.6)	.778	(14.4)
Maintenance	.306	(10.3)	.341	(9.6)	.372	(8.3)	.401	(7.4)
Total operating cost	1.136	(38.1)	1.512	(42.6)	2.216	(49.7)	2.988	(55.2)
Taxes and insurance	.312	(10.5)	.348	(9.7)	.381	(8.5)	.412	(7.6)
Capital recovery (0% int)	1.53	(51.3)	1.71	(47.9)	1.86	(41.7)	2.01	(37.2)
Total cost	2.98	(100.0)	3.57	(100.0)	4.46	(100.0)	5.41	(100.0)

TABLE 3.11   Total Cost to Purify Water by RO and Ion Exchange

	$ per 1000 Gallons (% of Total)							
	Case 1		Case 2		Case 3		Case 4	
RO electricity	.266	(8.6)	.271	(8.7)	.275	(8.6)	.280	(8.5)
RO chemicals	.097	(3.1)	.097	(3.1)	.097	(3.0)	.097	(2.9)
1 × electricity	.078	(2.5)	.078	(2.4)	.078	(2.4)	.078	(2.4)
Sulfuric acid	.003	(0.1)	.006	(0.2)	.014	(0.4)	.024	(0.7)
Caustic soda	.015	(0.5)	.032	(1.0)	.076	(2.4)	.130	(3.9)
Lime	0	(0.0)	.002	(0.1)	.006	(0.2)	.011	(0.3)
Subtotal	.459	(14.8)	.486	(15.5)	.546	(17.0)	.620	(18.8)
Resin replacement	.087	(2.8)	.087	(2.8)	.087	(2.7)	.087	(2.6)
Membrane replacement	.227	(7.3)	.227	(7.3)	.227	(7.1)	.227	(6.9)
Labor	.217	(7.0)	.224	(9.4)	.243	(9.3)	.267	(9.0)
Maintenance	.297	(9.6)	.297	(9.4)	.297	(9.3)	.297	(9.0)
Total operating cost	1.287	(41.6)	1.321	(42.2)	1.400	(43.7)	1.498	(45.3)
Taxes and insurance	.316	(10.2)	.316	(10.1)	.316	(9.9)	.316	(9.6)
Capital recovery (0% int)	1.49	(48.2)	1.49	(47.6)	1.49	(46.5)	1.49	(45.1)
Total cost	3.093	(100.0)	3.127	(100.0)	3.206	(100.0)	3.304	(100.0)

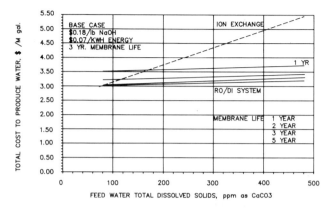

FIGURE 3.11.   **Comparison of IX and RO/IX desalination cost—effect of membrane life.**

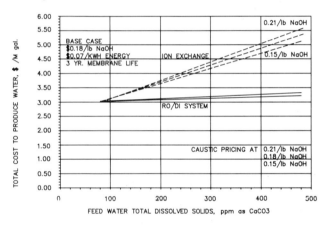

FIGURE 3.12.   **Comparison of IX and RO/IX desalination cost—effect of caustic price.**

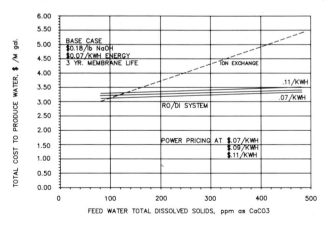

FIGURE 3.13.   **Comparison of IX and RO/IX desalination cost—effect of electricity cost.**

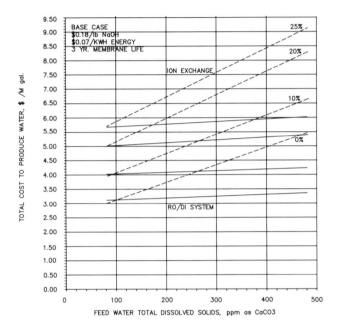

FIGURE 3.14.   Comparison of IX and RO/IX desalination cost—effect of interest rate.

SUMMARY

There are a variety of alternative treatment methods for the desalination of seawater. Each has a different cost structure. Total costs are dependent on feed quality, required product water quality, prevailing interest rates, and, of course, the desalination process. To compare these different processes properly, valid procedures must be used which accurately weigh the various investment and operating cost components. In most cases, IRR and NPV methods are suitable.

Seawater desalination is generally conducted by using multiple-effect evaporation or RO. The cost of either process varies within the range of approximately $7.00 to $10.00 per 1,000 gallons of water produced.

For desalination of brackish water supplies of 1,000 to 5,000 ppm TDS, RO and ED are most commonly used. Under exceptional circumstances, IX or distillation may be appropriate. The total cost of brackish water demineralization generally ranges from $4.00 to $7.00 per 1,000 gallons produced.

Demineralization of potable water, which generally has less than 500 ppm TDS, is ordinarily conducted by IX, with RO as pretreatment in some cases. Generally, RO pretreatment is cost-effective at TDS levels over 200 ppm. The total cost of demineralizing potable water generally varies from $1.00 to $5.00 per 1,000 gallons of water produced.

References

Aqua Chem a. *Theory of operation, multistage flash evaporator, once-through system,* Technical Bulletin No. 750–1007. Milwaukee: Water Technologies Division of Aqua Chem., Inc.

Aqua Chem b. *Process selection guide to seawater desalting,* Technical Bulletin No. 750–3550. Milwaukee: Water Technologies Division of Aqua Chem. Inc.

Birkett, J.D. 1988. Factors influencing the economics of reverse osmosis. Chap. 2 in *Reverse Osmosis Technology: Application for High Purity Water Production,* edited by S. Parekh. New York: Marcel Dekker.

Brigham, E.F., and L.C. Gapenski. 1985. Intermediate financial management, New York: CBS College Publishing.

Codina, O.J., and K.A. Curry. 1987. Side by side comparison of RO and EDR. *Ultrapure Water* 4(7):24–27.

Coulter, B.L. 1980. The application of reverse osmosis to Mexican water. In *Proceedings of 1st Mexican Water Treatment Conferences.* Mexico City, Mexico, February 20–22.

Epstein, A.C. 1978. Pretreatment of seawater for membrane processes. In *Proceedings of 39th International Water Conference,* Pittsburgh, Pennsylvania, October 31–November 2.

Hickman, J.C., and W.J. Conlon. 1984. Membrane processes more economical for potable water treatment than lime softening. Paper presented at the 12th Annual Conference of the National Water Supply Improvement Association, Orlando, Florida, May 13–18.

King, C.J. 1971. *Separation processes.* New York: McGraw-Hill.

Meller, F.H., editor. 1984. *Electrodialysis and electrodialysis technology.* Watertown, Mass.: Ionics Incorporated.

O'Brien, M., and T. LaTerra. 1981. Operating experience with six large reverse osmosis plants. In *Proceedings of International Water Conference,* Pittsburgh, Pennsylvania, October 25–27.

Polidoroff, C. 1987. Plant experience with temporary reverse osmosis makeup water systems. *Ultrapure Water* 4(3):15–21.

Polidoroff, C., E.L. Dubost, M. Humphreys, K.S. Jackson, and K.A. Curry. 1988. Seawater reverse osmosis: utility plant operating experience. *Ultrapure Water* 5(3):17–25.

Rice, D.B. 1991. Impact of reverse osmosis and electrodialysis reversal on the performance of ion exchange demineralizers. Paper presented at the 11th Annual Electric Utility Workshop, Urbana, Illinois, March 21, 22.

Smith, J.M., and H.C. Van Ness. 1959. *Introduction to chemical engineering thermodynamics.* New York: McGraw-Hill.

Vera, I. 1981. Operating experience of a reverse osmosis plant which converts sea water into boiler feed water. In *Proceedings of International Water Conference,* Pittsburgh, Pennsylvania, October 25–27.

Whipple, S.S., E.A. Ebach, and S.S. Beardsley. 1987. The economics of reverse osmosis and ion exchange. In *Proceedings of the First Annual Ultrapure Water Conference and Exposition,* Philadelphia, Pennsylvania, April 13–15.

4

Design Considerations for Reverse Osmosis Systems

Robert Bradley

Arrowhead Industrial Water, Inc.
A BFGoodrich Company

INTRODUCTION

In designing a reverse osmosis (RO) treatment system, a number of unit operations are placed in series with the RO system to adjust the water chemistry for optimal performance of the system. The most important considerations are the feed water chemistry entering the system and the quality of the finished water leaving the treatment process.

The feed water characteristics and the finished product quality requirements often dictate the type of membrane selected for the RO system. In turn, membrane type dictates the number of pretreatment unit operations required.

As in most processes involving water purification, there are a variety of treatment schemes that can provide acceptable results. The goal is to select the most economical and reliable system.

WATER SUPPLIES AND CHARACTERISTICS

Water supplies vary widely around the world, across the country, and even within local areas. Normally, water supplies contain both organic and inorganic contaminants. These may be dissolved, colloidal, or simply suspended (Kemmer 1988).

The sum of dissolved salts is referred to as *total dissolved solids* (TDS). Table 4.1 lists the normal naturally occurring ions that make up these dissolved salts.

The sum of the calcium, magnesium, dissolved iron, and dissolved manganese is referred to as hardness. The sum of the bicarbonate, carbonate, and phosphate is referred to as alkalinity.

104

TABLE 4.1   Naturally Occurring Ions in Water

Calcium	Ca	Carbonate	CO_3
Magnesium	Mg	Phosphate	PO_4
Sodium	Na	Chloride	Cl
Potassium	K	Sulfate	SO_4
Iron	Fe	Nitrate	NO_3
Manganese	Mn	Silica	SiO_2
Bicarbonate	HCO_3	Fluoride	F

The pH of the water is the negative log of the hydrogen ion concentration. The pH values between 6 and 9 may be considered neutral. Waters with pH values lower than 6 are somewhat acidic, and waters having a pH value over 9 are considered alkaline.

Generally, natural surface water supplies (lakes and rivers) have pH values between 7.5 and 8.2. Well or ground water supplies normally have pH values between 6.5 and 8.0.

In natural water supplies, pH is controlled or buffered by the ratio of dissolved carbon dioxide and dissolved bicarbonate ions. Most ground water supplies are stable with respect to the pH value and the amount of limestone (calcium carbonate) dissolved in the water. In other words, the dissolved calcium bicarbonate and carbon dioxide are in chemical equilibrium.

If the carbon dioxide concentration is reduced, the pH of the water will increase slightly and calcium bicarbonate will disassociate to calcium carbonate and carbon dioxide. Scale will begin to form at this point.

Although ground water supplies are normally higher in TDS than surface supplies, they generally contain a low amount of suspended solids (turbidity) and organic materials. However, there are exceptions. Most ground water supplies are well suited for RO treatment. The main points of concern, are high amounts of calcium and magnesium, or dissolved manganese and iron.

Surface supplies are a different issue altogether. In addition to dissolved salts, surface supplies contain large amounts of silt or suspended solids. Many surface supplies also contain organic materials, which may occur naturally or as a result of human intervention. Tannin and lignin resulting from the decomposition of vegetation are colloidal suspensions, but many compounds are dissolved. The concentration of these compounds usually varies seasonally (Briggs et al. 1977).

This variability is perhaps the most difficult aspect of treating surface supplies. On the other hand, multiple analyses of a deep well supply is more often a check of the chemist, not of the water chemistry. Municipalities with well supplies therefore usually produce very consistent water that is acceptable for RO treatment with little pretreatment. Municipal treatment plants treating surface water reduce the amount of suspended solids and turbidity to low levels. Chlorine is also added to deactivate biological growth. However, surface waters often maintain and transport some of their variable and difficult characteristics into the city mains.

Generally, the water supplies of large cities and cities with lake water sources are more consistent and treatable than those of smaller communities or those with river water sources. All cities' supplies, however, can contain corrosion products resulting from the distribution system. Larger cities often have older distribution systems and greater corrosion-related turbidity. Corrosion-related turbidity is also a variable problem. Fires, main flushing, and peak demand periods can knock deposits loose and cause sudden spikes in turbidity.

Many process engineers have said, "We can treat almost anything as long as it's consistent." An RO pretreatment system must be designed to function year round.

Testing Water Supplies

The first step in designing a system is to get a complete and current water analysis. The analysis should include all of the ionic constituents mentioned earlier, as well as turbidity, suspended solids, and total organic carbon (TOC). An empirical measurement, the Silt Density Index (SDI), should also be collected (ASTM Standard D-4189-82, 1987).

The pH of the supply should be measured at the source, as it may change after exposure to air. If the water is not from a well supply, the technician should note whether the sample was clear when it was drawn.

If the water is from any natural source and not a chlorinated municipal supply, a biological assay may be appropriate. If the water is from a surface source, government and EPA publications should be checked for seasonal extremes. One of the best sources for information is often the local city water department.

To understand the nature of the water supply, a day at the city water plant is well spent. Note the source of raw water, unit operations, and chemicals used, as well as variations in the finished water chemistry. Local water plant personnel are usually quite helpful in ascertaining this information.

Occasionally, a municipality has two water plants operating on different water supplies. Industries may receive a variable blend of the two sources. In such cases, a study of the water at the plant site may be required. Water supplies with turbidity values greater than 1 nephelo turbidity unit (NTU) generally require filtration prior to RO treatment.

Turbidity is measured by shining a light through a water supply and measuring the amount of light reflected off the particulate matter. The color, opacity, size, amounts, and shape of the particles affect the measurement. At best, the relationship between turbidity and suspended solids is a casual one.

Silt, colloids, bacteria, colloidal silica, organic molecules, and corrosion products can foul RO membranes. Many water clarification and filtration aids, such as cationic polymers and alum, may foul membranes as well. In order to predict the fouling tendency of a water supply, the SDI test was developed (Figure 4.1) (Permasep Engineering Manual 1982).

Generally, RO systems operating on feed water supplies with SDI values less than 1 run for years without problems, and those operating on supplies with SDI values less than 3 run for months without need of membrane cleaning. However, systems operating on supplies with values between 3 and 5 are cleaned regularly and are often considered problem systems. SDI values greater than 5 are not acceptable at this time.

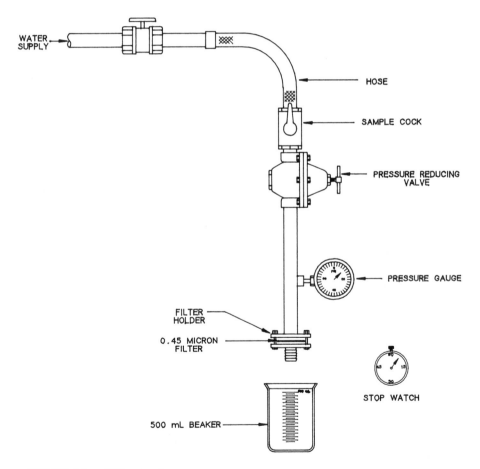

FIGURE 4.1.   SDI apparatus.

MATERIALS OF CONSTRUCTION

Iron corrosion products constitute one of the major foulants in membrane systems. As a result, RO pretreatment systems are normally made of noncorrosive materials. The most popular piping material is PVC, but fiberglass reinforced plastic (FRP) and stainless steel are also used.

Pretreatment pressure vessels are generally of carbon steel, lined with a 5 to 10 mil coating of baked phenolic epoxy. High-pressure feed pumps are usually stainless steel, but some use Noryl internal parts. RO pressure vessels are usually FRP or stainless steel.

The RO process rejects salts, but allows dissolved gas to pass. RO product water is low in alkalinity, has the same concentration of carbon dioxide as the feed water, and is enriched in oxygen. It also tends to be acidic and very corrosive. In most cases, the product piping is PVC, storage tanks are FRP, and pumps are stainless steel.

Highly brackish water or seawater will pit stainless steel because of its high chloride concentration. Type 316 stainless steel is much more tolerant than type 304 and is commonly used in these applications. Brass or aluminum bronzes are excellent in high-chloride waters, but only in a neutral pH environment. Cupro-nickels are superior to all of the aforementioned, but are generally not used because of their cost.

PRETREATMENT

At this point, the reader may suspect that few water supplies are suitable for RO treatment. Without proper pretreatment of water supplies, this is true. The science of water treatment, however, has progressed to the point where almost all water supplies can be adequately pretreated.

The RO process is an effective method of removing dissolved salts from water. The membrane will become fouled with anything that is not dissolved when entering the process or that precipitates during the process. Pretreatment equipment is designed to condition the supply water to preclude such fouling.

Suspended Solids and Silt Reduction

Suspended solids are normally both the first and the last impurities to be removed as part of the RO feed-water treatment. Media filters are usually used first. A cartridge filter, to stop any media transferred from previous devices, is the last treatment device to be used before the water enters the high-pressure RO pump. The following are some typical examples of suspended solids:

Mud and silt	Iron corrosion products
Organic colloids	Precipitated iron

Algae   Precipitated manganese
Bacteria   Precipitated hardness
Rocks   Aluminum hydroxide floc
Frogs   Beer cans
Silica/sand

Clarification

Under normal conditions, traditional open clarifiers are required only when the water supply comes directly from a lake or stream, and then only if the lake or stream is fairly turbid (greater than 50 NTU for at least part of the year). Water supplies such as the Great Lakes may be treated adequately by multimedia filtration with a coagulant feed.

Clarifiers have other uses, however. Because of their holding capacity, they can dampen out some variations in rapidly changing supplies, or provide contact time for chlorine in highly bioactive water supplies. Clarifiers can also be designed as cold lime softeners to reduce hardness or silica (American Society of Civil Engineers and American Water Works Association 1990).

Media Filters

The most popular means of removing suspended solids from feed water is multimedia filtration. Multimedia filters feature layered beds of anthracite coal, sand, finely crushed garnet, or other materials. A typical filter is shown in Figure 4.2. The top layer of the bed consists of the lightest and most coarsely graded material, whereas the heaviest and most finely graded material is found on the bottom. The principle is filtration in depth—larger particles are removed at the top layers, and smaller ones are removed deeper in the filter media.

In single-media filters, the finest granular material is backwashed to the top of the bed. Most of the filtration takes place in the top 2 inches of the bed, the remainder being support media. A layer of mud is formed. Although single-media filters are limited to filtration rates from 2 to 4 gpm/ft^2 of filter area, multimedia filters can hydraulically process flow rates up to 20 gpm/ft^2, but are generally limited to flow rates of 7.5 gpm/ft^2 in RO pretreatment applications because of the high water quality requirements.

Because suspended matter of a colloidal nature is either too small or electro-statically repelled from the media, filtration alone may not work. In these cases, a coagulant or flocculating chemical must be added to the water prior to filtration. Popular coagulants are ferric chloride, alum, and cationic polymers. The polymers are by far the most popular because they are effective at low dosages and do not substantially increase the solids loading on the filter media. On the other hand, cationic polymers are very effective foulants if they get onto

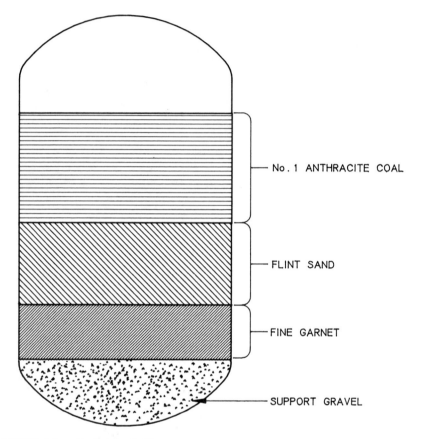

No . 1 ANTHRACITE COAL

FLINT SAND

FINE GARNET

SUPPORT GRAVEL

FIGURE 4.2.   Typical depth filter.

some of the most popular membranes currently being used. Very low amounts of polymer can blind these membranes, and sometimes they are difficult to remove. It is important to remember that great caution must be exercised when using polymers as a filtration aid.

Alum and sodium aluminate deserve special comment. Both are commonly used in municipal and industrial water treatment. Both precipitate at neutral pH's to form aluminum hydroxide. Figure 4.3 gives the solubility of aluminum as a function of pH. In Figure 4.3, the minimum solubility of aluminum is at a pH of 6.5 to 6.7. Aluminum salts are often used as a coagulant in lime softeners operating at a pH of 10. At this pH, aluminum is soluble to a level of 1 ppm. To reduce the feed-water pH, acid is frequently used as an RO pretreatment. When the pH of the feed water is lowered, the aluminum precipitates from solution and fouls the membranes. When used in this fashion, aluminum is one of the most

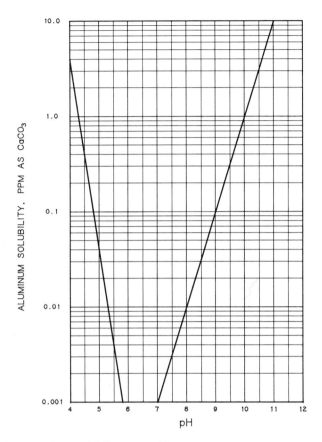

FIGURE 4.3.   Aluminum solubility versus pH.

common foulants found in membrane systems. For this reason, if the water supply contains aluminum, always feed acid as far ahead of the media filter as possible.

Ultrafilters

In recent years, ultrafiltration membranes have been introduced. These membranes are not nearly as tolerant of suspended solids as are media filters. They will, however, handle higher SDI values and turbidity levels that would blind an RO membrane. Ultrafilters are often polishing filters between a media filter and an RO unit. Ultrafilters typically remove particles as small as 0.01 microns, and so produce very high quality RO feed water. The principal disadvantage to using ultrafilters is their cost. They are expensive, and if they are improperly applied, a potential fouling problem simply occurs on a different set of membranes.

Cartridge Filters

Almost every RO system is equipped with a cartridge filter to prevent particles in the feed water, chemical feed equipment or pretreatment media from entering the RO system. Cartridge filters, normally the last pretreatment step, are constructed of stainless steel. The most widely used elements are polypropylene with polypropylene center cores. Cotton and other materials that can shed fibers are not acceptable. The most popular pore size is 5 micron nominal, but in some applications 5 micron elements are followed by 1.0 micron elements to provide better filtration in depth. They do not replace media filters when media filters are required, however.

Scale Control

As the feed water passes across the surface of the RO membrane from the feed to the exit, relatively pure water passes through the membrane to form product stream, and the feed water becomes ever more saline as it becomes the waste stream (frequently called brine or reject). As the salt concentration increases, some salts may supersaturate and begin to precipitate (Cowan 1976). If the precipitates coat the membrane, the result is scale that will eventually foul or damage the membrane. The following discussion illustrates this concentration effect.

The ratio of the RO product water to the feed water is called the *recovery (Y)*.

$$Y = \frac{\text{product flow rate}}{\text{feed flow rate}} \times 100\% \tag{1}$$

Assuming that the amount of any ion passing through the membrane is close to zero, then the concentration factor (CF) for any ion is given by:

$$CF = \frac{1}{1 - Y} \tag{2}$$

For example: If the calcium concentration in the feed water is 300 ppm (as $CaCO_3$), the feed flow is 100 gpm, and the product flow is 75 gpm; the CF will be 4, and the calcium concentration in the waste will be 1,200 ppm (as $CaCO_3$). Assume that as calcium sulfate is near saturation in the feed water, and that all other ions will concentrate in the same manner, calcium sulfate could be nearly 16 times saturation in the final elements before the reject is discharged.

In order of occurrence, the following compounds commonly form scales:

Calcium carbonate
Calcium sulfate
Silica complexes

Barium sulfate
Strontium sulfate
Calcium fluoride

The scale compounds not listed are hydroxides of aluminum, iron, or manganese. They are normally precipitated before contact with the membrane, and not crystallized on it.

Scales are almost never pure substances, but are collections of a number of different compounds. It is common for one compound to precipitate first and provide nucleation sites for other precipitates. As this occurs, silica is often complexed into the scale.

Considerations for the prevention of scaling are detailed in the following paragraphs.

Calcium Carbonate

By far the most common scale found in RO systems is calcium carbonate. Calcium carbonate, carbon dioxide, calcium, and bicarbonate ions form an equilibrium given by the following equation:

$$Ca^{+2} + 2\ HCO_3^- \rightarrow CaCO_3 + CO_2 + H_2O \qquad (3)$$

Precipitation of calcium carbonate is favored by increasing calcium or bicarbonate concentration, decreasing carbon dioxide or total dissolved solids, and increasing temperature or pH.

There are several indexes commonly used to quantify the calcium carbonate scaling potential. These include the Langelier Saturation Index (LSI), the Ryzmar, Stiff and Davis, and several others.

Common methods to prevent calcium carbonate scaling include the following:

1. Removal of the vast majority of the calcium with a sodium ion exchange (IX) water softener or a lime-softening clarifier
2. Removal of all or some of the bicarbonate alkalinity by feeding acid
3. Use of scale-control agents

The choice among these three methods is normally based on the membrane selection, the amount of hardness in the water, the final water specifications, and the economics of the treatment.

Cellulose acetate membranes require that the feed-water pH be less than 6.5 to reduce the effect of hydrolysis (degradation of the membrane). Acid is almost always fed to the RO feed water prior to its entry into a cellulose acetate (CA) system. In most parts of the world, sulfuric acid is much less expensive than other types. Feeding sulfuric acid, however, increases the sulfate content of the water, leading to the possible precipitation of other salts, including calcium, barium, and strontium sulfates (ASTM Standard D-4692-87 1987). Under most conditions whenever sulfuric acid is used, either to prevent calcium carbonate scaling or simply to reduce the feed-water pH, an antiscalant chemical is used as well.

Economics is always an important factor in system design. Both sulfuric acid and salt are inexpensive, but acid is fed on a stoichiometric basis to react with alkalinity in water. Water softeners regenerate with salt, but the salt used for regeneration is about three to five times the stoichiometric amount of the hardness removed. In high hardness supplies or large systems, acid with an antiscalant is usually more economical. Furthermore, it is not necessary to remove all of the alkalinity by feeding acid. Common practice is to feed a sufficient quantity to reduce the LSI in the waste stream to positive 1 and allow the antiscalant to prevent scaling at that level. Acid reacts with the bicarbonate alkalinity to produce carbon dioxide. Under normal conditions, the pH of the feed water is reduced to 5.7 prior to a CA RO system. Figure 4.4 is a graph of the relationship of total alkalinity (M)/carbon dioxide ratio to pH. The following example illustrates the calculations for determining changes to feed-water chemistry upon the addition of acid for pH adjustment.

If a water supply contained 400 ppm of total alkalinity and had a pH of 7.8, the ratio of M alkalinity to carbon dioxide should be 35 (see Figure 4.4). Therefore, the supply contains 11.4 ppm of carbon dioxide. If the pH is to be lowered to 5.7, the ratio should be 0.23.

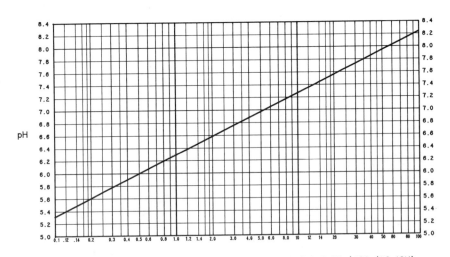

RATIO OF ALKALINITY TO CARBON DIOXIDE M ALK. (AS $CaCO_3$)/CO_2 (AS ION)

FIGURE 4.4.   Ratio of alkalinity to carbon dioxide.

One ppm of sulfuric acid (93 percent) will reduce M alkalinity by 0.95 ppm (expressed as $CaCO_3$) by converting it to carbon dioxide. The carbon dioxide content of the water will increase 0.84 ppm (as ion). The following equation can then be written:

$$R = \frac{M \text{ alk} - (0.95) \text{ X ppm } H_2SO_4}{CO_2 + (0.84) \text{ X ppm } H_2SO_4} \tag{4}$$

$$0.23 = \frac{400 - (0.95) \text{ X ppm } H_2SO_4}{11.4 + (0.84) \text{ X ppm } H_2SO_4} \tag{5}$$

$X = 347.6$ ppm 93% H_2SO_4
CO_2 after acid feed $= 11.4$ ppm $+ (0.84) (347.6$ ppm)
M alk after acid feed $= 400 - (0.95) (347.6)$
M alk $= 69.8$ ppm (as $CaCO_3$)

Naturally, the SO_4 ion content of the water will increase an amount equal to the M alkalinity reduction. This must be considered when calculating the calcium, barium, and strontium solubilities.

Because carbon dioxide is not rejected by the membrane, it passes through into the product. RO product water is always somewhat acidic, but product water from a system using acid feed may contain several hundred parts per million of carbon dioxide. In most situations, this is not acceptable. Carbon dioxide may be removed by a degasifier, but forced draft degasifiers introduce airborne dust and bacteria into the RO product water. Degasifier air intakes can be equipped with submicron filters, but they are still breeding grounds for bacteria. Vacuum degasifiers are expensive both to buy and to operate. In high-purity treatment systems, the RO product is always polished by a deionizer. If the RO product water is not degasified, however, the high CO_2 content will quickly exhaust anion resin. For this reason, RO units requiring acidified feed water are not popular for high-purity systems.

Reconsider the previously discussed water supply containing 400 ppm of alkalinity, with 11.4 ppm of CO_2. Suppose this supply contains 600 ppm of TDS, and the RO system uses a thin film composite (TFC) membrane. The product water from the system will contain approximately 25 ppm of TDS. However, the product will also contain 10 ppm of CO_2 (expressed as $CaCO_3$ equivalents). Assuming once again that the product will be deionized, the carbon dioxide will increase the load on the anion resin by 40 percent. To negate this effect, caustic may be fed to the water after softening, thus increasing the feed pH to between 8.3 and 8.5. The carbon dioxide will subsequently be converted to sodium bicarbonate, and it can then be removed by the membrane with no detectable effect on the product quality.

Softening has several drawbacks which should be considered:

1. As mentioned earlier, several times more salt is required than acid.
2. A significant quantity of spent regenerant brine must be discharged to waste, which can increase the chloride discharge from a facility to an environmentally unacceptable level.
3. Finally, sodium is not rejected by the membrane as well as are calcium and magnesium. RO units operating on soft water have slightly higher TDS levels in the product water than they would if softeners were not used.

Antiscalants and Dispersants

As previously discussed, acid is normally fed to reduce the LSI in the reject stream to a value of +1. At this level, an antiscalant is also used to prevent scaling (Amjad 1988). The first antiscalant used for this purpose was sodium polyphosphate, sometimes called sodium hexametaphosphate (SHMP). Scale inhibition with SHMP was at best a hit-or-miss proposition. Today, a number of organic scale inhibitors have been introduced that will inhibit scaling at LSI levels over 2. If the calcium and alkalinity levels are not very high and the pH of the water supply is less than 8.0, it may be possible to treat the water by RO without softeners or feeding acids. This approach has several advantages:

1. It eliminates the cost, environmental, and handling problems associated with acid or salt.
2. The product water quality is slightly better.

Some antiscalant chemicals also contain dispersants. These compounds keep the precipitates suspended and off the membrane. A dispersant is also useful in preventing fine silt, which was not effectively removed by filtration, from settling out on the membrane. Dispersants can even break up metal hydroxide floc. Antiscalants and dispersants are normally fed to the RO feed water after the cartridge filter, but before the high-pressure pump.

Silica

Silica deserves a special comment. The solubility of silica is a function of pH and temperature, as given in Figure 4.5. To date, there has not been an antiscalant or dispersant developed that can effectively prevent silica deposition. There are, however, many systems operating very successfully with supersaturated silica concentrations in their reject streams. Exceeding the saturation level in cold water (less than 50°F) is not a serious problem because the precipitation is very slow. However, exceeding 120 ppm, SiO_2 in any temperature presents a potential problem. Metal hydroxides (iron, aluminum, magnesium, etc.) absorb or complex silica and catalyze the precipitation. (ASTM Standard D-4993-87

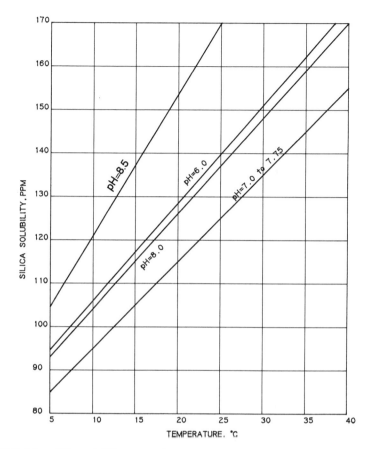

FIGURE 4.5.   Silica solubility versus temperature.

1987). Silica is difficult to clean from the membrane and should be treated with caution. The only economical pretreatment method to reduce dissolved silica is lime softening. Another option is to reduce the recovery of the RO system to keep the silica concentration in the brine below the saturation point.

Organics
Chlorinated municipal water supplies with low SDI values normally do not have levels of organic matter sufficient to cause a problem in an RO system. Municipal supplies with high SDIs or natural supplies, particularly if colored, should have a TOC check. A high TOC level is not a sure sign of a problem. Some organics such as humic acid are rejected by the membrane and will not foul it. Others, such as ethyl alcohol or acetone, will pass through the membrane with no

effect other than contaminating the product water. Some, however, such as tannic acid, will foul the membrane.

There are several ways to reduce organics in the feed water. Some species of low molecular weight may be volatile enough to be removed by degasification. Others that are nonpolar, with medium to high molecular weights, can be absorbed by activated carbon. Many polar, ionized compounds are anionic and can be coagulated with a cationic polymer and filtered out. Some weakly ionize and can be removed by an organic trap resin. In the extreme, some are of sufficient molecular weight to be filtered out with a dirty-water ultrafilter. No one process will address all organic problems. Water supplies with high SDI values and high TOC levels require pilot studies.

Oxidizing Agents

With the exception of some new polysulfone membranes that are used in limited applications, all RO membranes are damaged by oxidizing agents, mainly chlorine. Cellulose acetate (CA) is the most tolerant, but even it is damaged to some extent. CA has the added problem of being biodegradable. Given the two, chlorine and biodegradation, chlorine is the lesser evil. If the feed-water chlorine concentration is held between 0.2 and 0.5 ppm in cool water (less than 60°F), the membrane will last about 3 years (a normal life span for CA membranes) without a noticeable effect owing to chlorine. If unchlorinated feed water is used, bacteria have been known to eat through CA membranes in 2 weeks. In warm water (greater than 70°F) supplies, both the chlorine and the bacteria are more aggressive. This limits the application of CA membranes in southern climates. If the water supply contains more than 1 ppm of chlorine, it is best to remove all the chlorine and then add 0.2 to 0.5 ppm of chlorine back into the supply.

Asymmetric homogeneous polyaramid (PA) membranes are quickly destroyed by chlorine. TFC polyaramids are slightly more tolerant. One manufacturer limits chlorine exposure to 1,000 ppm—hours before failure is likely. Both TFC and PA membranes require complete chlorine removal prior to the water's entering the RO system.

The two most popular methods of removing chlorine are the addition of activated carbon and the addition of sodium bisulfite. If the water supply is cool (less than 60°F), the most reliable method for chlorine removal is to pass the water through a pressure vessel filled with activated carbon. The chlorine is absorbed by the carbon in the top 2 to 3 inches of the bed and reacts with the carbon and water to form carbon dioxide and hydrochloric acid. With warmer waters, the reaction is too quick. Because the chlorine is gone within the top few inches of carbon, the water is then free to grow bacteria. Carbon beds on warm water (greater than 60°F) supplies grow bacteria very quickly. TFC and PA membranes are not biodegradable but can and will be fouled by bio-growth. For this reason, carbon is primarily used in northern climates.

Sodium bisulfite will remove chlorine by the following reaction:

$$NaHSO_3 + Cl_2 + H_2O \rightarrow NaHSO_4 + 2HCl \tag{6}$$

The reaction takes a few seconds, so the sodium bisulfite is fed to the feed water ahead of the cartridge filter or softener. To ensure a complete reaction, at least two times the stoichiometric amount is used: 1 ppm Cl_2 requires 1.465 ppm $NaHSO_3$ (White 1972).

MEMBRANE SELECTION

As previously stated, the raw water chemistry and the finished water requirements normally dictate membrane selection. Once a membrane has been selected, it will dictate the pretreatment required. Table 4.2, on page 124, lists some of the membranes in use today, as well as their characteristics.

Spiral Wound Cellulose Acetate

CA membranes were the first commercial RO membranes to be developed. CA membranes marketed today are actually blends of cellulose acetate and cellulose triacetate (CTA).

CA on its own has rather poor chemical stability, and can hydrolyze faster than CTA. CTA, however, compacts quickly and loses a great deal of its flux during the first few months of operation. The CA-CTA blends have reasonable hydrolysis and compaction characteristics. The hydrolysis rate of a CA-CTA blend leads to a 3-year useful life when treating 50°F water with a pH of 5.7. Over the life of the membrane, the salt passage rate will double. A membrane rated at 96 percent salt rejection will drop to a rate of 92 percent at the end of its useful life. Hydrolysis is a function of temperature and pH. Figure 4.6 shows this relationship.

CA-CTA membranes also compact over their life. Operating at 400 psig on 50°F water, the typical compaction rate is at 12 to 20 percent loss of flux (capacity in gpd/ft^2) over the 3-year life, with half of that loss occurring during the first year. Like hydrolysis, compaction is a function of temperature and, in this case, pressure. Increasing pressure increases compaction fairly linearly. Increasing temperature increases compaction only slightly. Figures 4.7, 4.8 and 4.9 show how CA membranes age at various temperatures and pressures (UOP Fluid Systems 1988).

CA membranes are inexpensive and more tolerant of foulants than other membranes. They are also the only major commercial membranes that are chlorine tolerant. Because of this, CA is often preferred in waste water applications where the SDI is high and in potable water purification. CA does not reject

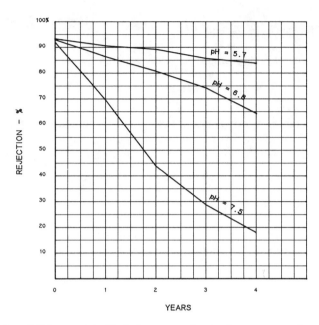

FIGURE 4.6. Cellulose acetate. Typical system: salt rejection versus membrane age. Temperature = 25°C.

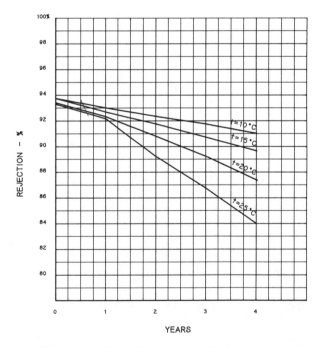

FIGURE 4.7. Cellulose acetate. Typical system: salt rejection versus membrane age.

120

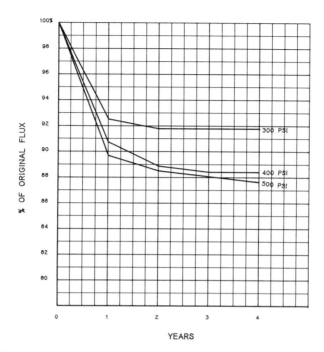

FIGURE 4.8. Cellulose acetate. Typical system flux decline (% of Original Flux) versus pressure. pH = 5.7. Temperature = 20°C.

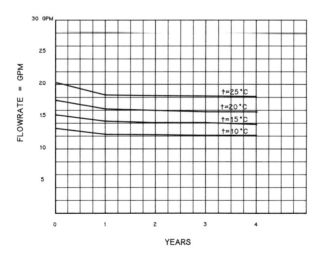

FIGURE 4.9. Cellulose acetate. Typical system flow rate versus membrane age. pH = 5.7. pressure = 360 psig.

salt as well as TFC membranes, and it also requires acid feed. Consequently, CA is not often used in high-purity applications where the product will be deionized. CA systems operate at 400 psig, resulting in the pumping cost's being substantially higher than in systems with TFC membranes, which operate at 250 psig. A final note: In warm water supplies, CA membranes frequently do not work out well because of the increased likelihood of biological attack. Both CA and TFC membranes are available in spiral wound construction (see Figure 4.10).

Hollow Fine Fiber Polyaramid

Another membrane developed early in the short history of RO is the hollow fine fiber (HFF) polyaramid. HFF membranes do not hydrolyze and are not biodegradable. They are, however, subject to compaction and have absolutely no tolerance for chlorine in feed water. Today, HFF membranes are used on deep well supplies in which the SDI is typically less than 1. These membranes are suited for high TDS supplies, and many are used on seawater wells and high TDS ground water in the Middle East. These elements are quite susceptible to fouling. SDIs below 3 are critical for maintaining the life of HFF membranes.

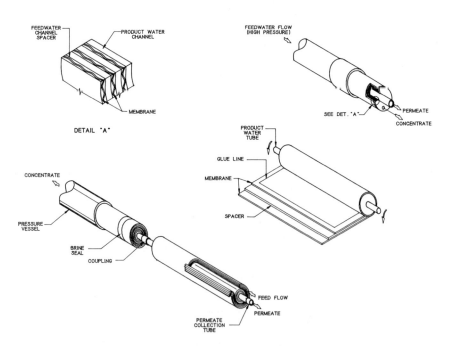

FIGURE 4.10.   Spiral wound RO membrane element construction.

Thin Film Composite

All RO membranes are asymmetrical, which means that from one side to the other the membrane is not of a consistent density. CA membranes are made by a process that allows one side of the membrane to be more dense with much finer pores. The layer of dense material is sometimes called the rejecting layer. The rejecting layer is always wound toward the feed water.

The thickness of the rejecting layer is often referred to as Tau (τ), and the flux rate at a given temperature and pressure is inversely proportional to it. With time, the continuous pressure on the rejecting layer forces it down into the less dense material below it, and Tau effectively increases. This process is known as compaction.

In the late 1970s, composite membranes were developed. A rejecting layer of a totally different material was cast onto a porous polysulfone support layer (see Figure 4.11). Because the two layers are of different media, composite mem-

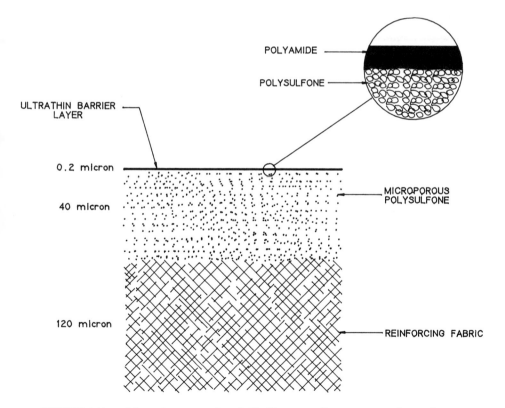

FIGURE 4.11.   Schematic cross-section of thin film composite reverse osmosis membrane.

branes do not truly compact. Some of the pores in the membrane may become fouled with dirt or scale as time passes, but cleaning may restore the flux rate. Because the rejecting layer is thinner than other types of membranes, composite membranes operate at higher flux levels and lower pressures. These membranes have other advantages as well:

1. They are not biodegradable.
2. They have higher salt rejection rates.
3. They reject silica, nitrate, and organics much better than other types of membranes.

These membranes are fairly inert chemically and will resist hydrolysis while operating in a pH range from 2 to 11. They may also be cleaned with much stronger solutions than can be used with other types of membranes.

Composite membranes are by far the first choice for high-purity systems. Because of their lower energy consumption, they are often preferred for any large RO system. A comparison of the three membranes discussed is given in Table 4.2.

For all of their advantages, composite membranes are not without their limitations. They are more tolerant of foulants than HFF membranes, but they will foul more readily than CA membranes. While operating, some composite membranes take on an anion charge on the surface, causing cationic foulants, such as aluminum and ferric hydroxides, along with cation coagulant polymers, to be attracted to it. Further, because of the high flux rates on composite mem-

TABLE 4.2   Membrane Characteristics and Recommended Operating Ranges

	CA	HFF	TFC
Salt rejection	90%–97%	90%–96%	96%–98%
Silica rejection	85%	85%	98%
Nitrate rejection	85%	85%	92%
Chlorine tolerance (ppm)	0.2–0.5	0	0
Maximum SDI	5	3	5[1]
Temperature range	32–95°F	32–95°F	32–113°F
Nominal organics rejection range	300+ mwt	300+ mwt	200+ mwt
Typical operating pressure, psig	400[2]	400[2]	250[2]
pH range	3.0–6.5	4–11	2–11
Flux rate, gpd/ft^2	12–16	2–4	15–20
Biological resistance	Poor	Good	Excellent
3rd year compaction	20%	20%	0
Hydrolysis	2× SP@3 yr	None	None

[1]Less than 3 strongly preferred
[2]Typical system values

branes, fouling occurs faster and is more noticeable than with other membranes. Because of these considerations in making a correct choice in membrane selection, the process engineer must often choose between a more complex pretreatment system and increased operating costs.

SYSTEM DESIGN PARAMETERS

After membrane and pretreatment selection has been made, the process engineer focuses attention on the membrane configuration. The client's flow and quality requirements must be satisfied without violating any of the membrane manufacturer's design parameters.

Temperature, Pressure, and Flux

A membrane is rated at a certain capacity at a given standard or test condition. At the test condition, the temperature, pressure, recovery, and feed water salinity are given, and the membrane is tested for flux and salt rejection. Beyond these conditions, the flux and salt rejection will vary by known relationships. The temperature at the test condition is 25°C. As the water temperature changes from 25°C, the viscosity also changes—increasing at lower temperatures and decreasing at higher temperatures. The flux will vary because of these viscosity effects according to the following relationships:

$$J = Jo\ 1.03^{t\ -\ 25} \tag{7}$$

Where

t = water temperature (°C)
Jo = flux at standard conditions
J = flux rate on the feed water at temperature (t)

The flux rate is proportional to the net applied driving force. The net applied driving force is the applied pressure less the difference in the osmotic pressure (feed minus product water).

$$J = Km\ (P - JC) \tag{8}$$

Where

J = flux
Km = permeability constant
P = difference in hydraulic pressure (average feed brine minus product)
JC = osmotic pressure difference (average feed brine minus product)

Km will be given by the membrane manufacturer or may be found by solving the equation at the standard test conditions. The osmotic pressures are a function of temperature, salinity, and water chemistry.

$$\pi = 1.21(T) \ (\Sigma \ Mi) \tag{9}$$

Where

T = water temperature, °K
M = summation of molalities of all ionic and nonionic materials
π = osmotic pressure, psi

Figure 4.12 gives the osmotic pressure for various concentrations of compounds in water.

From the preceding discussion, it can be understood that cooler water reduces the flux and increases the number of membranes necessary to provide the required flow. In some cases, however, the feed pressure can be increased to increase the flux and reduce the number of membranes required.

Salt passage is the converse of rejection:

$$
\begin{aligned}
\% \ SP &= (C_P/C_F) \times 100\% \\
\% \ REJ &= 100\% - \%SP \\
\% \text{ salt passage} &= 100\% - \% \text{ rejection} \\
&= (\text{Product Concentration} \times 100\%)/\text{Feed Concentration}
\end{aligned}
\tag{10}
$$

Where

$\% \ SP$ = salt passage as a percent
$\% \ REJ$ = salt rejection as a percent
C_P = concentration of the product
C_F = concentration of the feed

Flux is a function of temperature and net applied pressure. The rate at which salt passes through a membrane, however, is only a function of temperature and total salt concentration. It is a constant for the membrane in question. Therefore, raising the flux by increasing the net driving pressure can improve the product water quality.

Recovery and Membrane Array

As previously outlined, the recovery rate for an RO system is defined as follows:

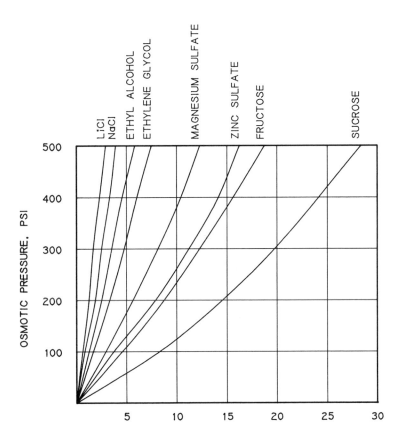

FIGURE 4.12.   Osmotic pressure of various compounds.

$$Y = \frac{\text{product flow rate}}{\text{feed flow rate}} \times 100\% \tag{1}$$

For large systems, a typical recovery rate is 75 percent. To accomplish this, the membranes must be staged. "Staging" occurs when the reject from a group of membranes (first stage) is gathered together and becomes feed water to a second, smaller group of membranes (second stage).

As an example, consider a group of eighteen TFC membranes arranged in three pressure vessels (see Figure 4.13). Membrane manufacturers have computer projection programs available to aid in this process. Figure 4.14 (FilmTec Corp 1989) is a computer printout for an RO system designed to produce 15

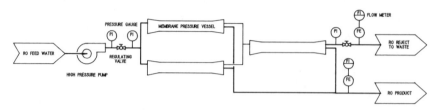

FIGURE 4.13.   Typical reverse osmosis system process and instrumentation diagram.

```
FEED:      20.00 GPM ,    500 MG/L,   25 DEG C
RECOVERY:  74.9 PERCENT

ARRAY:          1            2
NO.OF PV:       2            1
ELEMENT:    BW4040       BW4040
NO.EL/PV:       6            6
EL.TOTAL:      12            6

FOULING FACTOR:    0.85

                            FEED    REJECT   AVERAGE
PRESSURE(PSIG)             193.4    131.7     162.4
OSMOTIC PRESSURE(PSIG)       5.0     19.3      10.4
NDP(MEAN)=  151.9 PSIG
AVERAGE PERMEATE FLUX=  15.0 GFD        PERMEATE  FLOW=      14.98 GPM
```

ARRAY	EL.NO.	RECOVERY	PERMEATE		FEED		
			GPD	MG/L	GPM	MG/L	PRESS(PSIG)
1	1	0.098	1414	9	10.0	500	188
	2	0.105	1359	10	9.0	554	181
	3	0.113	1311	12	8.1	617	176
	4	0.123	1270	14	7.2	694	171
	5	0.136	1233	18	6.3	789	167
	6	0.153	1198	23	5.4	911	164
2	1	0.085	1119	27	9.2	1072	157
	2	0.088	1066	31	8.4	1168	151
	3	0.092	1018	38	7.7	1278	146
	4	0.097	974	45	7.0	1404	141
	5	0.103	933	55	6.3	1550	137
	6	0.110	893	68	5.6	1722	134

```
ARRAY:          TOTAL        1         2
REJECT GPM :                 9.2       5.0
REJECT MG/L:                1072      1926
PERM GPD :      21571      15569      6002
PERM MG/L:         22         14        43
```

FEED WATER IS WELL OR SOFTENED WATER (BW) SDI < 3

FILMTEC ASSUMES NO LIABILITY FOR RESULTS OBTAINED OR DAMAGES
INCURRED FROM THE APPLICATION OF THIS INFORMATION.   FILMTEC
RECOMMENDS THAT ANY FINAL DESIGN BE REVIEWED BY FILMTEC
APPLICATIONS ENGINEERING DEPARTMENT.

FIGURE 4.14.   Computer projection for 2:1—6 M array (acceptable design).

gallons of water per minute. Notice that the flow rate to the first membranes in the first stage is about equal to the flow rate in the second stage. This is known as natural array. Also note that the feed pressure to each membrane is slightly lower than the preceding one, the salinity is higher, and the product flow rate is lower.

Membrane manufacturers have design guidelines for the following parameters:

1. Maximum feed flow rate to any element
2. Minimum brine flow rate from any element
3. Maximum recovery rate for an individual element
4. Maximum flux rate for an element
5. Maximum average flux rate for a system
6. Maximum applied pressure

Manufacturers often alter these guidelines, based on the type of feed water and degree of pretreatment. Figure 4.15 is a printout for twelve elements in a 2:1 array to produce 12 gpm. Note the warning statement at the bottom of the printout. FilmTec recommends that the recovery for any one element not exceed 19 percent softened feed water with an SDI less than 3. The last element in the first stage has a recovery rate of 21 percent. By reducing the system recovery to 70 percent, as shown in Figure 4.16, the condition is corrected.

Figure 4.17 shows a system with twenty-four membranes in four vessels in a 2:1:1 array. The system is designed to produce 20 gpm. The printout does not indicate a problem. However, this is a poor design. Consider the following factors:

1. The array is a long, slender one, and the pressure drop from the feed to the reject end of the array is well over 100 psig.

2. Because these TFC membranes provide a high flux at low pressures, the first stage membranes are producing most of the water. The average flux rate is 15 gfd, which is conservative. However, the first membranes in the first array are producing 21.5 gfd, so fouling is likely to occur in the first membranes.

Figure 4.18 shows a printout for the same number of membranes arranged in a 3:2:1 array using four-element pressure vessels (4M). This time the pressure drop from the feed to the reject across the array is reduced to 65 psig. All the membranes are working together and the flux on the first membranes has been reduced to 17.5 gfd.

Figure 4.19 is a printout for a 100-gpm TFC system operating on unsoftened surface water, with an SDI between 3 and 5. Because FilmTec recommends that

```
FEED:       16.00 GPM ,    500 MG/L,  25 DEG C
RECOVERY:   75.2 PERCENT

ARRAY:           1          2
NO.OF PV:        2          1
ELEMENT:   BW4040      BW4040
NO.EL/PV:        4          4
EL.TOTAL:        8          4

FOULING FACTOR:    0.85

                          FEED    REJECT    AVERAGE
PRESSURE(PSIG)           211.7     177.2      193.6
OSMOTIC PRESSURE(PSIG)     5.0      19.5       10.6
NDP(MEAN)=  182.9 PSIG
AVERAGE PERMEATE FLUX=  18.0 GFD        PERMEATE  FLOW=     12.03 GPM

ARRAY EL.NO. RECOVERY    PERMEATE            FEED
                       ----------- -----------------------
                        GPD  MG/L   GPM    MG/L PRESS(PSIG)

   1      1     0.136   1565    9   8.0     500   207
          2     0.153   1525   11   6.9     577   202
          3     0.177   1489   14   5.9     680   199
          4     0.209   1454   20   4.8     823   196

   2      1     0.126   1380   24   7.6    1035   189
          2     0.139   1335   30   6.7    1181   185
          3     0.156   1291   40   5.7    1367   182
          4     0.179   1245   55   4.8    1613   179

ARRAY:       TOTAL        1          2
REJECT GPM :               7.6        4.0
REJECT MG/L:            1035       1952
PERM GPD :   17317     12066       5251
PERM MG/L:      20        13         37

FEED WATER IS WELL OR SOFTENED WATER (BW) SDI < 3


WARNING ! MAXIMUM RECOMMENDED ELEMENT RECOVERY HAS BEEN EXCEEDED

FILMTEC ASSUMES NO LIABILITY FOR RESULTS OBTAINED OR DAMAGES
INCURRED FROM THE APPLICATION OF THIS INFORMATION.  FILMTEC
RECOMMENDS THAT ANY FINAL DESIGN BE REVIEWED BY FILMTEC
APPLICATIONS ENGINEERING DEPARTMENT.
```

FIGURE 4.15.   Computer projection for 2:1—4 M array (maximum recovery exceeded—array 1, element 4).

```
FEED:      17.14 GPM .    500 MG/L,  25 DEG C
RECOVERY:  70.0 PERCENT

ARRAY:          1           2
NO.OF PV:       2           1
ELEMENT:     BW4040      BW4040
NO.EL/PV:       4           4
EL.TOTAL:       8           4

FOULING FACTOR:    0.85

                           FEED   REJECT   AVERAGE
PRESSURE(PSIG)            212.6    171.8     192.3
OSMOTIC PRESSURE(PSIG)     5.0     16.3       9.7
NDP(MEAN)=  182.6 PSIG
AVERAGE PERMEATE FLUX=  18.0 GFD      PERMEATE  FLOW=      12.00 GPM

ARRAY EL.NO. RECOVERY   PERMEATE            FEED
                       -----------  -----------------------
                       GPD   MG/L   GPM    MG/L PRESS(PSIG)

   1      1    0.127   1571    8    8.6    500   208
          2    0.142   1527   10    7.5    572   202
          3    0.161   1488   13    6.4    664   198
          4    0.187   1452   18    5.4    789   195

   2      1    0.109   1375   21    8.8    967   188
          2    0.118   1324   26    7.8   1083   183
          3    0.129   1278   32    6.9   1224   178
          4    0.143   1234   41    6.0   1400   175

ARRAY:          TOTAL        1        2
REJECT GPM :                8.8      5.1
REJECT MG/L:                967     1627
PERM GPD :      17286     12076     5210
PERM MG/L:         18        12       30

FEED WATER IS WELL OR SOFTENED WATER (BW) SDI < 3


FILMTEC ASSUMES NO LIABILITY FOR RESULTS OBTAINED OR DAMAGES
INCURRED FROM THE APPLICATION OF THIS INFORMATION.  FILMTEC
RECOMMENDS THAT ANY FINAL DESIGN BE REVIEWED BY FILMTEC
APPLICATIONS ENGINEERING DEPARTMENT.
```

FIGURE 4.16. Computer projection for 2:1—4 M array (acceptable design).

```
FEED:      26.67 GPM ,    500 MG/L,  25 DEG C
RECOVERY:  75.0 PERCENT

ARRAY:           1          2          3
NO.OF PV:        2          1          1
ELEMENT:    BW4040     BW4040     BW4040
NO.EL/PV:        6          6          6
EL.TOTAL:       12          6          6

FOULING FACTOR:   0.85

                         FEED    REJECT   AVERAGE
PRESSURE(PSIG)           234.8     86.2     163.2
OSMOTIC PRESSURE(PSIG)     5.0     19.3      11.1
NDP(MEAN)=  152.2 PSIG
AVERAGE PERMEATE FLUX=  15.0 GFD     PERMEATE  FLOW=     20.01 GPM
```

ARRAY	EL.NO.	RECOVERY	PERMEATE		FEED		
			GPD	MG/L	GPM	MG/L	PRESS(PSIG)
1	1	0.090	1724	8	13.3	500	230
	2	0.094	1636	9	12.1	549	218
	3	0.099	1561	10	11.0	604	209
	4	0.105	1496	12	9.9	669	201
	5	0.113	1440	14	8.9	746	194
	6	0.123	1391	18	7.9	839	188
2	1	0.064	1281	20	13.8	954	179
	2	0.064	1183	23	12.9	1018	167
	3	0.063	1095	27	12.1	1086	156
	4	0.062	1016	31	11.4	1157	146
	5	0.062	943	35	10.6	1231	137
	6	0.061	877	41	10.0	1310	129
3	1	0.058	778	49	9.4	1392	117
	2	0.057	723	57	8.8	1474	110
	3	0.056	672	66	8.3	1559	105
	4	0.055	625	76	7.9	1648	99
	5	0.054	582	88	7.4	1740	94
	6	0.053	541	102	7.0	1835	90

```
ARRAY:          TOTAL       1         2         3
REJECT GPM :                13.8      9.4       6.7
REJECT MG/L:                954      1392      1932
PERM GPD :      28813      18496     6395      3921
PERM MG/L:         23         12       28        71
```

FEED WATER IS WELL OR SOFTENED WATER (BW) SDI < 3

FILMTEC ASSUMES NO LIABILITY FOR RESULTS OBTAINED OR DAMAGES
INCURRED FROM THE APPLICATION OF THIS INFORMATION. FILMTEC
RECOMMENDS THAT ANY FINAL DESIGN BE REVIEWED BY FILMTEC
APPLICATIONS ENGINEERING DEPARTMENT.

FIGURE 4.17.   Computer projection for 2:1:1—6 M array (exceeding feed/brine pressure drop).

```
FEED:       26.67 GPM ,    500 MG/L,   25 DEG C
RECOVERY:   75.0 PERCENT

ARRAY:          1            2            3
NO.OF PV:       3            2            1
ELEMENT:    BW4040       BW4040       BW4040
NO.EL/PV:       4            4            4
EL.TOTAL:      12            8            4

FOULING FACTOR:    0.85

                            FEED    REJECT   AVERAGE
PRESSURE(PSIG)             190.5    125.2     162.5
OSMOTIC PRESSURE(PSIG)       5.0     19.3      10.4
NDP(MEAN)=   152.1 PSIG
AVERAGE PERMEATE FLUX=  15.0 GFD     PERMEATE  FLOW=      19.99 GPM

ARRAY EL.NO. RECOVERY     PERMEATE              FEED
                        ----------  -----------------------
                        GPD  MG/L    GPM    MG/L PRESS(PSIG)

    1      1    0.109    1396    9    8.9    500    185
           2    0.118    1350   10    7.9    560    180
           3    0.130    1310   13    7.0    634    175
           4    0.146    1274   16    6.1    727    172

    2      1    0.107    1198   19    7.8    848    164
           2    0.116    1157   23    6.9    948    159
           3    0.126    1118   29    6.1   1068    156
           4    0.140    1081   37    5.4   1219    153

    3      1    0.075    1001   43    9.2   1411    145
           2    0.077     946   50    8.5   1522    139
           3    0.079     895   59    7.9   1645    134
           4    0.081     848   71    7.3   1781    129

ARRAY:          TOTAL       1         2         3
REJECT GPM :               15.6       9.2       6.7
REJECT MG/L:                848      1411      1931
PERM GPD :      28789     15989      9109      3691
PERM MG/L:         22        12        27        55
```

FEED WATER IS WELL OR SOFTENED WATER (BW) SDI < 3

FILMTEC ASSUMES NO LIABILITY FOR RESULTS OBTAINED OR DAMAGES
INCURRED FROM THE APPLICATION OF THIS INFORMATION. FILMTEC
RECOMMENDS THAT ANY FINAL DESIGN BE REVIEWED BY FILMTEC
APPLICATIONS ENGINEERING DEPARTMENT.

FIGURE 4.18.   **Computer projection for 3:2:1—4 M array (acceptable design).**

```
FEED:     133.33 GPM ,     500 MG/L,   15 DEG C
RECOVERY:  74.9 PERCENT

ARRAY:          1          2          3
NO.OF PV:       3          2          1
ELEMENT:  BW8040     BW8040     BW8040
NO.EL/PV:       4          4          4
EL.TOTAL:      12          8          4

FOULING FACTOR:    0.85

                            FEED   REJECT   AVERAGE
PRESSURE(PSIG)             288.0    219.5     258.9
OSMOTIC PRESSURE(PSIG)       4.9     18.9      10.0
NDP(MEAN)=  249.0 PSIG
AVERAGE PERMEATE FLUX=  18.2 GFD        PERMEATE  FLOW=      99.91 GPM
```

ARRAY	EL.NO.	RECOVERY	PERMEATE GPD	MG/L	FEED GPM	MG/L	PRESS(PSIG)
1	1	0.103	6612	7	44.4	500	283
	2	0.113	6468	8	39.9	557	277
	3	0.125	6343	10	35.4	627	272
	4	0.140	6233	12	31.0	714	269
2	1	0.104	5996	14	39.9	828	261
	2	0.114	5862	16	35.8	923	256
	3	0.126	5740	20	31.7	1040	252
	4	0.141	5625	25	27.7	1187	249
3	1	0.078	5371	28	47.6	1377	241
	2	0.082	5191	32	43.9	1492	234
	3	0.087	5025	37	40.3	1622	229
	4	0.092	4872	44	36.8	1773	224

```
ARRAY:          TOTAL        1        2        3
REJECT GPM :                79.9     47.6     33.4
REJECT MG/L:                 828     1377     1948
PERM GPD :     143877      76969    46449    20459
PERM MG/L:         16          9       18       35
```

FEED WATER IS SURFACE WATER (BW) SDI 3 - 5

WARNING ! MAXIMUM RECOMMENDED PERMEATE FLUX HAS BEEN EXCEEDED

FILMTEC ASSUMES NO LIABILITY FOR RESULTS OBTAINED OR DAMAGES
INCURRED FROM THE APPLICATION OF THIS INFORMATION. FILMTEC
RECOMMENDS THAT ANY FINAL DESIGN BE REVIEWED BY FILMTEC
APPLICATIONS ENGINEERING DEPARTMENT.

FIGURE 4.19.   Computer projection for 3:2:1—4 M array (excessive average flux rate).

```
FEED:     133.33 GPM ,    500 MG/L,  15 DEG C
RECOVERY:  75.0 PERCENT

ARRAY:          1          2          3
NO.OF PV:       4          2          1
ELEMENT:  BW8040    BW8040     BW8040
NO.EL/PV:       4          4          4
EL.TOTAL:      16          8          4

FOULING FACTOR:    0.85

                        FEED    REJECT    AVERAGE
PRESSURE(PSIG)          244.6    186.3      223.6
OSMOTIC PRESSURE(PSIG)    4.9     18.9       10.0
NDP(MEAN)=  213.6 PSIG
AVERAGE PERMEATE FLUX= 15.6 GFD       PERMEATE  FLOW=      100.02 GPM

ARRAY EL.NO. RECOVERY     PERMEATE          FEED
                         ----------  -----------------------
                         GPD  MG/L    GPM    MG/L PRESS(PSIG)

  1     1     0.116     5591    8    33.3    500   240
        2     0.130     5495    9    29.5    565   236
        3     0.147     5408   11    25.6    648   233
        4     0.169     5325   14    21.9    757   231

  2     1     0.098     5115   17    36.4    908   224
        2     0.106     4996   20    32.8   1004   220
        3     0.116     4885   24    29.3   1121   217
        4     0.128     4780   29    25.9   1264   214

  3     1     0.070     4537   33    45.3   1446   207
        2     0.072     4370   37    42.1   1551   201
        3     0.075     4216   43    39.1   1669   195
        4     0.078     4072   50    36.1   1801   190

ARRAY:        TOTAL        1         2         3
REJECT GPM :              72.7      45.3      33.3
REJECT MG/L:              908      1446      1949
PERM GPD :    144026    87279     39552     17195
PERM MG/L:        17       11        22        40

FEED WATER IS SURFACE WATER (BW) SDI 3 - 5


WARNING ! MAXIMUM RECOMMENDED ELEMENT RECOVERY HAS BEEN EXCEEDED

FILMTEC ASSUMES NO LIABILITY FOR RESULTS OBTAINED OR DAMAGES
INCURRED FROM THE APPLICATION OF THIS INFORMATION.   FILMTEC
RECOMMENDS THAT ANY FINAL DESIGN BE REVIEWED BY FILMTEC
APPLICATIONS ENGINEERING DEPARTMENT.
```

FIGURE 4.20.   Computer projection for 3:2:1—4 M array (excessive recovery—stage 1, element 4).

```
FEED:    142.86 GPM ,    500 MG/L,   15 DEG C
RECOVERY:  69.9 PERCENT

ARRAY:          1            2            3
NO.OF PV:       4            2            1
ELEMENT:    BW8040       BW8040       BW8040
NO.EL/PV:       4            4            4
EL.TOTAL:      16            8            4

FOULING FACTOR:    0.85

                             FEED    REJECT    AVERAGE
PRESSURE(PSIG)              246.2    173.7      222.4
OSMOTIC PRESSURE(PSIG)       4.9      15.8        9.2
NDP(MEAN)=  213.2 PSIG
AVERAGE PERMEATE FLUX=  15.6 GFD       PERMEATE   FLOW=      99.80 GPM

ARRAY EL.NO. RECOVERY    PERMEATE            FEED
                         ----------  -----------------------
                         GPD   MG/L    GPM    MG/L PRESS(PSIG)

   1      1     0.109    5626    8    35.7     500    241
          2     0.121    5521    9    31.8     560    237
          3     0.135    5427   11    28.0     636    234
          4     0.153    5341   13    24.2     733    231

   2      1     0.087    5119   15    41.0     864    224
          2     0.092    4982   18    37.4     944    219
          3     0.099    4858   21    34.0    1039    215
          4     0.108    4744   25    30.6    1151    211

   3      1     0.057    4469   27    54.6    1287    203
          2     0.057    4254   31    51.5    1363    195
          3     0.058    4055   34    48.6    1444    187
          4     0.059    3873   39    45.7    1530    180

ARRAY:        TOTAL         1         2         3
REJECT GPM :              82.0      54.6      43.1
REJECT MG/L:              864      1287      1624
PERM GPD :    143715     87659     39405     16651
PERM MG/L:        15        10        19        33

FEED WATER IS SURFACE WATER (BW) SDI 3 - 5


FILMTEC ASSUMES NO LIABILITY FOR RESULTS OBTAINED OR DAMAGES
INCURRED FROM THE APPLICATION OF THIS INFORMATION.  FILMTEC
RECOMMENDS THAT ANY FINAL DESIGN BE REVIEWED BY FILMTEC
APPLICATIONS ENGINEERING DEPARTMENT.
```

FIGURE 4.21.   Computer projection for 4:2:1—4 M array (acceptable design).

the average flux rate be limited to 16 gfd on this type of water, the printout contains a warning. This problem may be corrected by adding one pressure vessel to the first stage, but a warning for excessive recovery again appears (see Figure 4.20). The problem occurs because the last element in the first stage is operating at 17 percent recovery. FilmTec recommends limiting the recovery to 16 percent on this type of water. By reducing the system recovery to 70 percent (see Figure 4.21), the problem is corrected.

INSTRUMENTATION AND RECORD KEEPING

In order to ensure the long-term operation of an RO system, a number of parameters need to be monitored.

Pretreatment

The feed water SDI is one of the important items to be checked. How often the test is run will vary, depending on the water source. In the case of a deep well, monthly checks are probably adequate. The SDI of a deep well should be low and constant. When the source is a river or a stream, running the SDI test once each 8-hour shift is appropriate.

If the RO unit is equipped with an acid feed system to lower the feed water pH, a pH monitor with high and low alarm settings is required. This instrument should be checked each shift and standardized on a regular basis.

If an antiscalant or dispersant is used, the chemical feed system must be checked regularly. The amount fed should be logged each shift and compared with the amount required for the quantity of water produced during that period.

Chlorine levels should be checked with a test kit each shift. Similarly, if sodium bisulfite is used to remove chlorine, its residual should be checked with a test kit each shift. Oxidation reduction potential (ORP) meters are available, but to date they are not sufficiently accurate at the low range required in this application.

Other readings to be monitored include media filter pressure drop and polymer addition, softener hardness and salt consumption, cartridge filter pressure drop, and cartridge consumption.

RO Unit Instrumentation

Flow, temperature, and pressure are interrelated and must be monitored to ensure the correct operation of the system. Most systems are equipped with high-pressure pumps capable of delivering 10 percent more pressure than the unit requires with the coolest water temperature expected.

A throttling valve is used on the pump discharge to adjust the feed pressure and maintain the product flow rate at design. Pressure gauges are used on the feed to the first stage and on the reject from the last stage. On start-up, the pressure drop between these two gauges is noted. An increase in this pressure drop at the same flow rate is an indicator of membrane fouling and should be investigated.

On each shift, the pressure, temperature, and flow is noted in the log, at which time trends should also be noted. Periodically, the data should be normalized, that is, corrected for changes in temperature, and compared with the original design.

A drop in flow rate at a constant pressure and temperature may indicate that the membranes are becoming fouled. They should be cleaned before 10 percent of the flux rate is lost. Otherwise, it may not be possible to clean them effectively. If cleaning is required more than every 2 months, the pretreatment may not be adequate.

The product and brine flows are monitored every shift. The ratio of these flows governs the recovery of the unit. It is critical to maintain the recovery rate for which the unit was designed. A high recovery will foul the membranes, whereas a recovery that is too low wastes water and may overload the pump.

In addition, the quality of the feed water and the product water (TDS) must be monitored. A drop in product water quality (increased TDS) can indicate membrane fouling, a bad O-ring seal, or bacterial or chlorine attack.

References

American Society of Civil Engineers and American Water Works Association. 1990. *Water treatment plant design* New York: McGraw Hill.

Amjad, Z. 1988. Mechanistic aspects of reverse osmosis mineral scale formation and inhibition. *Ultrapure Water Journal* 5(6):23–28.

ASTM Standard D-4189-82. 1987. *Standard test method for silt density index (SDI) of water,* Philadelphia: ASTM.

ASTM Standard D-4692-87. 1987. *Standard practice for calculation and adjustment of sulfate scaling salts ($CaSO_4$, $SrSO_4$ and $BaSO_4$) for reverse osmosis,* Philadelphia: ASTM.

ASTM Standard D-4993-89. 1990. *Standard practice for calculation and adjustment of silica (SiO_2) scaling for reverse osmosis,* Philadelphia: ASTM.

Briggs, J.C., and J.F. Ficke. 1977. *Quality of rivers of the United States, 1975 water year.* Based on the National Stream Quality Accounting Network (NASQAN). Reston, VA: United States Department of the Interior Geological Survey.

Cowan, J.C., and D.J. Weintritt. 1976. *Water formed scale deposits.* Houston: Gulf Publishing.

FilmTec Corp. 1988. FilmTec RO system design software.

Kemmer, F.N. 1988. *Nalco water handbook.* New York: McGraw Hill.

UOP Fluid Systems. 1988. ROPROV4 RO projection software.

White, G.C. 1972. *Handbook of chlorination.* New York: Van Nostrand Reinhold.

5

The Importance of Water Analysis for Reverse Osmosis Design and Operation

Joseph P. Hooley

Arrowhead Industrial Water, Inc.
A BFGoodrich Company

Gregory A. Pittner

Arrowhead Industrial Water, Inc.
A BFGoodrich Company

Zahid Amjad

The BFGoodrich Company
Brecksville Research Center

INTRODUCTION

There are numerous industrial uses for purified water. Most of these uses have historically been satisfied by ion exchange (IX), but more recently reverse osmosis (RO) has begun to play a major role. The success of RO is due not only to its far-reaching ability to remove particles and ions from water, but also to the availability of a large number of well-trained individuals to design, build, and successfully operate these systems. The advances in RO technology has made it economically competitive with more traditional methods for industrial water purification.

One of the foundations of this success is the ability and willingness of users to analyze source waters adequately, enabling the skillful design and operation of the RO systems. Proper water analysis can avoid misapplication of RO technology, diagnose problems in existing systems, and allow more accurate evaluation of the system performance.

The performance of RO systems is, to a large degree, controlled by the composition of the feed water. The close relationship between water composition and membrane performance is the basis of the requirement by some membrane manufacturers that a periodic water analysis be performed to maintain the membrane warranty (Himelstein and Goochee 1986).

This chapter addresses the types of analyses that are required, the water components of importance to the operation of the system, and the benefits likely to result from accurate water analysis.

WATER SOURCE INVESTIGATION

A water sample represents a glimpse of the quality of a water supply at a particular time. A single water sample, however, does not lead to an understanding of trends or of changes that may occur over time that could affect the operation of the system. Better understanding of the nature of a water supply can be obtained by collecting multiple samples, as well as by analyzing the probable reasons for water quality variation. Proper profiling of an existing or potential water source is an essential element in the effective design and operation of an RO system.

When an RO system is being designed to operate using a potable water supply from a large municipality, one can be reasonably certain that variations in the quality of the water will remain within the limits imposed by the EPA for potable water supplies. One may also expect that the pH will be high enough to require the addition of acid and/or antiscalant to avoid calcium carbonate precipitation. This is because most supplies are pH adjusted to avoid corrosion of the distribution mains. However, it is important to consider the original source, or sources tributary to the municipal potable water system. Is the water from a surface supply, such as a lake or a stream, which tends to contain high particulate levels? Bacterial particles tend to be present in surface supplies, although nonviable and thus not culturable under the conditions present in a drinking water supply. Well sources tend to be quite uniform in both particulate and ionic quality, as well as very low in particulate matter. Groundwater sources tend to be higher in hardness (calcium and magnesium salts) and alkalinity. If the water source is not potable, or is perhaps an untreated surface or waste supply, much greater emphasis must be placed on the site investigation to determine the possible range over which the quality of the water can vary. Under these circumstances, multiple water samples should be taken over the period of time during which the quality of the water is expected to vary.

Sample analysis can be expensive, thus any techniques that allow more information to be derived from each analysis, or that reduce the number of analyses, should be employed. A proper site investigation is an integral part of designing a sampling program.

SAMPLING

Sampling is an integral part of the overall water analysis program. Two basic requirements should be met in sample procurement (Hamilton 1978). First, samples should accurately represent the water that is being evaluated. This may require a composite sample taken over a period of 24 hours or longer. Second, samples must be of adequate size to ensure that all proper analyses may be performed. Based on the analysis requirements in ASTM Designation: D 4195-88 *Standard Guide for Water Analysis for Reverse Osmosis Applications,* two liters should be adequate for most situations (Fazio et al. 1990).

Prior to taking the first sample, the following questions should be considered carefully (Himelstein and Goochee 1986):

1. What information must be obtained from the water sample?
2. What kinds of analysis procedures will be employed?
3. Are sampling points available that adequately ensure that a representative sample is obtained?
4. Which analyses are best performed on site?
5. What delay will occur in transporting the samples to the laboratory, and what preservation methods are required?
6. What volume of sample is required?
7. What container is best suited for the analyses being performed?

Through the careful consideration of these questions, an effective sampling program can be designed. A complete understanding of the required information will enable one to obtain just the proper amount of information without excessive analysis or the exclusion of valuable data. Through knowledge of the specific analytical steps to be performed, proper sample volumes can be taken, and compatible and effective preservatives can be employed. Through an understanding of the program scope, the sampling can be coordinated with transport to a qualified laboratory.

Suitable sample-collection containers can be critical to ensuring an accurate water analysis. One must consider the possibility of contamination being caused by a container. Figure 5.1 illustrates degradation of a lid caused by a highly caustic sample, releasing high levels of aluminum into the sample and thus destroying its validity. Plastic or glass sample containers are often used. Figure 5.2 shows the kind of sample kit typically available from laboratories specializing in the analysis of water samples for RO systems.

Sample containers must be cleaned prior to use to avoid contamination. Glass

FIGURE 5.1.   Degradation of sample lid caused by caustic sample.

bottles are most often cleaned with a solution of sulfuric acid and sodium dichromate, which is especially effective in avoiding organic contamination of the bottle (Clesceri et al. 1989). Plastic containers may be scrubbed with laboratory detergents or rinsed with concentrated hydrochloric acid, followed by a very thorough rinse with deionized water. Containers used to sample for biological activity must be autoclaved. Table 5.1 gives a list of the proper sample containers and preservation measures based on the constituents being analyzed (Environmental Protection Agency 1979).

Sampling point selection within the system is a major concern. New systems should have sample ports positioned to allow representative samples to be taken easily and should avoid dead areas within the system. If proper care is not taken, the sampling point can often be a source of contamination. The best sampling points are locations where conditions encourage a homogeneous mixing of the water (Hamilton 1978). Prior to sample collection, the sample port should be adequately flushed. The sample container should also be rinsed several times with the water to be sampled, unless the preservative in the container must remain with the sample. Finally, always take feed, product, and brine samples simultaneously. After sample collection, proper logging of the information, including the following, is required (Fazio et al. 1990):

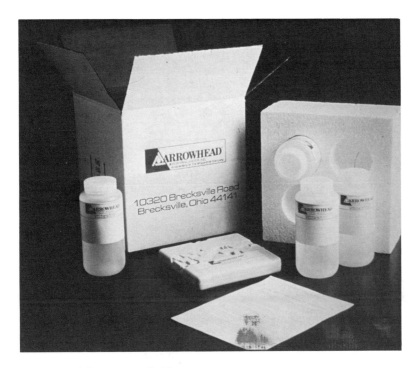

FIGURE 5.2.   RO water sample kit.

1. Name of the sampler
2. Sample point
3. Collection time and date
4. Preservation, if any
5. Water temperature at the time of collection
6. Any analyses done on site, such as for dissolved oxygen or carbon dioxide concentrations

Other data may be useful to include with the sample log; for example, information concerning the operation of the RO pretreatment at the time the samples were taken.

TYPES OF ANALYSES

The cost of routine analytical testing is nominal and generally insignificant in comparison with the cost of shutdown, maintenance, cleaning, or membrane replacement (Himelstein and Amjad 1985). By employment of analyses developed specifically for RO systems, the design and operation of the system may be tailored to ensure maximum efficiency. Often, the true value of a water

TABLE 5.1   Standard Practices for Container, Preservation, Holding of Samples

Physical Properties	Container	Preservation	Holding Time
Conductance	P, G	Cool, 4°C	24 hours
pH	P, G	Determined on site	6 hours
Residue	P, G	Cool, 4°C	7 days
Turbidity	P, G	Cool, 4°C	7 days
Temperature	P, G	Determined on site	None
Inorganics			
Alkalinity	P, G	Cool, 4°C	24 hours
Chloride	P, G	None required	7 days
Chlorine	P, G	Determined on site	None
Fluoride	P, G	Cool, 4°C H_2SO_4 to pH <2	24 hours
Silica	P	Cool, 4°C	7 days
Sulfate	P, G	Cool, 4°C	7 days
Sulfide	P, G	2 ml zinc acetate	24 hours
Sulfite	P, G	Determined on site	None
Metals	P, G	HNO_3 to pH <2	6 months
Organics			
BOD	P, G	Cool, 4°C	24 hours
Oil and grease	G	Cool, 4°C H_2SO_4 or HCl pH <2	24 hours
Organic carbon	P, G	Cool, 4°C H_2SO_4 or HCl to pH <2	24 hours

P = plastic
G = glass

From Himelstein and Goochee, 1986. Reprinted by permission.

analysis is uncovered only after severe problems develop. A brief overview of the types of water analyses and their utility follows.

Feed-water Analysis

Feed-water analysis serves to characterize key contaminants present in the feed water to the RO system (Himelstein and Amjad 1985). This analysis is useful for evaluating and profiling potential water sources, as well as for tracking changing water conditions and their effects on operating systems. The data generated are useful for determining pretreatment requirements, in addition to parameters of the system design such as flux rate and recovery. The ASTM Designation: D 4195-88 "Standard Guide for Water Analysis for Reverse-Osmosis Applications" provides a list of the basic minimum constituents that should be analyzed prior to RO system design. A discussion of the constituents given in this standard follows.

Calcium

Calcium is always present as a divalent ion and forms insoluble or slightly soluble salts with common ions such as carbonate and sulfate. Both can be adequately controlled through crystal growth inhibitors or, if the precipitate is allowed to form, can readily be cleaned from the RO membrane. Calcium concentration, in conjunction with alkalinity and sulfate concentrations, often establishes the upper limit to the water recovery of the RO system.

Magnesium

Magnesium forms sparingly soluble salts such as magnesium silicate and, under high pH conditions, magnesium hydroxide. Both are uncommon in RO systems.

Sodium

Because the sodium ion is monovalent, it forms relatively soluble salts with most anions, including bicarbonate, sulfate, and chloride, and thus seldom presents a fouling issue. However, sodium is the cation that passes most readily through the RO membrane, and will thus be present at the highest concentrations in the permeate and generally control overall rejection.

Strontium

Strontium is a divalent ion found in 20 percent or less of water supplies. It forms a salt with sulfate that is soluble to the level of less than 1 ppm, requiring the feed of a crystal growth inhibitor whenever strontium is present.

Barium

Barium is also a divalent ion, which forms a low solubility compound with sulfate. This must be handled in a manner similar to that used with strontium. Barium and strontium tend to occur together.

Aluminum

Aluminum-based compounds such as sodium aluminate, aluminum sulfate, etc. are common coagulants used in potable water treatment. Under proper circumstances, the aluminum precipitate formed in the water treatment plant is completely filtered out, and thus is not present in the finished water. Time-to-time fluctuations in pH at the water treatment plant, however, cause excessive amounts of aluminum to pass into the distribution system, usually in a dissolved form. Aluminum is generally found as the hydroxide, which causes particulate fouling of the RO membrane.

Manganese

Manganese is usually present at a level of less than 0.3 ppm in public water supplies. Private well water supplies often contain higher manganese levels, often in a dissolved form that precipitates on exposure to the air through reaction with oxygen. A well water supply containing manganese should be pretreated for manganese removal, or steps should be taken to eliminate contact with air or oxidants to assure that the manganese remains soluble.

Iron

Iron, like manganese, is generally present in the dissolved ferrous form, but can oxidize to the ferric state and precipitate as the hydroxide. Municipal potable water supplies generally contain less than 0.3 ppm of iron, but subsurface supplies may contain levels of iron in excess of 10 to 20 ppm. Prior to use in an RO system, a water supply containing iron should be pretreated to remove the iron, or steps should be taken to avoid contact of the water with air or oxidizing substances such as chlorine.

Potassium

Although chemically similar to sodium, potassium is not likely to be present in appreciable quantities in a water supply. No operating or fouling problems are caused by potassium.

Bicarbonate

Bicarbonate (HCO_3^{-1}) is an ionized form of carbon dioxide (CO_2), resulting from the chemical combination of dissolved CO_2 with the hydroxide ion in water. A portion of the bicarbonate present in the feed water can be converted to carbonate as a result of pH changes owing to concentration of salts on the concentrate side of the membrane. Such cases require the addition of acid or a crystal growth inhibitor to avoid calcium carbonate fouling.

Phosphate

Phosphate forms a low solubility salt with calcium. Thus, waters containing phosphate generally require the addition of an antiscalant.

Silica

Silica very often limits the extent to which water can be purified with RO. Although the true solubility level of silica is affected by numerous additions, under the conditions of time, temperature, and pH normally found in an RO unit, the maximum silica concentration is customarily given as 150 ppm. Because many water supplies contain silica of more than 20 ppm, this is often the factor limiting recovery. Moreover, it is extremely unwise to allow silica to precipitate in an RO unit because precipitated silica is virtually impossible to remove from the elements.

Total Dissolved Solids

Total dissolved solids (TDS) is a measure of the total weight of impurities found in a water supply. The measure is too general to predict key operational features of a unit, but does permit a quick, rough estimate of permeate quality.

Carbonate

Carbonate (CO_3^{-2}) forms a sparingly soluble salt with calcium which, as mentioned above, can readily foul RO membranes. Calcium carbonate can easily be removed from a membrane by dropping the pH below 5, but conditions within the RO unit should be such that fouling by this material is avoided.

Sulfate

Sulfate (SO_4^{-2}) forms a sparingly soluble salt with calcium, strontium, or barium. Sulfate does not usually limit the cycles of concentration, unless one or more of these three cations are present at high levels.

Chloride

Chloride is relatively safe, having no negative effects on the life of the membranes, nor generating insoluble salts. Because it passes through RO membranes more easily than most anions, it is one of the predominant permeate ions.

Nitrate

Nitrate (NO_3^-) is similar to chloride in that it is not aggressive to the membrane, nor does it tend to form insoluble salts.

Fluoride

Fluoride has no negative effects on the membrane. It forms insoluble salts with divalent metal ions such as calcium, magnesium, barium, and strontium. The precipitate of these insoluble salts must be handled in a manner similar to that used for other insoluble salts.

Hydrogen Sulfide

Hydrogen sulfide is not present in chlorinated water supplies because it is quickly converted to sulfate. It is, however, associated with some subsurface water supplies that have been depleted of oxygen. Under the right conditions, hydrogen sulfide will form sulfur particles in the water and contribute to RO fouling. Sulfide ion, under the right conditions, may form insoluble metal sulfides.

Free Chlorine

Free chlorine, being oxidizing in nature, can oxidize various materials in the water, and/or attack the RO membrane itself. Thin film composite membranes made from polyamides are sensitive to chlorine, which triggers an oxidation reaction that cleaves the polymer. Chlorine cannot, therefore, be tolerated in the polyamide RO systems. However, it is beneficial for cellulose acetate (CA) membrane systems that are subject to bacterial attack.

Oxygen

Oxygen does not play a direct role in the fouling or operation of RO units, but is indicative of the character and source of the water supply. Waters void of oxygen are likely to contain soluble iron, manganese, and hydrogen sulfide, often in appreciable quantities. Upon addition of oxygen to these waters, precipitates are likely to form.

Carbon Dioxide

Carbon dioxide does not play a major role in RO fouling. However, it does pass readily through any RO membrane, equilibrating on both sides. Under some circumstances, it is the major dissolved constituent in the permeate.

TOC

TOC (total organic carbons) is a measure of the organic, carbonaceous material in water. Under normal circumstances, TOC concentrations will be less than 15 ppm, ranging in most water supplies from 2 to 6 ppm. Under unusual circumstances or in waste water streams, the TOC level can be 50 ppm or higher. Organic compounds that are greater than 200 molecular weight units are well removed with RO. Compounds of lower molecular weight, such as trihalomethanes, are removed only to 40 percent. TOC is not, under most circumstances, a key parameter in the design of RO systems. It may, however, be the predominate contaminant to be removed by the RO system, or to contribute to organic or bacterial fouling of the membrane.

pH

The pH, or the measure of the concentration of hydrogen ions in the water, determines the percentage of inorganic carbon that is in the form of carbon dioxide, bicarbonate, or carbonate. pH determines the extent to which carbon dioxide will appear in the permeate water, or whether calcium carbonate is likely to precipitate.

Temperature

Temperature is important in determining the pressure drop through the membranes at the intended flux rate. It may also be important in determining the rate at which salts will precipitate in the membrane, and thus the extent to which these salts could become a major fouling problem.

Turbidity

Turbidity is a measure of the extent to which light is scattered by particulates in the water. Municipal water supplies normally have a turbidity of less than 1 formazin turbidity unit (FTU), indicating a content of less than approximately 0.5 ppm of suspended solids. Waters with a turbidity in excess of 1 FTU should be adequately pretreated to avoid excessive particulate fouling.

SDI

SDI (Silt Density Index) is a measure of the tendency of water-borne particles to foul the RO membrane. It is not a direct measure of the particle concentration, which is more properly measured by turbidity.

The SDI value is derived from the time required to filter a standard volume of water through a membrane at a constant pressure. It is a measure of the rate at which the membrane becomes plugged with the feed-water source under standard conditions. The SDI value is calculated using the following formula (Fazio et al. 1990):

$$SDI = (1 - T_i/T_f)/T_t \times 100 \tag{1}$$

Where

T_i = initial time required to collect 500 ml.
T_t = total test time, usually 15 minutes.
T_f = total time required to collect 500 ml after the total test time.

SDI is often the method of choice for measuring the colloidal fouling potential of RO feed waters. Caution should be used when interpreting SDI values and care given to conducting the SDI test to prevent misleading results. Not all individuals proficient in RO applications agree regarding the correlation of SDI values and the tendency for RO membrane fouling. An improved correlation will result when SDI changes are monitored over time and the response by a particular RO unit to the SDI of a particular feed water is known.

Some laboratories offer SDI pad digestion to analyze the material retained on the surface of the membrane to identify the particulate matter present. Consultants, design engineers, and system operators may use this information for system adjustments and evaluation of pretreatment. An example of such an SDI pad digestion is shown in Table 5.2. The analysis gives the weight of each foulant material left on the pad, and its percentage of the total amount of material on the pad. The identification of chemical compounds making up the particulate matter can be derived from such information.

Table 5.3 is a feed-water analysis result for a typical RO system. It has a majority of the constituents suggested by ASTM, with a few additions that provide useful information. The first of those additions is information regarding bacterial activity. This provides the design engineer or operator with warnings concerning potential problems stemming from such activity. The second addition to the water analysis is a projection of fouling and scaling tendencies based on the sparingly soluble nature of various salts. These projections are not absolute, but provide guidelines useful in optimizing the performance of the system.

TABLE 5.2   SDI Pad Digestion

Client Code: John Customer   Arrowhead Industrial Water, Inc.

Date: 01/03/90   A BFGoodrich Company

Component	As ppm Lab Solution	As ugm Pad	Component % of Total
Aluminum	1.58	31.6	3%
Barium	0.04	0.8	0%
Boron	2.8	56	5%
Calcium	10.56	211.2	18%
Chromium	<DL	0	0%
Copper	1.83	36.6	3%
Iron	2.42	48.4	4%
Lead	<DL	0	0%
Magnesium	2.4	48	4%
Manganese	0.37	7.4	1%
Phosphorus	8.84	176.8	15%
Potassium	<DL	0	0%
Silicon	2.51	50.2	4%
Sodium	20.96	419.2	36%
Strontium	0.07	1.4	0%
Zinc	2.17	43.4	4%
Alkalinity	<DL	0	0%
Sulfate	<DL	0	0%
T.O.C.	2.01	40.2	3%
pH*	6.00/5.50	—	—
TOTAL:		1171	100%

All concentrations are ppm's unless noted.
<DL = less than detection limit
ppm values are normalized to SDI pad/20 grams solution
*Initial pH water/final pH water after 24 hour soak

A complete feed-water analysis also yields a cation-anion balance. The balance shown in Table 5.3 is accurate to within 93 percent, indicating that only 93 percent of the required anions have been detected in the analysis. Cation-anion balance is required to maintain the electrical neutrality of the water. If there were more cation charges present in the water than anion charges, the water would have a net positive charge and would tend to spark to ground. Therefore, one law of analysis is that the cations and anions must balance. A water sample that is balanced to the extent of 93 percent must be inaccurate in the measure of anions, the measure of cations, or both. Typically, if anions are missing, chloride is usually added. Similarly, if cations were missing, sodium is added.

Table 5.3 also shows the nonionic quality parameters, including alkalinity,

TABLE 5.3   Feedwater Analysis

Project Code:	John Customer	Arrowhead Industrial Water, Inc.
Collected:	01/02/90	A BFGoodrich Company
Received:	01/03/90	
Analyzed:	01/03/90	By: JPH

Component	Feed ppm
Alkalinity[1]	36.0
Aluminum	<DL
Barium	<DL
Boron	0.63
Calcium	10.75
Chloride	189.9
Chromium	<DL
Conductivity[2]	746.0
Copper	<DL
Fluoride	<DL
Iron	<DL
Lead	<DL
Magnesium	0.7
Manganese	0.02
Nitrate	<DL
Phosphate	<DL
Phosphorus	<DL
Potassium	<DL
Silicon	0.85
Sodium	110.5
Strontium	0.16
Sulfate	7.8
Zinc	0.851
Ionic Strength	0.0066
Ion Balance	93%
pH	8.20
Temperature (°C)	20.0

Foulants	Saturation of Brine (%)
Calcium Carbonate Scaling Index	0.75
Calcium Sulfate	0.3
Calcium Fluoride	0.0
Barium Sulfate	0.0
Strontium Sulfate	0.4
Silica	5.7

Projection of brine % of saturation at 75% recovery.

TABLE 5.3   *(continued)*

Additional Components	Feed
Alkalinity, total (ppm CaCO$_3$)	36.0
Alkalinity, bicarbonate (ppm CaCO$_3$)	33.4
Alkalinity, carbonate (ppm CaCO$_3$)	2.6
Carbon, total organic (ppm)	1.5
Carbon dioxide (ppm as ion, calculated)	0.3
Chloride, total (ppm Cl$_2$)	0.7
Hardness, total (ppm CaCO$_3$)	29.9
Osmotic pressure (psi)	4.2
Silica, total (ppm SiO$_2$, calculated/Si)	1.8
Specific gravity (g/ml)	1.0
TDS (ppm as NaCl by conductivity)	369.6
Turbidity (NTU)	1.5

Bacterial Analysis

Bacterial Type	Feed
Aerobic	
Yeast and mold (colonies per ml)	0
Standard plate count (colonies per ml)	0
Total coliform (colonies per 100 ml)	0

TNTC indicates too numerous to count

Anaerobic	
Sulfide generating bacteria (degree of infection)	0
Degree of infection: 3 = heavy infection	
2 = moderate infection	
1 = slight infection	
0 = no infection	

<DL = less than detection limit
Ratio = specific ion recovery/overall recovery
% rejection is based on feed-brine average concentration.
[1]Alkalinity in ppm CaCO$_3$
[2]Conductivity in umhos/cm

hardness, and osmotic pressure. Alkalinity is the measure of the chemical constituents of the water that react with hydrogen ion and thus resist a decrease in pH. Those constituents include carbonate, bicarbonate, and hydroxide. Under normal circumstances, the hydroxide contribution to the alkalinity is almost zero, thus the total alkalinity is often simply the sum of the bicarbonate and carbonate ions.

Osmotic pressure offsets the pump pressure in the RO unit, reducing the force pushing water through the membranes. The osmotic pressure is approximately 11 pounds per 1,000 ppm, and thus does not become a key factor in the design of an RO unit until the TDS of the feed water is higher than several thousand ppm.

System Analysis

System analysis is a tool for providing summary performance data on the operating RO system. It is performed by conducting complete analyses on the feed, product, and concentrate streams. As such, this approach not only demonstrates the concentrations of various components in the feed, product, and concentrate, but when properly conducted, also computes parameters that can be useful for predicting the causes of performance decline, such as various saturation indices. A thorough system analysis provides information concerning the rejection and recovery of individual constituents and calculates the scaling potential of the brine stream.

Table 5.4 provides an example of a typical system analysis. The principal difference between this analysis and the feed-water analysis shown in Table 5.3, is that fouling projections are based on actual operating data, whereas feed-water projections are not. This type of analysis also serves as a check of actual versus design flows of the operating system. System operation may change over a period of time, and a historical perspective is an excellent point from which to investigate the performance characteristics of an RO system.

Whenever possible, quarterly system analyses should be performed to reduce operating cost, assure integrity of membrane performance guarantees, and help to avoid premature membrane replacement.

The system analysis given in Table 5.4 provides information condensed into a simple and usable form. Numerous computer calculations that lead to a better understanding of the operation of the RO unit are printed on the two-page report, enabling the operator to readily understand and interpret significant trends. The printout in Table 5.4 includes the rejection of each ion, with the calculation based on the feed brine average. For example, the rejection of alkalinity is given as 87.4 percent. This is computed by taking the average of 48 and 142, which is 95, and dividing it into the product alkalinity level of 12, then subtracting the result from 1, and expressing the number as a percentage. The recovery, in percentage, is calculated by taking the difference of the brine alkalinity and the feed alkalinity, and dividing that result by the difference between the brine and product alkalinities. Thus the recovery is equal to 94 divided by 130 or 72.3 percent. The recovery ratio is simply the ratio of the recovery of alkalinity, divided by the overall recovery, and indicates the relative performance of the RO system with respect to each ion. The following is a summary of the formulae:

TABLE 5.4   System Analysis

Project Code: John Customer   Arrowhead Industrial Water, Inc.
Collected: 01/02/90   A BFGoodrich Company
Received: 01/03/90
Analyzed: 01/03/90   By: JPH

Component	Feed ppm	Product ppm	Brine ppm	Rejection %	Recovery %	Recovery Ratio
Alkalinity[1]	48.0	12.0	142.0	87.4	72.3	0.977
Aluminum	<DL	<DL	<DL	NA	NA	NA
Barium	0.011	<DL	0.042	100.0	73.8	0.997
Boron	0.33	0.33	0.39	8.3	100.0	1.351
Calcium	58.48	10.21	195.00	91.9	73.9	0.998
Chloride	514.4	121.4	1771.9	89.4	76.2	1.030
Chromium	<DL	<DL	<DL	NA	NA	NA
Conductivity[2]	2121.0	487.0	5670.0	87.5	68.5	0.925
Copper	<DL	<DL	<DL	NA	NA	NA
Fluoride	2.3	<DL	8.5	100.0	72.9	0.986
Iron	<DL	<DL	<DL	NA	NA	NA
Lead	<DL	<DL	<DL	NA	NA	NA
Magnesium	86.1	11.8	265.0	93.3	70.7	0.955
Manganese	0.03	0.01	0.07	80.0	66.7	0.901
Nitrate	<DL	<DL	<DL	NA	NA	NA
Phosphate	<DL	<DL	<DL	NA	NA	NA
Phosphorus	<DL	<DL	<DL	NA	NA	NA
Potassium	10.3	<DL	34.8	100.0	70.4	0.951
Silicon	12.21	3.33	35.00	85.9	72.0	0.972
Sodium	257.6	64.4	797.0	87.8	73.6	0.995
Strontium	9.57	1.43	31.00	93.0	72.5	0.979
Sulfate	286.3	27.7	944.0	95.5	71.8	0.970
Zinc	<DL	<DL	<DL	NA	NA	NA

Ionic strength	0.0297	0.0053	0.0961
Ion balance	99%	98%	93%
pH	5.75	5.10	6.20
Temperature (°C)	26.8		

Foulants	Saturation of Brine (%)
Calcium carbonate scaling index	−0.88
Calcium sulfate	15.7
Calcium fluoride	2434.3
Barium sulfate	413.9
Strontium sulfate	194.5
Silica	64.7

If the % of saturation exceeds 100 then antiscalant is required!
Projection of brine % of saturation at 75% recovery

TABLE 5.4   *(continued)*

Additional Components	Feed	Product	Brine
Alkalinity, total (ppm CaCO₃)	48.0	12.0	142.0
Alkalinity, bicarbonate (ppm CaCO₃)	48.0	12.0	142.0
Alkalinity, carbonate (ppm CaCO₃)	0.0	0.0	0.0
Carbon, total organic (ppm)	2.1	0.5	6.3
Carbon dioxide (ppm as ion, calculated)	178.9	209.0	182.0
Chloride, total (ppm Cl₂)	<DL	<DL	0.1
Hardness, total (ppm CaCO₃)	511.8	75.8	1614.6
Osmotic pressure (psi)	12.6	2.8	35.2
Silica, total (ppm SiO₂, calculated/Si)	26.1	7.1	74.9
Specific gravity (g/ml)	1.0	1.0	1.0
TDS (ppm AS NaCl by conductivity)	1086.4	238.6	3037.2
Turbidity (NTU)	0.8	0.5	1.0

Bacterial Analysis

Bacterial Type	Feed	Product	Brine
Aerobic			
Yeast and mold (colonies per ml)	0	0	0
Standard plate count (colonies per ml)	0	0	0
Total coliform (colonies per 100 ml)	0	0	0
Anaerobic			
Sulfide generating bacteria (degree of infection)	0	0	0

Degree of infection: 3 = heavy infection
2 = moderate infection
1 = slight infection
0 = no infection

<DL = less than detection limit
NA = not analyzed
Ratio = specific ion recovery/overall recovery
% Rejection is based on feed-brine average concentration.
[1]Alkalinity in ppm CaCO₃
[2]Conductivity in umhos/cm

$$\text{Rejection } (\%) = \left[1 - \frac{(\text{product}) \times 2}{(\text{brine} + \text{feed})}\right] \times 100\% \qquad (2)$$

$$\text{Recovery } (\%) = \left[\frac{(\text{brine} - \text{feed})}{(\text{brine} - \text{feed})}\right] \times 100\% \qquad (3)$$

$$\text{Recovery ratio} = \frac{\text{individual ion recovery (\%)}}{\text{overall recovery (\%)}} \qquad (4)$$

Where

Overall recovery

$$= \frac{[(\text{total brine ions (Meq/L)}) - \text{total feed ions (Meq/L))}] \times 100\%}{[(\text{total brine ions (Meq/L)}) - \text{total product ions (Meq/L))}]}$$

Below the list of ions the pH is given for the feed, product, and brine. No chemicals are added within the RO, but there is generally an increase in pH of the brine and a decrease of pH in the product, both as compared with the feed. This results from the fact that a higher percentage of the carbon dioxide passes through the membrane to the product than does the bicarbonate. The ratio of carbon dioxide and bicarbonate determines pH, with a higher ratio resulting in a lower pH. This causes the low pH of the product and the relatively higher pH of the concentrate.

The saturation of the brine is also given in Table 5.4. At a glance one can see the potential for precipitation of various sparingly soluble salts. It is not uncommon to operate an RO unit at above the solubility of many salts without precipitation. This is due to the fact that most salts tend to supersaturate readily and thus remain soluble for at least as much time as it takes for the brine to pass out of the RO unit.

In Table 5.4, the feed, product, and brine are analyzed with respect to the ionic balance. The match between the ions given as milligrams per liter is important in assuring that the analysis has successfully detected the presence of all ions. An ion balance that is not within 95 percent would be suspect.

Cartridge Filter Analysis

In many cases, membrane fouling is due to inadequate pretreatment of the feed water. Its cause, however, cannot always be diagnosed by feed-water or system analysis. These tests are most useful in calculating scaling potential based directly on dissolved ions. However, the state of the art has not reached the point at which feed-water analyses can accurately predict colloidal or particulate fouling.

Cartridge filters are normally in operation for a period lasting from several days to as long as months (Himelstein and Goochee 1986). This provides a long-term accumulation of particulate matter from the feed-water source, which can be converted into information through proper analysis work.

Once the cartridge filter is removed from the system, a close examination can provide useful information and insight into fouling problems. Visual examination through an optical microscope is the first step in revealing the nature of foulants. Table 5.5 provides a complete analysis of a cartridge filter wherein both

TABLE 5.5   Cartridge Filter Digestion
Client Code: John Customer   Arrowhead Industrial Water, Inc.
Date: 01/03/90   A BFGoodrich Company

Component	Outer Fibers (ppm)	Inner Fibers (ppm)	Component % of Total	% Change Inner versus Outer
Aluminum	13.97	8.47	11	−39
Barium	0.15	0.08	0	−47
Boron	0.76	0.76	1	0
Calcium	17.88	11.71	15	−35
Chromium	<DL	<DL	0	0
Copper	0.75	0.39	1	−48
Iron	20.56	9.6	17	−53
Lead	<DL	<DL	0	0
Magnesium	3.22	2.09	3	−35
Manganese	0.26	0.13	0	−50
Phosphorus	2.66	<DL	2	−100
Potassium	<DL	<DL	0	0
Silicon	18.98	12.12	16	−36
Sodium	2.89	2.78	2	−4
Strontium	0.2	0.14	0	−30
Zinc	0.68	0.59	1	−13
Alkalinity	25.9	21.96	21	−15
Sulfate	4.78	4.2	4	−12
TOC	8.78	5.68	7	−35
pH*	6.00/6.90	6.00/6.80	—	—

All concentrations are ppm's unless noted.
<DL = less than detection limit
ppm values are normalized to 1 gram filter/40 grams solution
*Initial pH water/final pH water after 24 hour soak

inner and outer fibers have been separately analyzed. This can be useful in identifying possible sources of particulate contamination, such as microbial slime, silt, corrosion products, and silicates.

The particulates that tend to adhere to a cartridge filter that precedes an RO system correlate well with the particles that tend to foul the RO unit. A complete diagnosis and digestion of the cartridge filter thus provides an excellent opportunity to gain insight into the types of foulants that lie within the membrane. From a complete analysis, one can derive the chemical composition of the particles. In addition, the relative distribution of these particles toward the outer and inner portions of the filter cartridge is important in gaining insight into the types of particles. For instance, assume that the silica concentration on the outer portions of the fibers is approximately 19 ppm, whereas the silica concentration on the inner fibers is 12 ppm. This would indicate that there is a portion of the silica present as relatively large particles, which are screened out on the surface of the cartridge filter. Iron, present at a level twice as high on the outside of the cartridge filter, is also being removed by the same mechanism. The calcium level differential is also high, indicating a large number of calcium-containing particles.

Membrane Autopsy

A more drastic analytical method for troubleshooting performance difficulties in RO systems is membrane autopsy. It is occasionally used to help diagnose problems that routine analytical methods are unable to pinpoint. Some types of chemical, physical, or membrane degradation can be identified only by this method.

Autopsy involves varying degrees of analytical complexity, depending on the nature of the problem. The analytical procedures can range from simple visual inspection of the surface of the membrane to complex X-ray and infrared analysis. Autopsies are normally performed in two distinct phases (Amjad et al. 1988).

Nondestructive phase
Visual inspection for physical damage and fouling loading
Distilled water soak and analysis for mineral content
Performance test to determine membrane condition relative to original specifications
Evaluation of membrane performance following cleaning treatments

If performance is not acceptable following these procedures, the next phase must be carried out.

Destructive Phase
The membrane is opened and a representative sample of the fouled area is taken.
Foulant material is digested to enable positive identification.
Scanning electron microscopy is used to identify fouled and control samples.
Energy dispersive X-ray analyses (EDX) of fouled and controlled samples are performed to determine elemental composition of the foulant.
Biological examination is performed to characterize the type of biofoulant material that may be present.
Infrared spectroscopy is performed on fouled and control samples to determine the presence and identity of organic materials, as well as the chemical degradation of membranes (i.e., hydrolysis of acetate group).
Optical microscopy is used to discern features in the range of 1 to 100 μm.

COST OF WATER ANALYSIS

The value of feed-water source analysis is illustrated by the cost associated with the design and operation of RO systems. The total water cost, including operating and capital recovery costs, for RO systems using brackish water, can range between $1.00 to $1.50 per 1,000 gallons of produced water (Parekh 1991). A typical 400 gpm system can cost in excess of $300,000 per year. Poor design based on improper water analysis can lead to unscheduled downtime, increase in chemical cleaning costs, premature membrane replacement, and increased labor cost, all of which affect the economic attractiveness of RO systems and operating efficiency.

The cost of a water analysis can vary, based on the number of constituents to be analyzed and the laboratory chosen to perform the analysis. Membrane manufacturers, service companies, and chemical supply companies often offer supporting water analysis. Environmental laboratories also perform water analyses. The cost to perform an individual water analysis to generate the information listed by the ASTM (provided at the end of this paragraph) can be $250 to $400 per sample. This is a minor cost to ensure proper design and operation of an RO system.

Calcium (Ca)	Sulfate (SO_4)
Magnesium (Mg)	Chloride (Cl)
Sodium (Na)	Nitrate (NO_3)
Strontium (Sr)	Fluoride (F)
Barium (Ba)	Hydrogen sulfide (H_2S)
Aluminum (Al) (total and dissolved)	Free chlorine (Cl_2)
Manganese (Mn) (total and dissolved)	Oxygen (O_2)
Iron (Fe) (total and dissolved)	Carbon dioxide (CO_2)
Potassium (K)	TOC (organic content)
Bicarbonate (HCO_3)	pH
Phosphate (PO_4) (total)	Temperature
Silica (SiO_2) (total and dissolved)	Turbidity
Total dissolved solids (TDS)	SDI (Silt Density Index)
Carbonate (CO_3)	

Note: If analysis is complete, the total cations and total anions should balance within ±2 percent.

WHO BENEFITS FROM WATER ANALYSIS

Companies often have internal engineering capabilities or use professional engineering firms for RO system design. Whatever the source of the design, an accurate water analysis is essential in evaluating a feed water for an RO system. Proper profiling of a potential feed water depends on the water source itself.

Companies frequently have little choice in their feed-water source. Different sources of water offer a wide range for potential problems with RO systems. The engineer has to be aware of not only the difference between various water sources and their impact on the design, but also of the seasonal variation within single water sources. Temperature changes, turnover, and intrusions or upsets into water sources all have an effect on the operation and design of an RO system. An accurate water analysis taken over time to profile proper- ly the changes in the feed water greatly improves the system design and op- eration.

Computer models are presently used by engineers and operators to project fouling tendencies based on water chemistry. Water analyses that accurately represent the feed water are imperative to ensure proper use of this valuable design tool. Evaluations of pretreatment considerations are often based on these projections. Inaccurate analysis or misinformation can be more dangerous than a lack of information.

Pilot testing is often employed to increase the reliability of the design by simulating actual operating conditions. Analytical support through water analysis is essential at this stage. Pilot testing can aid in predicting problem areas before large expenditures are committed.

Membrane manufacturers supply water analysis services to support their sales and provide technical support to their customers. The guarantees given by most manufacturers are usually dependent on a routine water analysis and accurate record keeping. Failed or shorter membrane life because of fouling, scaling or physical degradation can often be traced back to problems in the feed water. The original water analysis performed may no longer be valid, owing to changes in the water. Customers feel that membranes constitute the key piece in the RO system and, when performance is poor, often blame it on the membrane itself. In such situations, an accurate water analysis can often aid in diagnosing the cause of poor membrane performance. Often it is not a failure of the membrane, but a design problem.

Operators of the system can benefit greatly from accurate water analysis by tracking the performance of their system and identifying downward trends. Steps to correct such trends can be implemented before costly downtime or membrane failure occurs. Membrane replacement can be costly. Periodic water analysis can also alert the operator to changes in the feed-water composition and its possible impact on the RO system. Pretreatment considerations can be adjusted accord- ingly.

RO consultants' true value becomes apparent when there are design or operational problems or changes in the feed water. They provide a valuable service in troubleshooting and diagnosing problems. Based on experience and reliable information, adjustments can be made to improve the performance of troubled RO systems. Water analysis is an important diagnostic tool. It allows consultants to understand how the system is presently operating in comparison with design performance standards. Evaluating the historical operational data base, including periodic feed-water and systems analysis, is helpful in determining whether the problem is a slow decline in system performance or a drastic upset.

SUMMARY

As sources of quality feed water become scarce, the demands on RO as a technology become complex and demanding. The composition of the feed water is the limiting factor for both RO design and operation. The pretreatment considerations and changing water conditions must be included in any design and operational programs. Accurate water analysis is essential to ensure successful application of RO as a water treatment program.

Two basic requirements must be fulfilled to ensure proper analysis. The sample must accurately represent the water being evaluated and must be of sufficient quantity for the analyses being performed. A properly designed sampling program can ensure that the information generated will be meaningful. Avoid contamination by choosing representative sample points. Sample containers can often be a source of contamination. The sample bottles shown in Figure 5.3 can offer only questionable results at best. Proper sample containers and preservation techniques should be coordinated with the particular analyses.

FIGURE 5.3.   Improper sample containers.

For proper feed-water characterization, certain constituents, as described earlier, should be measured. Beyond those characterizing the feed water, other types of analyses, such as cartridge filter, SDI pad digestions, and membrane autopsies provide valuable information in troubleshooting existing systems.

The capital and operating costs associated with RO systems demand that feed water be accurately characterized. RO systems are quickly becoming a technology of choice and, as water quality decreases, the need for accurate information for design is magnified.

References

Amjad, Z., J. Isner, and R. Williams, 1988. The role of analytical techniques in solving reverse osmosis fouling problems. *Ultrapure Water* 5(5):20–26.

Clesceri, A.E., A.E. Greenberg, and R.R. Trusse II, editors. 1989. *Standard methods for examination of water and wastewater.* 17th ed. Washington, D.C.: The American Public Health Association, American Water Work Association and American Water Pollution Control Federation.

Environmental Protection Agency. 1979. *Methods for chemical analysis of water and wastes.* Pub. No. 600/4-79-020, Cincinnati, Ohio.

Fazio, P.C., M. Gorman, E.L. Gutman, C.T. Hsia, G. Johns, and J. Kramer, editors. 1990. *Annual book of ASTM standards.* Section II, Vols. 11.01 and 11.02, *Water and environmental technology,* Philadelphia: ASTM.

Hamilton, C.E. 1978. *Manual on water.* ASTM Special Technical Pub. 442A, Dow Chemical Company.

Himelstein, W. and Z. Amjad. 1985. The role of water analysis, scale control and cleaning agents in reverse osmosis. *Ultrapure Water* 2(2):32–36.

Himelstein, W., and J. Goochee. 1986. Understanding water chemistry: The need for improved analysis in reverse osmosis plant operations. In *Proceedings of National Water Supply Improvement Association Conference,* Washington, D.C.

Parekh, B. 1991. Get your process water to come clean. *Chemical Engineering* 98(1):75–85.

6

Mechanistic Aspects of Reverse Osmosis Membrane Biofouling and Prevention

Hans-Curt Flemming, Ph.D.

Institut für Siedlungswasserbau der Universität Stuttgart

INTRODUCTION

Fouling is a term generally used to describe the undesirable formation of deposits on surfaces. This occurs when rejected solids are not transported from the surface of the membrane back to the bulk stream. As a result, dissolved salts, suspended solids and microorganisms accumulate at the membrane surface. Fouling can be caused by the following processes:

1. Inorganic deposits (scaling)
2. Organic molecules adsorption (organic fouling)
3. Particulate deposition (colloidal fouling)
4. Microbial adhesion and growth (biofouling)

In the reverse osmosis (RO) operation, all of these processes tend to decrease the performance of the permeators. The different types of fouling frequently occur at the same time, influencing each other. Unfortunately, the interactions between the various types of fouling are poorly understood; biofouling is not always easy to recognize and is usually considered last.

RO technology had to contend with biofouling quite early (Kissinger 1970; McDonough and Hargrove 1972). In many cases, a shotgun approach may have replaced systematic research, and an operator was content to solve the actual problem rather than doing expensive, difficult, and time-consuming investigations with unpredictable results. The success of RO technology, however, demanded more systematic research on biofouling in order to develop more rational strategies for prevention and cleaning.

Biofouling is the result of the complex interaction between the membrane material, fluid parameters (such as dissolved substances, flow velocity, pressure, etc.), and microorganisms. Biofouling is basically a problem of biofilm growth. It can really be understood only if the implications of the biofilm mode of bacterial growth (Costerton et al. 1987) and its dynamics (Characklis and Marshall 1990) are considered.

BIOFOULING IN RO WATER SYSTEMS

Cases and Causes

Reports of biofouling problems in RO technology come from various fields (Eisenberg and Middlebrooks 1984), such as high-purity water production (Dlouhy and Marquardt 1984; Rechen 1985; Martyak 1988; Flemming and Schaule 1989), food technology (Kissinger 1970; McDonough and Hargrove 1972), seawater desalination (Winters and Isquith 1979; Winters et al. 1983; Kaakinen and Moody 1985; Winters 1987; Ebrahim and Malik 1987; Applegate et al. 1989; Ahmed and Alansari 1989), domestic water purification units (Payment 1989), and waste water reclamation (Winfield 1979a; Argo and Ridgway 1982; Ridgway et al. 1983a,b; 1984a,b; 1985, 1986; Whittaker et al. 1984).

The main source of microbial contamination is the feed water. Surface waters with high numbers of microorganisms are bound to lead to microbial problems (Applegate et al. 1989; Nagel 1990). RO systems are characterized by large membrane surface areas, which increases the likelihood of adhering bacteria. In addition, RO membranes experience a vertical vector that transports cells toward the rejecting surface and increases the probability of contact with suspended microorganisms. The composition of the microflora forming the biofilm may be quite different from the bulk microflora composition (Dott and Schoenen 1985; Ridgway 1987). Practically every surface in both natural and technical environments are colonized by bacteria. From the total spectrum of microorganisms, those preferring the membrane material best will be found most commonly adhering to the membrane surface.

Pretreatment steps can also be sources of biofouling. A common cause of membrane fouling is the overfeeding of flocculants used to assist in the removal of suspended solids (Graham et al. 1989). This provides a suitable habitat for microbial growth. Ahmed and Alansari (1989) report that the conditioning agent sodium hexametaphosphate (SHMP) is both a source of microorganisms (requiring sanitation) and a source of nutrients for the biofilm (hydrolysis of SHMP yields orthophosphates, a good food source). Sodium thiosulfate, used to neutralize chlorine, can also be utilized as a nutrient source by bacteria (Winters and Isquith 1979). Applegate and colleagues (1989) found that biofouling was favored by the chlorination of raw seawater; chlorine degrades humic acids into biologically assimilable material, which supports the growth of the biofilm. One would expect ozonation to produce similar reactions (Gilbert 1988). Oil and other hydrocarbons are not usually present in significant quantities in feed water but can enter the RO system through leaks or spills. Even in small quantities, oil serves as a nutrient for accelerated microbial growth or can agglomerate other suspended particles into a sticky mass (Himelstein and Amjad 1985).

The piping and treatment system prior to the RO unit offers many surfaces suitable for biofilms, such as ion exchangers (Flemming 1987), granulated activated carbon filters (Geldreich et al. 1985; Mittelman 1987; LeChevallier and McFeters 1990), degasifiers (Mittelman 1987) and holding tanks. All of these biofilms can release microorganisms that colonize other surfaces of the system, including the membranes. In one instance, algal growth was reported in a system using fiberglass reinforced plastic piping (Ebrahim and Malik 1987). It was determined that light penetrating the pipe walls was sufficient to induce this growth.

It is important to realize that the biofilm mode of growth enables microorganisms to survive and multiply even in extremely low nutrient habitats (Geesey 1987; Morita 1988) such as ultrapure water systems (Rechen 1985). Adhesion is considered to be a microbial strategy for starvation survival (Marshall 1988). When adhering to a surface, bacteria can scavenge and accumulate the necessary nutrients. Moreover, protection against biocides is provided (Costerton and Lashen 1984; Costerton et al. 1987). Bacteria may respond to nutrient limitation by turning into ultramicrobacteria of less than 0.1 μm in diameter (Geesey 1987; Morita 1988). ZoBell and Grant (1943) showed that *E. coli*, *Staphylococcus citreus*, *Bacillus megatherium*, *Proteus vulgaris*, and *Lactobacillus lactis* multiplied in solutions containing 0.1 mg/l of glucose, which represents 40 μg/l carbon. More recently, *Ps. aeruginosa* has been shown to grow in tap water at 25 μg/l total carbon supplied by a variety of compounds (Geesey 1987; van der Koij and Hijnen 1988). Thus it is not surprising that problems associated with biofilms occur even in high-purity industrial waters containing <5 to 100 μg/l total organic carbon (TOC).

Effects of Biofouling on Performance of RO Units

Normally, biofouling is a comparatively slow process and is, in many cases, undiscovered and associated problems are related to other causes for a long period of time. A biofoulant may manifest itself in many ways, depending on the length of time it has been developing. Ridgway (1988) lists the following signs and symptoms:

1. A gradual decline in the rate of water transport per unit area (i.e., a membrane flux decline)

2. A gradual increase in the transmembrane and differential pressure (i.e., feed pressure and [delta P])
3. A gradual increase in the permeability of the membranes to dissolved materials (i.e., decreased mineral rejection)

The cumulative effects of membrane biofouling are the following:

1. Increased cleaning and maintenance costs of the RO system
2. A noticeable deterioration of product water quality (waters may have to be posttreated)
3. A significantly reduced membrane lifetime

Combined, these factors result in substantially increased operating and maintenance costs for an RO system and tend to become irreversible if allowed to go unchecked. Membranes can account for as much as 20 percent of the installed costs of a typical brackish water plant and 30 percent of a seawater plant (Graham et al. 1989).

The biofouling process may result in secondary mechanical deformation ("telescoping") of spiral wound modules. Biofouling of membrane surfaces is invariably accompanied by some degree of mineral deposition (inorganic scaling); however, the types and concentrations of the inorganic constituents associated with RO biofilms may very greatly, depending on the specific chemical and microbiological properties of the feed water (Ridgway 1988).

It is not yet well understood by which molecular mechanisms bacteria and their extracellular material reduce the membrane flux. The relationship between biofilm thickness and membrane permeability has not been fully evaluated (Ridgway 1988). The membrane flux decline is often biphasic: a rapid decline phase lasting for a few weeks that asymptotically approaches some equilibrium value that may be only 60 to 80 percent of the original permeate production (Ridgway 1988; Nagel 1990).

The increase of the delta-p results not only from the attachment of the biofilm to the membrane surface, but also from the colonization of the Vexar spacer. Certain effects will be noted before gross accumulations of slime. Rapid increases in frictional flow resistance have been reported for water moving at 2 m/sec, when the film thickness exceeds the thickness of the viscous sublayer (Characklis 1982). With inorganic deposits, flow is reduced owing to increased surface roughness and constriction. Biofouling, however, causes flow losses owing not only to constriction of the passage, but also to such factors as absorption of kinetic energy from the flowing water by its rippled elastomeric surface contour. To illustrate, one case is reported in which a 130 μm film in a half-inch tube caused a 110 percent increase in pressure drop. This amount of pure constriction can account only for a 10 percent increase in the pressure drop (Characklis 1982). Biofilms affect flow in at least three ways (Ridgway 1988):

1. Reduce the cross-sectional area available for flow
2. Increase the roughness of the surface
3. Increase drag by virtue of their viscoelastic properties

The fouling of the ends of spiral wound elements within a system can also result in a higher membrane delta-p. The uneven flow distribution (channeling) tends to increase the likelihood of further inorganic scale formation because turbulent mixing in proximity to the membrane is significantly reduced (Ridgway 1988).

Ridgway (1988) proposed that two primary mechanisms are instrumental in diminishing the mineral rejection properties of a biofouled RO membrane. First, the biofilm can be considered as a more or less dense gel layer. Dissolved minerals tend to accumulate in this layer and increase concentration polarization. The gel layer will tend to reduce turbulent flow very near the membrane surface and extend the viscous boundary layer beyond the normal range that would be characteristic of a clean membrane. When this occurs, there will be a greater opportunity for dissolved minerals to accumulate in close proximity to the membrane surface (trapped in the biofilm—eventually bound by exopolymers) above the concentration found in the bulk fluid phase; thus, the biofilm increases concentration polarization.

The second mechanism is either direct or indirect promotion of the decomposition of the RO membrane polymer. There are reports suggesting that bacteria attached to cellulose acetate (CA) RO membrane may produce enzymes or other substances that can attack and hydrolyze the membrane directly. However, enzyme-catalyzed hydrolytic cleavage of the fully substituted CA polymer has not yet been unequivocally demonstrated under controlled laboratory conditions. Therefore, the true significance of this potential mechanism of RO membrane degradation is not clear at this time. Ho et al. (1983) reported that bacteria and fungi isolated from fouled RO membranes were unable to degrade CA membranes and/or free polymer in vitro. Some of the fungal isolates were, however, capable of enzymatically hydrolyzing unsubstituted cellulose. Significant hydrolysis of the membrane polymer was reportedly evident after 4 weeks of incubation at 25°C. Yet no direct evidence was presented that linked the observed membrane degradation with enzymatic activity of any kind. It is possible that nonenzymatic processes, such as localized pH changes owing to metabolic activity of the adherent bacteria, could have indirectly caused an acceleration of hydrolysis (Ridgway 1988). Few, if any, reports exist concerning biodeterioration of polyamide or polysulfone type membranes.

Theoretically, RO membranes do not allow bacterial penetration. However, product water may be microbially contaminated (Jacobs 1981). The occurrence of microorganisms on the permeate side may be due to at least four different mechanisms:

1. Penetration through loose O-rings or other fittings (Whittaker et al. 1984)
2. Microscopic imperfections in the membrane material, through which the microorganisms may grow or may be transported by pressure changes (Christian and Meltzer 1986; Wallhäusser 1983)
3. Retrograde growth of microorganisms from contaminated outlets, contaminating the water from biofilms on surfaces of the permeate piping system
4. Microbial contaminations of the feed-water side resulting from manufacture or from operation.

Identification of Biofouling

It is generally difficult (sometimes nearly impossible) to differentiate between the potential causes of RO performance decline from only the water and solute transport data of an RO plant, because the performance responses to various types of membrane fouling and mechanical deformation are usually similar. Frequently, the problem is called "biofouling" when there is no other explanation. Changes in salt rejection, flux, and delta-p are associated with biofouling.

Indirect evidence of biofouling suggests the efficiency of cleaning and bactericidal actions as monitored by the increase of a unit's performance. The occurrence of microbial slimes in accessible parts of the system makes it very likely that the membrane surfaces bear biofilms as well. A very simple field method giving an indication for the biological nature of a deposit is to scratch a small portion from the surface, burn it on a lighter's flame, and check for the smell of burnt hair. Other simple testing procedures using specific dyes are under current development (Flemming and Schaule, in preparation). Slimy consistence, odor, and results of visual inspection provide good information to the trained observer. If macroscopically detectable slimes cover the system's walls, however, biofilm development has proceeded quite far already.

A biofilm is characterized by the following features:

• High water content (70 to 95 percent)
• High content of organic matter (70 to 95 percent)
• High numbers of colony-forming units (CFU) and cells (microscopic counts)
• High content of carbohydrates and proteins
• High content of adenosine triphosphate (ATP)
• Low content of inorganic matter

Appropriate sampling of biofilms (Geesey and Mittelman 1987) is the prerequisite for the reliable identification of a biofouling situation.

The situation may become complicated by mutual interactions between different types of fouling. If a primary foulant such as iron oxide or a biofilm is allowed to accumulate on the membrane, it will prevent the migration of other ions, calcium and sulfate ions, for example, back into the bulk flow stream, resulting in a supersaturated condition and the formation of gypsum scale beneath the primary foulant. Either the iron scale or biofilm would have been easily cleaned, the gypsum scale is not. In severe cases, upstream fouling with silt can cause zones of low flow in downstream portions of elements. This causes the membrane to act like a depth filter, rather than a cross-flow filter, and can lead to secondary fouling by causing precipitation of insoluble scales such as the sulfates of barium, calcium, or strontium; calcium fluoride; or silica, as the concentration of these species exceeds saturation in areas of low flow across the membrane surface (Graham et al. 1989).

Microorganisms Occurring in RO Water Systems

Observations indicate that a wide variety of physiological types of microorganisms are capable of adhereing to and colonizing RO membrane surfaces, creating a highly complex microbial biofilm. Table 6.1 gives an overview of identified species:

Mycobacteria are frequently found in the biofilms of fouled RO membranes (DuMoulin et al. 1988; Ridgway et al. 1983b). *Mycobacteria* are considered to be opportunistic pathogens, frequently found in opportunistic infections of HIV patients (DuMoulin et al. 1988). Winters and Isquith (1979) report that many bacteria that colonize, form slime, and become numerically important in the microfouling syndrome, do not form colonies on usual plating media.

TABLE 6.1   Microbial Species Isolated and Identified from Reverse Osmosis Membranes

Species	*Reference*
Fungi	
Fusarium	Ho et al. 1983
Penicillium	Ho et al. 1983
Trichoderma	Ho et al. 1983
Bacteria	
Acinetobacter	Ho et al. 1983; Ridgway et al. 1983a; Flemming & Schaule, 1988a; Payment 1989
Alcaligenes	Payment 1989
Arthrobacter	Ho et al. 1983
Bacillus	Ho et al. 1983; Ridgway et al. 1983a
Chromobacterium	Payment 1989
Flavobacterium	Ho et al. 1983; Ridgway et al., 1983a
Lactobacillus	Ho et al. 1983
Micrococcus	Ho et al. 1983
Mycobacterium	Ridgway et al. 1983a; DuMoulin et al., 1988
Pseudomonas	Ho et al. 1983; Ridgway et al. 1983a; Flemming & Schaule 1988a; Martyak 1988; Payment 1989
Staphylococcus	Flemming & Schaule 1988a

MECHANISTIC ASPECTS OF MEMBRANE FOULING

The biofouling process can be divided roughly into four phases (Winters and Isquith 1979; Winfield 1979; Flemming and Schaule 1988a). These processes are schematically depicted in Figure 6.1.

The Conditioning Film

The first step in biofilm formation, prior to microbial adhesion, is the irreversible adsorption of macromolecules, which leads to a "conditioning film" (e.g., humic substances, lipopolysaccharides, and other products of microbial turnover) on the membrane surface. These particles can mask the original surface properties (Loeb and Neihof 1975; Baier 1980; Corpe 1980) and cause a slightly negative surface charge. This phase is completed within seconds to minutes after immersion of a surface into an aqueous system. A macromolecular conditioning film on a solid substratum presents a new set of surface characteristics to the bulk liquid phase. Thus, the electrostatic charge (Loeb and Neihof 1975) and the critical

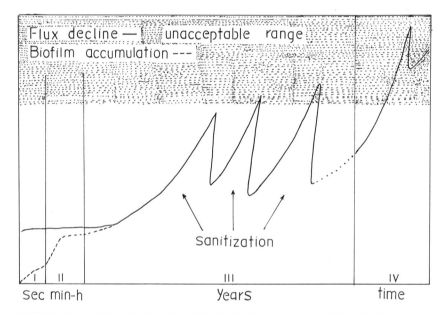

FIGURE 6.1.   Schematic depiction of the biofouling process in an RO unit. Phase I: formation of a conditioning film. Phase II: primary adhesion. Phase III: biofilm development, sanitization measures. Phase IV: irreversible blocking of the module. Biofilm accumulation: flux decline.

surface tension (an indication of surface free energy [Baier 1980]) may change with the accumulation of a conditioning layer. Accordingly, a variety of protein conditioning layers, on surfaces such as polystyrene and glass, may both increase and inhibit the attachment of various bacterial species (Fletcher 1976; Pringle and Fletcher 1986).

Primary Microbial Adhesion of Microorganisms to Membranes

Bacteria are transported to the membrane surface because of the following effects:

- Hydrodynamic forces (flow, turbulence, flux)
- Motility (chemotaxis)
- Brownian motion and diffusion

Under nonstatic conditions, a laminar boundary layer builds up, the thickness of which depends on flow velocity, viscosity, and wall roughness. It can be assessed between 10 and 50 μm under normal RO operating conditions, which exceeds the thickness of individual bacteria significantly. This explains how biofilms can exist in high-velocity systems (Characklis 1981). The shear forces of the turbulent flow do not reach through the viscous sublayer and affect a bacterial monolayer.

The microbial cell has to penetrate this sublayer before attaining contact to the surface. Hydrodynamic forces are considered to be responsible for transporting the cells to the boundary layer, whereas Brownian motion, motility, and/or diffusion seem to be the mechanisms to cross the boundary layer (Characklis 1981). The "sticking efficiency" of microbial cells, however, decreases with increasing flow velocities (Characklis 1990). It has been shown in experiments of biofouling of heat exchanger surfaces that increased flow slows down the primary adhesion rate.

In the RO process, the flux as a transport vector vertical to the membrane must be considered to be a strong force assisting the cell to penetrate the viscous sublayer (Flemming et al. 1992). However, there has not yet been established a clear relationship between the flow velocity, the flux, and the deposition of bacteria.

Probstein et al. (1981) found that the rate of colloid deposition on the RO membranes (and, hence, the flux decline kinetics) was largely independent of fluid shear (i.e., brine flow velocity) within the range tested. However, equilibrium foulant thickness was found to be an inverse function of the brine flow velocity, regardless of whether flow was laminar or turbulent. These findings suggest that fluid velocities typically encountered in most RO systems do little to impede the initial rate of colloidal (or microbial) foulant deposition, although they can significantly limit the overall thickness of the fouling layer (Ridgway 1988).

Primary adhesion is a heterogeneous process, including three phases:

1. *The microorganisms* (semisolid phase), governed by factors such as species composition of microflora, cell number in the bulk, viability, nutrient status, growth phase, hydrophobicity, surface charge, and extracellular polymeric substances (EPS)

2. *The fluid* (liquid phase), with factors such as temperature, pH, dissolved organic and inorganic substances, viscosity, surface tension, and hydrodynamic parameters (shear forces, flux, turbulence)

3. *The membrane surface* (solid phase), influencing adhesion via chemical composition, hydrophobicity, surface charge, conditioning film, and "biological affinity"

Role of Microorganisms
The number of cells in suspension influences the number of adhering cells. Bryers and Characklis (1981) described the rate of initial biofilm formation in flowing systems using a first-order expression. The rate constant is shown to be a linear function of the biomass concentration, the Reynolds number (R_e), and the biomass growth rate. According to this theory, particle flux from the bulk fluid is expected to increase with increasing R_e, but experimental results show that the rate of biofilm accumulation decreases as turbulence increases, suggesting that particle flux from the bulk solution is only one of the mechanisms contributing to biofilm accumulation. Flemming and Schaule (1988b) found a linear correlation between the number of *Pseudomonas diminuta* cells in suspension and that on a polysulfone membrane surface (until the surface was completely covered). However, this was true only in a concentration range from above 10^6 cells/ml. Below this limit, the adsorption rate was significantly lower. Ridgway and colleagues (1986) found a linear correlation from zero up to 10^6 cells/ml with mycobacteria adhering to CA membranes in a similar assay, when the slope turned into a plateau below complete covering of the surface.

The growth phase of the bacteria may influence their adhesion behavior as well (Fletcher 1980; McEldowney and Fletcher, 1986b; Little et al. 1986). Experiments with *Ps. diminuta* and *Staphylococcus warneri* adhering to polysulfone membranes, however, showed similar adhesion rates in both log phase and stationary phase (Flemming and Schaule, unpublished). The nutrient status of the cells has been reported to be of major influence. Starvation is considered to alter the cell surface and to increase the adhesion of marine microorganisms (Dawson et al. 1981; Kjelleberg and Hermansson 1984; Little et al. 1986). Adhesion experiments with *Pseudomonas fluorescens* on various membrane materials consistently showed much higher adhesion rates of the starving cells (Flemming and Schaule, in preparation). The same experiments with *Pseudomonas vesicularis*, however, resulted in a significantly slower adhesion rate of starving cells (Flemming and Schaule 1988a,b). Thus, various organisms respond to starvation with different surface alterations, leading to either a more or less adhesive cell surface. It is probable that under nutrient-limited conditions, more species tend to adhere than under nutrient-rich conditions.

The surface charge is very likely to influence primary microbial adhesion. Most bacteria are slightly negatively charged (Baier 1980). Therefore, they have to overcome a repulsion barrier when they attach to slightly negatively charged membrane surfaces. The surface charge of a cell may show a wide range under different physiological conditions, as indicated by adhesion to anion exchange resins (Hogt et al. 1985). Localized positively charged groups on the cell surface may play a more essential role than the overall negatively charged surface (Stenström 1989). In nature, surfaces of quite different charges are colonized by bacteria, indicating that there will always be some colonizing species among the bulk microflora in touch with the surface immersed into water. The role of microbial surface charge in RO biofouling has not yet been fully investigated.

Microbial cell hydrophobicity plays an important role in the hypothesis on bacterial adhesion (Absalom et al. 1983; Hogt et al. 1985; Marshall 1985, 1986; Loosdrecht et al. 1987, 1990; Stenström 1989; Vanhaecke et al. 1990). Hydrophobic interactions are considered to be the driving force for the primary adhesion of *mycobacteria* to CA membranes (Ridgway et al. 1983b; Ridgway 1987). Nutrient conditions could change the hydrophobicity of fast-adhering bacterial strains considerably (Kjelleberg and Hermansson 1984; Little et al. 1986; Flemming and Schaule 1991).

Experiments with differently hydrophobic mutants of fast-adhering *Pseudomonas diminuta* (between 15 and 73 percent), as compared with strongly hydrophobic *Rhodococcus* and strongly hydrophilic *Pseudomonas fluorescens*, showed that cell hydrophobicity does not play a crucial role in the adhesion to the relatively hydrophobic polysulfone membrane materials (see Figure 6.2).

This finding is consistent with results in other systems. Hogt et al. (1985) showed that the adhesion of staphylococci to fluorinated polyethylene-propylene films was not related to the relative surface charge or the hydrophobicity of the bacteria.

EPS are the first cell envelope structures to encounter a solid surface in primary contact. Thus, they are of great interest for the understanding of the adhesion process. Many but not all of the EPS consist of polysaccharides, built up by oligosaccharide repeating units which frequently contain sugar acids, giving an anionic character to the whole molecule (Sutherland 1983; Whitfield 1988). Proteins, glycoproteins, and glycolipids can also be among the EPS. It is

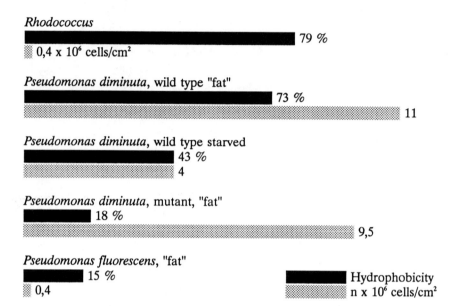

FIGURE 6.2.   Role of cell hydrophobicity in adherence of various bacterial strains with different hydrophobicity to polysulfone membranes. "Fat"—harvested from log phase. "Starved"—after 6 days in tap water.

proposed that additional EPS production in response to the adsorption is involved in the development of the surface films, but possibly not in the initial adhesion process. In those strains that do produce polysaccharides, the cells that attach develop into microcolonies (Allison and Sutherland 1987).

However, it is safe to assume that EPS may play an important role in adhesion through the following functions:

1. Determining the chemistry, and thus the potential adhesive interactions, of the cell surface
2. Interacting with the surface and its conditioning film
3. Eventually acting as a "polymer bridge" between cell and surface

Clearly, the potential for adhesive interactions varies with different organisms and their environmental conditions, and, accordingly, attachment ability differs enormously. However, not only do various bacterial strains differ in their ability to attach to solid surfaces, but the attachment of a given bacterium may also vary with environmental conditions.

Influence of Fluid Parameters

Primary adhesion has been reported to be influenced by temperature in several adhesion systems (Characklis 1982; Applegate and Erkenbrecher 1987; Applegate et al. 1989). As expected, rising temperatures (within a physiological range) result in increasing adhesion rates (Fletcher 1980; Pedersen 1982; Donlan and Pipes 1988). However, experiments with primary adhesion of *Pseudomonas diminuta* and *Staph. warneri* to various membranes showed almost no temperature dependence in a range between 10°C and 30°C (Flemming and Schaule 1990). These results indicate that temperature may not always be a significantly influencing factor in primary adhesion systems. This may mean that even in cold water many bacteria may adhere to a surface. Once temperature is rising, they may grow rapidly.

McEldowney and Fletcher (1986b) and Applegate et al. (1989) report a significant influence of pH on bacterial attachment. However, other authors found that pH changes in a range between 4 and 9 were not correlated with alterations of the adhesion behavior (Donlan and Pipes 1988; Stenström 1989). This was found to be true with CA membranes (Ridgway et al. 1985) as well as with other membrane materials (Flemming and Schaule 1991).

The electrolyte concentration influences the electrostatic double layer too, and it has been found to influence the initial adhesion, leading to a higher rate with increasing concentrations from zero to 0.1 M (Marshall et al. 1971; Fletcher 1980; Stanley 1983; Ridgway et al. 1984b; Gordon and Millero 1984; Loosdrecht et al. 1987, 1989, 1990a,b). Again, this may not be true for all systems, probably depending on the individual adhesion mechanism. In other experiments, Ridgway et al. (1985), Donlan and Pipes (1988), Stenström (1989), and Schaule et al. (1992) found no significant correlation between electrolyte concentration and initial adhesion rate.

The dissolved organic substances can be expected to influence initial adhesion in two ways: (1) by adsorbing to the surface of the membrane and formation of a conditioning film and (2) via adsorption to the microbial cell surface. Polycations from pretreatment have been reported to enhance biofouling (Winters 1987), which can be expected because of the electrostatic attraction to the slightly negative overall charge of microbial cells. Adsorbed organic substances can be utilized as a nutrient source, e.g., humic acid degradation products after chlorine treatment of raw seawater, which can induce a chemotactic response of the cells. Precoating of surfaces with various proteins led to enhanced and reduced adhesion rates (McEldowney and Fletcher 1986a,b; Pringle and Fletcher 1986). Thus, not surprisingly, the nature of the preadsorbed substances governs the adhesion process. Ridgway and colleagues (1986) report that the effect of quaternary amines on bacterial adhesion is dependent on the chain length of the aliphatic part, thus indicating that the molecular basis of the detergent might be more important than the surface active properties.

Influence of the Membrane Surface

Although up to now there is no known material that cannot be colonized by microorganisms, differences in "biological affinity" have been noted (Dempsey 1981; Exner et al. 1983; Dott and Schoenen 1985; Little et al. 1986; McEldowney and Fletcher 1986b; Pedersen 1990). Experiments on the colonization rate of different materials used as piping materials showed different biological affinities (Schoenen et al. 1988). This was also observed with membrane materials by Ridgway (1987), who found a prefouled CA membrane yielding the lowest amount of bacteria. Some authors (Ridgway 1987; Flemming and Schaule 1988a,b) found a significantly lower bacterial affinity of poly(ether urea) membrane material, as compared with polyamide and polysulfone material, which could be confirmed with pure and mixed cultures. The explanation of this effect is not yet established. The correlation of the low biological affinity to a low magnitude of surface charge (zeta potential) of this membrane in comparison with polyamide membrane material is offered as an explanation (Light et al. 1988); however, this overall surface parameter has not been systematically investigated in comparable systems.

Other overall surface properties such as hydrophobicity and, derived from measurement of surface hydrophobicity, surface energy have been widely accepted as important for bacterial adhesion. It is generally recognized that the majority of microorganisms show a preference for hydrophobic (low-energy) surfaces (Baier 1980; Fletcher and Marshall 1982; Absalom et al. 1983; Pringle and Fletcher 1986; Marshall 1986). Stenström (1989) proposed that an assay of the measurement of hydrophobic interaction would measure the adhesion potential to low-energy surfaces.

The results obtained with various membranes, however, showed only marginal differences in the hydrophobicity of the materials, whereas significant differences in biological affinity could be observed (Flemming and Schaule 1990). Thus, hydrophobicity of the membrane surface cannot be the factor discriminating high or low biological affinity. Efforts to change the membrane surface hydrophobicity in order to prevent microbial adhesion are, therefore, very likely to be ineffective.

Surface roughness is acknowledged to be an important factor in biofouling of heat exchanger surfaces (Characklis and Cooksey 1983). It significantly influences transport rate and microbial cell attachment for several reasons:

1. It increases the convective transport (mass transport) near the surface.
2. It provides more "shelter" from shear forces for small particles.
3. It increases the surface area for attachment.

If the dimension of the roughness is much smaller than the dimension of the microorganisms, however, this factor tends to be negligible. Vanhaecke and colleagues (1990) report that even electropolished metal surfaces were colonized after only 30 seconds of contact to *Pseudomonas* strains. Membrane surfaces usually are very smooth and have been shown to be colonized within a very few minutes (Flemming and Schaule 1988a). Biofilms can be expected to display a rougher surface than the original membranes. For the biofouling process of the total unit, however, other parts of the system have to be considered as well, which may offer much rougher surfaces and are thus colonized very well.

Ridgway et al. (1983b) assumed the existence of a limited number of adhesion sites because of adhesion kinetics of the saturation type which conform closely to the Langmuir adsorption isotherm. He calculated a cell number of 3×10^6 per cm^2, which was substantially less than the theoretical adsorption capacity of the RO membrane surface based on geometric and spacial considerations. Similar binding site numbers and adsorption kinetics have been reported for the adhesion of other kinds of bacteria to a wide variety of solid substrates (Daniels 1980; Celesk and London 1980; Belas and Colwell 1982; Marshall 1985). Absalom et al. (1983) found a maximal coverage of primary adhesion of 12 to 18 percent; Fletcher (1977) reported 45 percent coverage. However, no information is currently available concerning the physiochemical nature, distribution, or specificity of these adsorption sites on RO membrane materials. Flemming and Schaule (1988b) observed complete covering of the surface of nonstarving cells, whereas starving cells covered only 10 to 15 percent of the available membrane surface. This indicates that the adhesion pattern may be due rather to cell surface properties than to the substratum.

Kinetics and Models for the Mechanism of Primary Adhesion

Primary adhesion of microorganisms to RO membrane materials occurs within a short time (see Figure 6.3). The highest rate of attachment is during the first hour, slowly giving way to a plateau phase after 4 to 6 hours. Thus, the primary biofilm development that may lead to biofouling begins early in the module's lifetime. Interestingly, dead cells adhere as fast as living cells. Fletcher (1980) discriminates between "active" and "passive" adhesion. Active adhesion requires physiological activity of the cell in response to contact to a surface. One can conclude that mixed population will always contain strains of dead cells that adhere as well as living cells (Fletcher 1980; Argo and Ridgway 1982; Stanley 1983; Costerton and Lashen 1984; Whittaker et al. 1984; Allison and Sutherland 1987; Flemming and Schaule 1988a). This means for the RO practice that it is not enough to kill bacteria; they also have to be removed, otherwise many of them attach to the surfaces to be cleaned, providing both substrate and substratum for subsequent bacteria from the feed water.

A two-step model for the initial bacterial adhesion was proposed (Marshall et

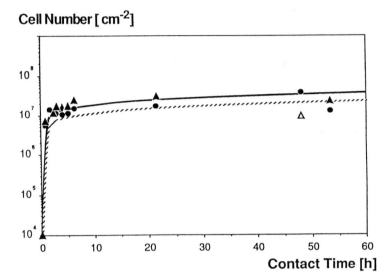

FIGURE 6.3.   Adhesion rate of living and dead *Pseudomonas diminuta* cells to polyethersulfone membrane material during the first hours of contact. ●—● = living cells, ▲—▲ = killed cells.

al. 1971). The first step includes reversible adhesion, which is an instantaneous attraction by long-range forces holding bacteria near a surface. They continue to exhibit Brownian motion and can be readily removed from the surface by the shearing effect of a water jet or by violent rotational movements of mobile bacteria. Some mobile bacteria are able to overcome the attractive forces, often becoming reversibly attached at other sites and utilizing nutrients adsorbed on the surface (Marshall 1986). Irreversible adhesion has been defined as a time-dependent firm adhesion whereby bacteria no longer exhibit Brownian motion and cannot be removed by washing.

Using some simplifications, the DLVO* theory, dealing with the interactions between suspended colloids and solid surfaces, can be applied to the initial adhesion of microorganisms as well (Marshall et al. 1971; Rutter and Vincent 1988). The DLVO theory proposes that the interaction between two particles is made up of two additive components. These are (1) an attractive component resulting from van der Waals forces and (2) a repulsive component (if both particles bear the same charge sign), which is due to the overlap between the electrical double layers associated with the charge groups on the two particles. There are two particle separation distances at which attraction may occur. These are called the primary minimum, at small interparticle distances where attractive

---

*Named after their authors; Derjaguin, Landau, Vermeer, and Overbeek.

forces are strong and governed by short-range forces, and the secondary minimum, at relatively large interparticle distances where attractive forces are weaker. There is also a separation by a repulsion barrier at intermediate distances. This theory fits for a number of combinations of microorganisms and surfaces (Marshall et al. 1971; Fletcher 1976, 1980), but it includes simplifications such as the absence of interactions between the particles (stickiness), the assumption of a perfect round shape, and the total reversibility of the process. Thus, it is not surprising that in spite of its merits in defining and quantifying the adhesion process, the theory describes only special cases.

The thermodynamic approach is based on the assumption that bacterial adhesion will occur when the thermodynamic function, i.e., the free energy of adhesion, is minimized. Unlike the DLVO theory, this approach assumes that the effect of the electrical charges and receptor ligand interactions may be neglected. The thermodynamic model may be expressed as follows:

$$\text{delta } F_{adh} = \text{gamma}_{BS} - \text{gamma}_{BL} - \text{gamma}_{SL} \tag{1}$$

where delta F_{adh} is the change in free energy associated with bacterial adhesion, and gamma_{BS}, gamma_{BL}, and gamma_{SL} are the bacterium-substratum, bacterium-liquid, and substratum-liquid interfacial tensions, respectively (Marshall 1986). Values for the parameters can be derived from contact angle measurements and from an equation-of-state approach (Absalom et al. 1983). The adhesion is predicted to be correlated to the surface tension of the bacteria in a way that depends on the relative value of the bacterial surface tension to the surface tension of the suspending medium.

There are, however, discrepancies between the theoretical model and observed adhesion, which may arise because of the fact that the electrostatic interactions are neglected. Indeed, when the bacterial surface tension is equal to the surface tension of the suspending liquid, zero adhesion is predicted—but does not occur. It was shown that in this case the level of adhesion was constant because of electrostatic interactions (Defriese and Gekas 1988).

The threadlike structure of the EPS may also help to bridge the repulsion barrier ("polymer bridging" [Loosdrecht et al. 1990]) and change the microenvironment drastically. This is probably the main reason for the divergences of experimental results from the predictions of these models. When EPS are located at or near to the primary attraction minimum, they become subjected to the short-range forces. For any particular EPS, it is likely that different combinations of these short-range forces will operate at different substratum surfaces. However, EPS are not "sticky" per se. Brown and colleagues (1977) report the enhanced production of EPS without enhanced adhesion.

Marshall (1986) states that the actual modes of interaction of different EPS and various substrata are very poorly understood. Many microbial cells have surface appendages such as flagellae, fimbriae, and pili. They will influence processes occurring at the first contact of cells with the substrate. Present evidence supports the importance of projections in the adsorption process and suggests that cells may "probe" the substratum in some cases, searching for a local inhomogeneity suitable for colonization (Pethica 1980; Characklis 1990). Fast-adhering *Pseudomonas diminuta* has a flagella, whereas fast-adhering *Staph. warneri* has no cellular appendages (Flemming and Schaule, unpublished).

The foregoing summarized results of various investigations indicate clearly that microbial adhesion cannot be adequately explained in terms of a single parameter, but depends on a complex mixture of factors that will vary with the particular microorganism, its surrounding environment, and the surface of attachment. McEldowney and Fletcher (1987) found no particular species combinations or conditions (i.e., surface composition or sequence of attachment) that consistently influenced attachment or detachment. Adhesion varied with bacterial species, substratum, and electrolyte type and concentration, with no apparent correlation between adhesion and electrolyte valence or concentration. (McEldowney and Fletcher 1986b). Moreover, there is evidence that a given microorganism can use different adhesion strategies, depending on the surface to be colonized (Paul and Jeffrey 1985).

Thus, it must be considered that microbial adhesion occurs in a very wide variety of conditions; for example, ionic strengths ranging from nearly pure water to saline solutions in excess of 3 M NaCl, pH from low to fairly high values, temperatures from supercooled to more than 120°F. Adhesion sometimes occurs in media of relatively low or high dielectric constant and with many undefined chemicals present. Various organisms adhere to various surfaces under various conditions, whereas overall parameters such as hydrophobicity, surface charge, and surface energy do not give a consistent pattern. It is unlikely that a single mechanism can explain microbial adhesion. Hydrophobicity, for example, can be caused by very different molecular structures. Some of them seem to act adhesively and some do not (see Figure 6.2).

Considering this situation, the molecular basis for various adhesion systems seems to be the key to an understanding of the adhesion process, which differs from one adhesion system to another. In addition, the exact cell surface component and the partner of the membrane surface is still unknown—be it the membrane material itself or some conditioning material. More fundamental research in this area is definitely needed.

Inhibition of Primary Adhesion

A number of substances are reported to inhibit the primary adhesion of microorganisms to various surfaces. In many cases, detergents are effective (McEldowney and Fletcher 1986b). Humphries and colleagues (1986, 1987) found that most of the tested linear ethoxylated surfactants had an antiadhesive effect on hydrophobic, but not hydrophilic, surfaces. Marshall and Blainey (1988) report the antiadhesive action of blocks of polymers of ethylene oxide and propylene oxide to be effective inhibitors against the adhesion of marine microorganisms to polystyrene (hydrophobic), but not to glass surfaces. The best results were achieved with surfactants of the procetyl series (Humphries et al. 1987). Ridgway (1988) and Flemming and Schaule (1988b) found that the Triton-X series (nonionic polyoxyethylene surfactants) inhibited bacterial adhesion to various membrane materials, but not with all combinations of bacterial strains and membrane materials. However, not all detergents are inhibitive, thus indicating that interactions other than solely hydrophobicity must be involved in the adhesion system. Similar results can be obtained with the "impregnation" of polysulfone surfaces with Tween 20 or Pluronic 64 (Schaule et al. 1990). This approach of influencing the conditioning film by adding antiadhesive substances seems to be interesting mainly for closed systems such as heat exchangers, rather than for RO systems. There are experiments in progress to investigate the adsorption rate for these substances and the durability of their inhibiting effect in combinations of mixed cultures (Flemming and Schaule, in preparation).

Interestingly, *Staph. warneri* cells or their EPS, preadsorbed on various membrane materials, inhibit significantly the subsequent adhesion of *Ps. diminuta*. If both strains are present at the same time in the adhesion assay, there is no mutual influence. There is no influence on adhesion when *Ps. diminuta* or its EPS is preadsorbed (Schaule et al., in preparation).

Detachment of Primary Biofilm

Whittaker and colleagues (1984) succeeded in removing biofilms with 6M urea/SDS and with cleaning agents. It is much more difficult to remove cells once attached than to inhibit their adhesion. Triton-X 100 has proven to be an agent that inhibits the adhesion of *Ps. diminuta* to polysulfone to 99 percent (Flemming and Schaule 1988b). It could not, however, remove more than 50 percent of *Ps. diminuta* already attached to the same surface (Flemming and Schaule 1989). Prolonged contact of the cells with the surface decreased the amount of detached cells, indicating that an older biofilm is more difficult to remove than a fresh one. Only a few days of contact can change the detachment rate drastically, as it has been found in detachment assays using ultrasonic energy (Flemming and Schaule 1988b; Zips et al., in press). The consequences for maintenance of an RO plant are obvious.

Biofilm Growth and Development

Once irreversibly attached, cells may grow and proliferate into microcolonies, excreting EPS (Allison and Sutherland 1987), colonizing free surface areas, and forming a multilayered biofilm. The production of an acidic exopolysaccharide matrix by sessile bacteria has been noted in freshwater systems, marine systems, flowing streams, and even in soils (Geesey 1982). The matrix is seen to be a highly hydrated mass of polyanionic polymers (Sutherland 1983). In nature, bacteria live predominantly in adherent biofilms in which they are protected from antibacterial agents by the ion exchange capacity and the physical blocking properties (diffusion barrier) of their enveloping polysaccharides (Costerton and Lashen 1984). They are almost ubiquitous in natural systems and may constitute the basic "fabric" of adherent biofilms.

Microbial adhesion continues during the biofouling process with cells arriving with the feed water. Ridgway and colleagues (1984b) found, after a few months, a microflora present that was significantly different from the biofilm formed by the pioneering organisms. They isolated and identified a total of 336 bacteria from various stages of RO membrane biofouling, ranging from 2 to 215 days. Surprisingly, all 200 isolates recovered during the initial 43 days of membrane operation fell within the acid-fast genus *Mycobacterium*. After 57 days, a small number of gram-positive organisms were also detected (2.3 percent), but following 215 days of operation the mycobacteria, which had predominated earlier, had been completely replaced by a much more diverse microbial assemblage in which *Acinetobacter* (53 percent) and *Pseudomonas* (24 percent) organisms were most prevalent. The remainder of the population sampled after 215 days was comprised of *Shigella, Alcaligenes, Klebsiella, Flavobacterium,* and a small group of gram-positive and unidentified microbes.

How long membrane material properties influence adhesion after being masked by a biofilm is an open question. Newly arriving cells, however, do not encounter a free membrane surface, but a more heterogeneous surface of an already preformed biofilm. This explains why biofilms show clear temporal diversities. The production of such surface biopolymers (i.e., bacterial glycocalyx) is correlated with substantially increased resistance to many chemical biocides and antibiotic substances (Costerton et al. 1987). Interestingly, Flemming and Schaule (unpublished) found a microflora from an irreversibly blocked polyamide membrane to significantly prefer this material, in primary adhesion assays, in comparison with other membrane materials. Particles may be adsorbed and buried in the EPS slime matrix between the microorganisms. At this point, the biofilm begins to develop its spatial heterogeneity (Hamilton 1987). Higher temperatures allow the microorganisms to become more active metabolically, leading to enhanced proliferation. However, the major portion of the biofilm usually consists of EPS, exceeding microbial mass up to more than 60 percent.

The biofilm is responsible for interferences with plant performance. The slope in Figure 6.1 indicates a flux decline, this answered by cleaning measures and subsequent flux declines. Phase III characterizes the module's lifetime. This phase is generally governed by (1) the nutrient situation, (2) the temperature, and

(3) the shear stress. It has to be considered that water with even a low nutrient level may provide a nutrient-saturated environment for a biofilm if the volumetric flow rate is high. The shear stress will, in many cases, be the limiting factor for biofilm growth in RO systems.

Structure and Composition of Fouling Biofilms:
Results of Module Autopsies

If all sanitizing methods fail to restore an acceptable level of performance, resulting in uneconomical operation, the module is irreversibly blocked (see Figure 6.1, Phase IV). An autopsy of the module may now be carried out (Himelstein and Amjad 1985; Ridgway et al. 1986; Flemming and Schaule 1989). The diagnosis in this case will often be "biofouling." Prerequisites for an appropriate module autopsy are information about the membrane type, the module type, the composition of the feed water and conditioning chemicals, the "history" of the operation (such as the slope of delta-p, cleaning chemicals and measures, etc.), and the solution in which the module has been stored. A typical analysis includes the determination of the content of water, dry matter, total organic carbon (TOC), chemical oxygen demand (COD), residue at 600°C, metals, silica, sulfate, carbohydrate, protein, and ATP, as well as colony and cell number per cm^2 (Ridgway et al. 1985; Flemming and Schaule, 1989).

The autopsies of irreversibly fouled modules have given a clear picture of typically biofouled modules. The feed-water surfaces of the membranes were often coated with a gray-black (Ridgway et al. 1983a) or brown (Flemming and Schaule 1989) mucilaginous substance, exhibiting a regular crisscross pattern which is congruent with the pattern of the overlying plastic feed-water channel spacer (Vexar) that separates adjacent membrane envelopes in the RO element. The fouling substance had a slippery, gel-like texture and could be readily scraped or washed from the membrane surfaces. The biofilm also exhibited a distinct laminar construction in the edgewise orientation. Usually, three to five such layers were evident within the biofilm, and each layer was on the order of 3 to 5 μm thick. Organisms of various morphological types were scattered throughout the entire biofouling layer. These bacteria were evidently firmly secured within the biofilm matrix by extracellular fibrillar secretions (Ridgway et al. 1983a).

The total number of bacteria per gram (wet weight) of fouling material on both fouled CA and polyamide membranes varied from approximately 5×10^7 to 5×10^8. These values corresponded to viable cell counts of 4.2×10^5 to 5.6×10^6 (Ridgway et al. 1983a; Flemming and Schaule 1989). Figures 6.4 and 6.5 show scanning electron microscope (SEM) pictures of the surfaces of a fouled membrane, indicating a biofilm of about 10 to 30 μm thickness with bacterial cells imbedded in a slime matrix.

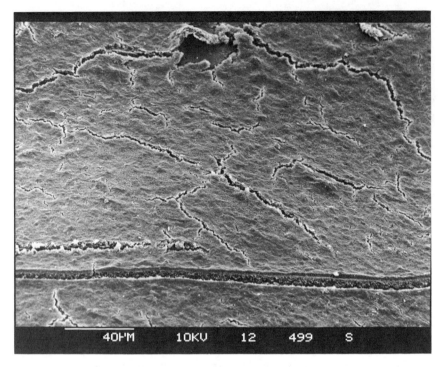

FIGURE 6.4.   Biofouled polyamide membrane surface.

Approximately 75 to 90 percent of the dehydrated materials are volatile at 600°C, suggesting a primarily organic composition. A significant proportion of this organic fraction was composed of protein (up to 30 percent wt/wt) and carbohydrate (up to 17 percent wt/wt), confirming the biological derivation of the fouling layer. In addition, significant quantities of ATP were also detected, indicating the presence of metabolically active microorganisms within the biofilm (Ridgway et al. 1984b). The feed-water surfaces of the membranes were observed by SEM to have been extensively colonized by predominantly rod-shaped bacteria measuring approximately 0.5×1.0 μm. When viewed edgewise by SEM (see Figure 6.6), the biofilm exhibited a distinctive multilayered construction, 10 to 20 μm in estimated total thickness, throughout which rod-shaped bacteria could be identified (Ridgway 1988; Flemming and Schaule 1989).

Bacteria were also observed attached to the exposed surfaces of the straight-woven, polyester support fibers (Texlon fibers) located on the permeate side of the RO membranes (Ridgway 1988). Large regions of the Texlon fibers were entirely void of adherent bacteria. Most of the attached bacteria were localized

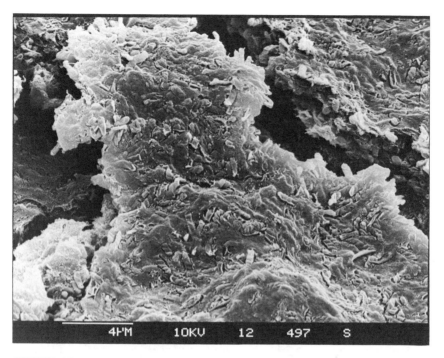

FIGURE 6.5.   Magnification of Figure 6.4; bacteria are embedded in a slime matrix.

within discrete microcolonies that had evidently arisen from the repeated growth and multiplication of a single pioneering organism. (Ridgway 1988). Figure 6.7 shows individual bacteria beginning to colonize a membrane surface. In the thin-sectioned samples, the biofilm was found to be composed primarily of autolysed bacteria, which were largely or completely devoid of their cytoplasmatic contents. Interspersed among such lytic cell debris were apparently morphologically intact bacterial cells, many of which clearly exhibited an extracellular capsule or slime layer (glycocalyx) having the characteristic fibrillar appearance (Ridgway 1988; Flemming and Schaule 1989).

Although bacterial growth is the most pervasive and troublesome of the microbial foulants encountered in membrane systems, it by no means represents the only genera. Algae, protozoa, and fungi are also prevalent in natural water systems. Surface waters in particular are breeding grounds for a menagerie of microorganisms. Figure 6.8 shows a fungus found on a membrane surface. Previous biofouling proved a good substrate for growth. In addition to causing the degradation of the membrane itself, fungal growth tends to disrupt the flow

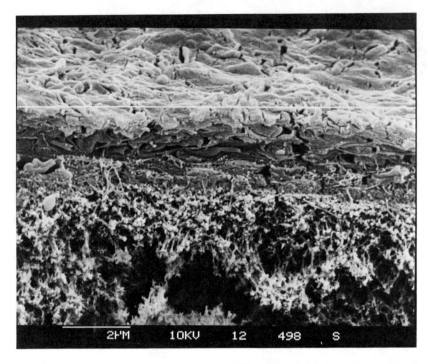

FIGURE 6.6.   Magnification of the fissure as visible in Figure 6.4; layered structure of the deposit. The white line indicates the upper edge of the biofilm.

characteristics within the membrane. Moreover, the filamentous stolons entrain other particles, further fouling the membrane.

Another microbiofoulant is algae, which can be found in a variety of waters and soils. When unable to find sufficient nutrients, algae will conglomerate, forming a slime outercoating. In this way, these unicellular microorganisms can utilize dead neighbors for additional sustenance. The slimey secretions of the cell mass can lead to further fouling.

CONTROL AND PREVENTION OF BIOFOULING

Assessment of the Biofouling Potential

The first prerequisite for effective control and prevention of biofouling is to be aware that a biofouling problem can occur. Many biofouling problems come to maturity only when no notice is taken for a long period of time.

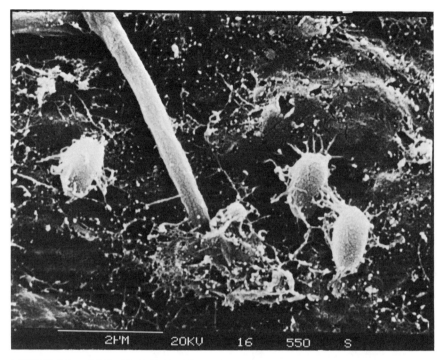

FIGURE 6.7.   **SEM of in individual bacteria beginning to colonize the surface of a biofouling layer on an aromatic polyamide composite membrane.**

There are a number of factors that can increase the risk of biofouling:

1. *Plant design:* Extended piping system, access of light, dead ends, fissures, armatures with dead zones, nondisinfected holding tanks

2. *Feed-water characteristics:* High temperature ($>25°C$), high amounts of organic and inorganic nutrients, large numbers of cells ($>10^4$ colony-forming units per ml), high SDI

3. *Operational characteristics:* Infrequent monitoring of performance characteristics, use of microbially contaminated pretreatment chemicals, relatively slow cross-flow, long storage periods

Prediction of biofouling potential is very desirable in order to prepare for eventual biofouling problems. A regular water analysis, checking salt content, TOC, COD, pH, and conductivity, normally gives no clues to the biofouling potential. If biofouling is expected, the microbial content of the water is

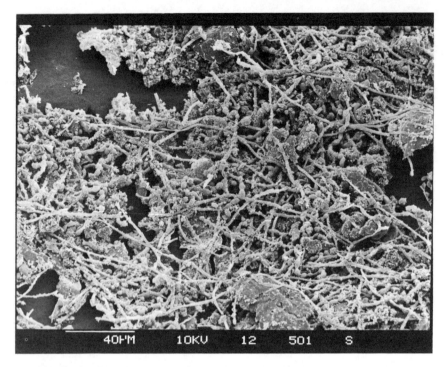

FIGURE 6.8. Mold is the most common fungus found in biofouled membranes. The fibers will cause increased fluid friction resistance.

checked, usually with cultivation methods. George and colleagues (1988) investigated the microbial content of three water samples, drawn at three different sites of an RO plant. The numbers gained by cultivation methods were always the lowest, indicating cells that were able to grow under the given laboratory conditions. However, epifluorescence microscopy (EM), not discriminating between living and dead cells, indicates the maximal number of cells present in the water sample. INT (iodonitrotetrazolium violet) is a fluorescent dye activated by microbial activity, indicating living cells in direct EM. There is no reliable correlation between cultivation and the microscopical methods. It is recommended to monitor the bacteriological quality of the raw water (before chlorination), the RO feed, the brine, and the product water in order to assess the biofouling potential of a feed water. Applegate and Erkenbrecher (1987) recommend the following microscopic and culturing techniques for the RO biomonitoring program:

- Acridine orange direct count
- INT (living cell number) direct microscopic technique

• Standard serial-dilution plating
• Membrane filtration
• Anaerobe enumeration with adequate nutrient broth

Details of the methods are given by Applegate and Erkenbrecher (1987) and (including biofilm sampling and analysis) by Geesey and Mittelman (1987).

It must be emphasized that cell numbers in the bulk water phase tell nothing about biofilms. All that is measured is living or total bacteria in the circulating water. However, it is the attached biofilm, rather than circulating microorganisms, that causes the deterioration in plant performance. There is no correlation between colony numbers in the bulk and the amount of biofilm bacteria (Dott and Schoenen 1985; Costerton et al. 1987). Cells may be released from biofilms irregularly and in response to environmental alterations of pH, ionic strength, temperature, shear stress, and cleaning agent action. The bacterial numbers of a typical slime on an RO surface are between 10^6 and 10^8 per cm^2 (Whittaker et al. 1984; Flemming and Schaule 1989). This exceeds the number of cells per millileter in tested waters in most cases by some orders of magnitude.

The Silt Density Index (SDI) is the only widely accepted test for fouling prediction in the RO industry (Potts et al. 1981). A correlation of SDI data with biofouling is not yet established. However, a high SDI always has to be considered as an alarm sign, whereas a low SDI does not necessarily indicate the absence of a fouling potential (Nagel 1990).

Pretreatment of the Feed Water

It is obvious that effective pretreatment results in less membrane cleaning. Himelstein and Amjad (1985) state that "in most cases, membrane fouling is due to inadequate pretreatment of the feed." But what is adequate? For example, an usual pretreatment scheme may include the following (Nagel 1990):

1. Addition of a biocide, preferably chlorine or its compounds (Belfort 1977; Eisenberg and Middlebrooks 1984; Mindler and Epstein 1986). This leads to the necessity of dechlorination prior to chlorine-sensitive membranes.
2. Flocculation, precipitation, and sedimentation.
3. Adsorption with granulated activated carbon.
4. Cartridge or micron filters.
5. Addition of a scale inhibitor.

Chlorination has some drawbacks; because it does not sterilize the water (i.e., kill all microorganisms), some bacteria will always survive. After chlorination, the surviving bacteria can grow (LeChevallier and McFeters 1990), resulting in biofouling. At higher water temperatures (25°C to 35°C) aftergrowth and biofouling can occur rapidly (Applegate et al. 1989). In addition, chlorine (as well as ozone) may break down humic acids and make them biologically available, thus providing additional nutrients. Applegate and collegues (1989) recommend chloramine, because it does not degrade humic acid, and no increase in AOC (assimilable organic carbon) would occur in the RO pretreatment system. This would significantly inhibit aftergrowth and reduce biofouling. There is greater attachment and significant aftergrowth because of the generation of AOC from humic acid. The chloramine-treated bacteria clearly showed less attachment and reached a maximum of only 10^2 to $10^4/cm^2$ after 4 days (Applegate et al. 1989). These authors recommend running pretreatment at low pH (6.0) in the presence of larger concentrations of humic substances, particularly when the water temperature is above 25°C. Even at that pH, however, a significant amount (40 percent) of humic acid degradation (nutrient formation) occurs.

It is very important to check the microbiological quality of the pretreatment chemicals used for other purposes than the prevention of biofouling. SHMP, used as a scaling inhibitor, has been reported to contain high numbers of bacteria and to provide orthophosphate as an inorganic nutrient for the fouling biofilm (Himelstein and Amjad, 1985; Ahmed and Alansari 1989). Therefore, SHMP had to be used with sodium metabisulphate (SMBS) for sanitization. SMBS is used for dechlorination prior to the RO membrane, and it has some biostatistical properties that are utilized either by shock dosage (~500 mg/l) or continuous dosage (20 to 50 mg/l) in some instances.

Activated carbon, used for the adsorption of undesired organics that would interfere with the RO process, offers a large amount of surface area for colonizing microorganisms and is often the source of microbial contaminations in water treatment systems (Graham et al. 1989; LeChevallier and McFeters 1990).

Prefiltration of the RO feed water is often utilized for pretreatment. However, most standard RO prefilters possess nominal or absolute pore sizes ranging from 5 to 25 μm in diameter, which is, of course, far too large to efficiently retain bacteria and prevent their entrance into the RO module. Prefilters will not prevent the development of biofilms in the water system prior to the membranes.

Perhaps the most frequently employed alternate method of disinfection for the control of biofouling in RO systems is ultraviolet (UV) irradiation of the feed-water stream. However, many types of bacteria that are commonly found in domestic water supplies are known to be extremely resistant to UV irradiation (Ridgway 1987). Once a biofilm is established in a piping system, UV irradiation will probably not be able to kill and remove these cells. Bacteria that survive UV irradiation would still be capable of rapid growth and proliferation on the RO membrane surface. Opaque biofilms, e.g., with entrapped colloids, will absorb most of the energy in the upper layers and the deeper biofilm will not be affected by the much-decreased UV irradiation.

Presterilization of the feed water has been shown not to affect biofilm development (Pedersen 1982) in nonsterile systems. Any type of treatment intended for biofilm control has to reach the biofilm to be successful.

A special case of "pretreatment" may be the prevention of sunlight intrusion into piping, filters, and clearwells. Typical fiberglass-reinforced plastic (FRP) permits sufficient light penetration to promote algae growth, which increases biofouling potential (Ebrahim and Malik 1987). If the FRP piping is painted, the algae will not grow.

Monitoring of Biofouling

Monitoring systems capable of directly detecting biofouling are not used in the RO practice. In most cases, biofouling is recognized by indirect effects on the unit's performance: flux decline, decrease of salt rejection, or an increase of the feed delta-p/brine. However, these are nonspecific reactions of the system, and without further information they cannot be simply correlated to biofouling, as indicated in phases I and II of Figure 6.1.

Therefore, it is often impossible to conclude that a certain increment of flux decline is the result of a certain amount of microbial fouling on the membrane surface. Ridgway (1988) reports that within the first 3 days of operation, approximately 2×10^5 bacteria per cm^2 (measured as total CFU) were recovered from the surfaces of CA membranes, although no visible evidence of biofouling could be observed by the unaided eye. Within 1 or 2 weeks of initial operation, the fouling bacteria produced a nearly confluent lawn (approximately one cell layer in thickness) extending over the entire membrane surface (Ridgway 1988). Flemming et al. (1992) observed this effect even after a few days. Thus, biofilm formation occurs before plant performance parameters respond, as schematically drawn in Figure 6.9. Therefore, monitoring should be carried out as directly as possible, i.e., as close to the biofilm as possible.

Some observers have noted that accumulations of biomass in the prefilter generally precede any significant biofouling on the membrane, and this can be confirmed by soaking a portion of the filter in sterile water and then running microbial analyses of the resulting extract (Graham et al. 1989). If there are indications of biofouling in a system, the analysis of acid, caustic, and distilled water extracts of the prefilter can be of considerable help in determining what may be fouling the system. Thus, it is important to keep in mind that biofouling usually does not occur on the membranes only. Inspection of the prefilters may help to discover biofouling in an early phase. The visual detection of slime in the system, however, indicates an already advanced phase of biofouling.

A reasonable approach to control biofouling is the use of "sacrificial elements" as biofouling monitoring systems. Winters and Isquith (1979) and Ridgway et al. (1984) used mini RO probes which were integrated on-line in the system and removed from time to time to check biofouling directly. Winters and

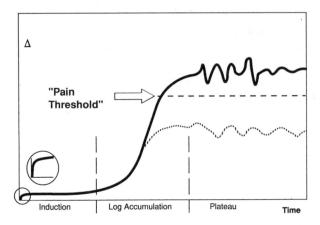

FIGURE 6.9.   Biofilm development above and below the "pain threshold." Δ = parameter for thickness, resistance, mass, etc.; the indent indicates primary adhesion as shown in Figure 6.3.

Isquith (1979) proposed an easy colorimetric test, based on the specific reaction of the dye Alcian blue with extracellular acidic polysaccharides (Allison and Sutherland 1984). Practical experiences of this approach are not reported. Once removed, the biofilm on the RO probe can be investigated in different ways, including determination of ATP content.

Although it is not often used in RO biofouling monitoring, the "Robbins device" is another tool for monitoring biofilm development (Ruseska et al. 1982; Geesey and Mittelman 1987). Principally, it consists of a removable part of the water system's surface which can be separated and microbiologically examined. This system is not suitable for the membranes themselves, but for representative surfaces in their proximity.

Other—and indirect—biofilm monitoring systems (BFM) can be adapted from different applications, such as biofouling control in heat exchangers. They are based on measuring of the increase of the drag resistance (Johnson and Howells 1981; Kozelski 1983; Donlan and Pipes 1988). Pressure drop in the BFM (and hence the friction factor) can be correlated with development of biofilm in the system. Characklis et al. (1986) described a monitoring device that continuously monitors the deposition of microbial biomass on surfaces indirectly by measuring changes in heat transfer and fluid frictional resistance. The micro-computer-controlled system simulates conditions by maintaining the appropriate process fluid velocity (or pressure drop) and heat flux (or wall temperature) to any metal surface on which deposition is occurring. The correlation of the data gained with this system to RO membranes has yet to be established.

Direct-process parameters such as the transmembrane differential pressure, the percent salt rejection, and the normalized product flow rate are the most critical operating parameters and, as a minimum, should be monitored daily (Himelstein and Amjad 1985). At times, recognition of even the potential for foulant accumulation should alert the operator to institute a cleaning procedure (Himelstein and Amjad 1985; Ebrahim and Malik 1987; Graham et al. 1989). The "pain threshold" (Figure 6.9) is reached when:

1. A 10 percent decrease in productivity while operating at constant pressure and temperature
2. A 10 percent increase in net driving pressure required to maintain constant product flow at constant temperature
3. An increase of 15 to 30 percent in the differential pressure between feed and reject during routine cleaning after 3 months operation of the system under adverse feed-water conditions, or 6 months operation of the system under normal raw-water conditions

An effective biofouling monitoring system should provide the following:

1. Monitoring of biofilm formation on representative surfaces
2. Monitoring of the microbial quality of feed water and all added pretreatment chemicals
3. Monitoring of plant performance data in order to evaluate correlations that allow the prediction and recognition of biofouling from these data, and permitting the establishment of the individual fouling characteristics of an RO plant.

Sanitization of the Biofoulant

It is a common fallacy that in a system suffering from biodeterioration, the problem is solved by killing all the microorganisms present. In many cases, it is the presence of the biofilm rather than its physiological activity that causes the problem. A biologically inactive biofilm may continue to cause biofouling effects (Argo and Ridgway 1982). Thus, removal of the biofilm is crucial for effective sanitization. In addition, dead ends, corners, and fissures may bear biofilms that are not reached by biocides, which can act as an inoculum for the cleaned system, inducing rapid bacterial aftergrowth, as is frequently reported in the practice (Ridgway 1988; Payment 1989; Nagel 1990). The spectrum of organisms that survive treatment processes includes spore formers, acid-fast bacilli, pigmented organisms, disinfectant-resistant bacterial strains, various yeasts, fungi, and *actinomycetes*. As previously mentioned, it is important to keep in mind that biofilms protect their members against biocides (Costerton 1984; Costerton and Lashen 1984; Exner et al. 1987; Costerton et al. 1987;

LeChevallier et al. 1988a,b; Nichols 1989). Starvation has been shown to increase microbial resistance against biocides (Matin and Harakeh 1990). Mycobacteria, which are frequently found in the biofilms of RO systems (Ridgway et al. 1983a; DuMoulin et al. 1988) are significantly more resistant to biocides when they are embedded in biofilms (Schulze-Röbbecke and Fischeder 1989). Data on effectiveness of biocides are usually gained with suspended cultures, which are 50 to 500 times more sensitive than biofilms of the same species. Biofilms have been found even on the walls of disinfectant piping systems (Exner et al. 1983)!

Sanitization of the feed water is part of the overall sanitization program. It should include a decrease of the cell numbers and, even more important, a decrease of nutrients. Characklis (1990) has shown that the adhesion of new cells is almost negligible as compared with biofilm growth caused by nutrients.

The effectiveness of biocides depends on a number of intrinsic and environmental factors, such as the following:

• Kind of biocide
• Biocide concentration
• Biocide demand (which can be decreased by prior cleaning)
• Interference with other dissolved substances
• pH
• Temperature
• Contact time
• Types of organisms present
• Physiological state of microorganisms
• Presence of biofilms

As a general rule, the higher temperature, the longer the contact time; and the higher the concentration of the disinfectant, the greater will be the degree of disinfection. Considering the fact that a biofilm is increasingly difficult to remove as it ages (Flemming and Schaule 1989), quick reaction may save a great deal of effort.

It should be emphasized that biocides are toxic materials and that every biocide added to the system can cause environmental problems. Restrictive environmental legislation may prohibit the application of an agent in effective concentrations without special sewage water purification (Nagel 1990).

If a biocide is added, the following additional processes must also be considered.

1. Live bacteria in suspension react with biocide. Result: dead suspended bacteria (which may attach as well as the living ones).
2. Live biofilm reacts with biocide. Result: dead biofilm (which remains on the surface, providing both substratum and substrate for subsequent cells and masking the original surface properties of the substratum).

3. Deposition of dead bacteria. Result: dead biofilm deposits.
4. Detachment of dead biofilm. Result: dead bacterial suspension.
5. Reaction of dead suspended bacteria (EPS) with oxidizing biocide. Result: long-chain organic molecules are reduced to smaller molecules and form a clear solution.
6. Bacteria may be injured and survive (McFeters et al. 1986), resulting in an inoculum for the cleaned system.

The chlorination/dechlorination system can be considered as the state-of-the-art technique for biofouling control and sanitization, but failures have been reported (Nagel 1990) when massive biofouling occurred despite chlorination (Argo and Ridgway 1982; Applegate et al. 1989). A high dose of combined chlorine was continuously added to the RO feed water, and viable organisms could no longer be isolated from the membrane surfaces. However, measurement of the biochemical parameters, as well as inspection of the membrane surfaces by SEM, confirmed that the chlorine-inactivated microbes continued to adhere to and rapidly foul the RO membrane surfaces (Ridgway 1988). Nagy and colleagues (1982) reported bacterial levels in the Los Angeles aqueduct system with more than 10^4 CFU/cm^2 in the presence of a residual of 1 to 2 mg/l of free chlorine. It was necessary to maintain a residual of 3 to 5 mg/l chlorine to reduce biofilms by more than 99.9 percent. Bacteria from a chlorinated system were more resistant to both the combined and free forms of chlorine than those from the unchlorinated system, suggesting that there may be selection of microorganisms for more chlorine-resistant microorganisms in chlorinated systems. The most resistant organisms were able to survive a 2-minute exposure to 10 mg/l free chlorine. These included gram-positive spore-forming bacilli, actinomycetes and some micrococci (Ridgway and Olson 1982). Chlorine reacts with the EPS in which the biofilm bacteria are embedded and penetrates a biofilm much slower than a chloramine (LeChevallier et al. 1988b). Moreover, if H_2S (hydrogen sulfide) is present in the water to be treated, chlorine will lead to the precipitation of elemental sulfur, which is extremely difficult to remove (Ahmed and Alansari 1989).

Hypochlorite (Miller and Bott 1981) is used as a cheap and effective biocide, which detaches the biofilm from surfaces (Costerton et al. 1987). However, membranes sensitive to oxidation cannot be treated this way.

Another choice of biocide is formaldehyde (Applegate and Erkenbrecher, 1987). Advantages of formaldehyde are its low cost and its broad antibiotic spectrum. Routine formaldehyde disinfection of unit piping, however, does not appear to eliminate the mycobacterial population appreciably. The resistance of mycobacteria to formaldehyde has been studied, and its efficacy as a disinfectant has been questioned. Some strains have been shown to survive 2 percent formaldehyde disinfection up to 96 hours (DuMoulin et al. 1988). Formaldehyde reacts with proteins and is used in preparative microscopy as a fixing agent. This fixing property keeps the biofilm in stasis instead of detaching it (Exner et al. 1987), leading to strong blocking of membranes in some instances (Nagel 1990). Formaldehyde has further disadvantages in that it is volatile, has a pungent, irritating odor, and also leads to the mutation of adaptive strains of bacteria and fungi. Moreover, it is extremely toxic and suspected to promote cancer.

Regular treatment with ozone (once a week) is reported to maintain sterility in high-purity water production (Nagel 1990). There are reports that ozone is more effective than chlorine, at equal concentrations, in controlling biofouling in low salinity waters. In seawater, ozone oxidizes the bromide ion. Bromine's effectiveness is approximately the same as chlorine. Ozone may also react with organics in water to form epoxides, which pose an environmental problem (Characklis 1990). It also degrades humic acids (Gilbert 1988) and turns them into assimilable compounds.

The merits of peracetic acid (PAcA) lie in its strong biocidal action even in low concentrations (0.02 percent) against biofilms (Flemming 1984; Nagel 1990; Flemming and Schaule, in preparation), although it raises the TOC to a certain extent.

Hydrogen peroxide needs either long contact times (a few hours), high concentrations (>3%) or high temperatures (>40°C) in order to act properly. It is used for membranes in food technology (Lintner and Bragulla 1988) but is restricted to those that are not sensitive to oxidizing agents.

Although there was no approval from the manufacturers, 2,2-dibrom-3-nitrilopropionamide (Labozid) has also been successfully applied to control biofouling in a large RO plant (Nagel 1990). Lintner and Bragulla (1988) recommend, preferably in this order, hydrogen peroxide, hypochlorite, peracetic acid, formaldehyde, and sodium metabisulphite for the disinfection of membranes.

Cleaners generally fall into two broad categories. Acidic cleaners are used to remove metal oxides, calcium carbonate, and other inorganics. Depending on the type of membrane, neutral-to-caustic materials are the choice for biofilms, collodial silts, and most naturally occurring organic foulants. There are reports from the practice that alkaline conditions were quite effective in the control and removal of microbial slimes (Nagel 1990).

Most proprietary cleaners are formulated from a wide range of chelating agents, surfactants, detergent builders, antideposition aids, enzymes, and biocides to provide the user with effective, safe and, easy-to-use products (Graham et al. 1989). The type of cleaner required depends on the type of membrane and the nature of the foulant. The best method of foulant identification is extensive analysis of the foulant gathered from a destructive autopsy. Before resorting to this measure, however, a first step may be to obtain a thorough chemical and biological analysis of the current feed water to see how it compares with the water the system was designed to process. Knowledge of the source of the feed water alone often is enough to make a good guess about the type of fouling likely to occur (Graham et al. 1989).

Whittacker et al. (1984) investigated both the biocidal and the biofilm removal efficiency of various chemical cleaning agents. They found that some cleaners only killed the biofilm without removing it, and vice versa. Removal of the biofilm, however, is much more important for the restoration of a plant's performance than merely killing the microorganisms. Biz® bleach has proven to be an effective cleaner, but it has the drawback of limited use owing to its oxidizing properties. Ultrasil U-53, a specially designed membrane cleaner, was found to be among the best agents to clean biologically fouled membranes (Flemming and Schaule 1989). The mechanical stress of the shear forces caused by the cleaning solution is of great importance (Lintner and Bragulla 1988).

Whittacker et al. (1984) and Yanagi and Mori (1984) report, however, that mechanical cleaning gave the best results in terms of biofilm removal. This method is obviously limited to mechanically accessible membranes and has to be checked carefully because of the possible presence of precipitated crystals with sharp edges that may cut the membrane while mechanically moved along the module (Belfort 1977; Nagel 1990).

It has been proposed that dead cells can be removed more easily than living cells (Argo and Ridgway 1982). Experiments dealing with the removal of primary biofilms from various membrane materials, however, showed clearly that dead cells adhered as strongly as living cells (Flemming and Schaule 1988a). To minimize the effect of biofouling, it is probably best to clean the RO membrane at frequent and regular intervals (e.g., at a minimum of every 30 days) before an extensive biofilm has had an opportunity to develop (Trägardh 1989). Membrane cleaning solutions become noticeably less effective as a function of biofilm age (Whittacker et al. 1984; Ridgway 1987; Flemming and Schaule 1988b).

The quality of the makeup water can also be crucial for the success of a cleaning measure. The quality of the cleaning water has been formulated as follows (Lintner 1989): Fe < 0.05 ppm, Mn < 0.02 ppm, chlorine < 0.1 ppm, silica < 40 ppm, and cell number < 1,000/ml.

Evaluation of Sanitization Effectiveness

The evaluation of the effectiveness of sanitizing measures is the Achilles heel of the whole procedure. In almost all cases, the restoration of better plant performance is the measure by which effectiveness of sanitization is assessed. However, numerous investigators and plant operators have observed a rapid resumption of biofouling immediately following biocide treatment and have termed this phenomenon "regrowth" or "aftergrowth." Characklis (1990) proposes *recovery* as a more appropriate term, inasmuch as growth may be only one of the processes contributing to reestablishment of the biofilm. Thus, recovery may be due to one or all of the following causes:

1. The remaining biofilm contains enough viable organisms to preclude any lag phase in biofilm accumulation, as observed on clean heat exchanger tubes. Thus, biofilm accumulation after a shock treatment is more rapid than on a clean surface.
2. The residual biofilm imparts an increased relative roughness to the surface and thus enhances transport and sorption of microbial cells and other compounds to the surface. The roughness of the deposits may provide a "stickier" surface.
3. Chlorine, for example, reacts preferentially with the EPS and removes it. It does not reach the biofilm cells, thus the cells are left more exposed to the nutrients when chlorination ceases.
4. EPS is rapidly created by surviving organisms as a protective response to irritation by chlorine.
5. There is a selection of organisms less susceptible to the biocide which proliferate between successive biocide applications.

Flemming and Schaule (1989) found that the membrane cleaners applied to a mature biofilm on a membrane surface could not remove more than 80 percent. Although this would lead to a significant improvement of the system, the 20 percent biofilm remainders act both as substrate and substratum for subsequent cells, making the effect of the sanitization as transient as it is observed in practice (Whittaker et al., 1984). As previously noted, cell numbers in the water phase are not indicative for the biofilm situation. The usual result of a disinfection is first, a few hours with colony numbers of zero per ml, then increasing numbers. Recall, also, that dead cells adhere if they are not removed (Flemming and Schaule 1988a; Ridgway 1988).

Thus, the use of monitoring techniques, as discussed in a previous section, is appropriate to evaluate the effectiveness of sanitizing measures (Whittaker et al. 1984). Some representative parts of the system must be accessible in order to be investigated for the biofilm situation. It does not seem necessary that these be the membrane surfaces themselves, although these would provide the most direct approach. Biofouling is a process occurring not only on the membranes but on all surfaces of the system in contact with water. Inspection of piping walls closely before and after the modules can tell a lot. Prefilters, which undergo the same procedures as the membranes themselves and are often more accessible, can be checked. The enumeration of the cells in the water alone, however, is not a suitable way of learning about the biofilm situation. While all cells in suspension may be dead, biofilm cells can still be alive, proliferating as soon as the biocide has disappeared.

DO WE HAVE TO LIVE WITH BIOFOULING?

What is called "biofouling" is nothing but a biofilm that starts to interfere with the demands of a technical process. Thus, it is an operational definition. Characklis et al. (1986) showed that the development of a biofilm did not interfere with drag resistance or heat transfer in a heat exchange system until it exceeded the thickness of the viscous sublayer. "No biofouling" does not necessarily mean "no biofilm"—it means only "no interference by the biofilm" (see Figure 6.9). A biofilm developing on a surface will produce an effect that exceeds a tolerance level—be it a pressure drop, an increased fluid frictional resistance, or a flux decline of product deterioration—and signals a problem.

Biofilms do not grow indefinitely. After a log phase, they will sooner or later reach a plateau phase, which can be described roughly as a dynamic equilibrium: the increase of biofilm mass per square centimeter is equal to the decrease. The simple equation is as follows:

$$\text{delta } M_{adhesion} = \text{delta } M_{growth} = \text{delta } M_{detachment} + \text{delta } M_{lysis} \qquad (2)$$

Delta $M_{adhesion}$ depends on the number of cells in the bulk, their nutrient status, the "biological affinity" of the membrane material, the fluid velocity, the sticking efficiency, the flux, and the motility.

Delta M_{growth} depends on the nutrient content of the feed water, its fluid velocity (low nutrient concentrations can be compensated by large volumes passing the biofilm, which can scavenge the nutrients), the temperature, and the composition of species present in the biofilm (having different temperature optima and nutrient requirements).

Delta $M_{detachment}$ is mainly governed by the tangential shear forces (detaching) and the flux, the surface structure (smooth or rippled), and the inner strength and elasticity of the biofilm, which regulates the response to shear forces. This measurement will increase with biofilm thickness and shear forces (Characklis 1990).

Delta M_{lysis} is dependent on the cell age, the biofilm age, the growth conditions, and the presence and action of lysing biocides.

For nonfiltration systems such as heat exchangers (Characklis 1990), it can be presumed that $M_{adhesion}$ will be the most important factor during the log phase of biofilm accumulation, whereas the equilibrium level of the biofilm is probably governed mainly by M_{growth} and $M_{detachment}$. Thus, the nutrient content of the feed water may be of greater importance than the cell number. However, the influence of the flux as a vector increasing the probability of contact of cells to the wall must be considered for the RO process, thus increasing the $M_{adhesion}$ factor in a yet-not-quantified way.

Until now, this correlation has not been acknowledged expressively, although it is used more or less unknowingly in a pragmatic approach, to design an RO plant in a special way if biofouling is expected. A higher cross-flow shall provide higher shear forces and, thus, a thinner biofilm. Exceeding membrane surface shall absorb a 20 to 40 percent flux decline, a level at which a stable situation is expected (Ridgway 1988; Nagel 1990). This formulation is addressed by the fouling factor. This factor, however, does not discriminate between the different causes of fouling but indicates their cumulative effect. It is described by a simple formula:

$$P_{prod.} = f \cdot P_{theor.} \tag{3}$$

$P_{prod.}$ is the product rate as produced in the operating plant. $P_{theor.}$ is the calculated product rate, and f is the fouling factor, which can range between 1 (no fouling) and 0 (total fouling). The fouling factor is an arbitrary value, which stays constant when the system is in equilibrium. Figure 6.10 shows a practical example of the process of reaching a fouling equilibrium with biofouling as a major factor, with practically no scaling and a low SDI. In the first few days, there is a strong flux decline. Cleaning with a proprietary membrane cleaner (U-53) is not effective. In this period, $M_{attachment}$ may be the most important factor. After a flux decline of almost 50 percent, a shock treatment with NaOH (pH 10) is carried out and repeated daily, and equilibrium is reached, where $M_{detachment}$ is enhanced by the action of NaOH (Nagel 1990). The resulting fouling factor is 0.75.

Most certainly, we do have to live with biofilms. Whether they result in biofouling depends on the awareness of the problem and on the "threshold."

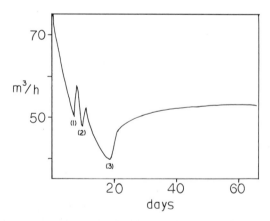

FIGURE 6.10.   Permeate flow decline in an RO plant (thin film composite membrane) with high biofouling potential: (1) and (2), treatment with a membrane cleaner; (3), beginning of daily treatment with NaOH (pH 10).

There is no hope for a universal biofouling inhibitor or sanitizer. General principles, as discussed in this chapter, are devices to find the way of appropriate individual treatment of each single biofouling problem. Further research on microbial adhesion and on the dynamics of biofilm development will help to explain the underlying processes and to live with biofilms that are below the threshold of biofouling.

ACKNOWLEDGMENTS

The technical assistance of Hanna Rentschler and the additional research work of Gabriela Schaule are gratefully acknowledged, as are the fruitful discussions with Rudolf Wagner. This work has been supported by the Willy Hager Stiftüng.

References

Absalom D.R., F.V. Lamberti, Z. Policova, W. Zingg, C.J. van Oss, and W.A. Neumann. 1983. Surface thermodynamics of bacterial adhesion. *Applied Environmental Microbiology* 46:90–97.

Ahmed, S.P., and M.S. Alansari. 1989. Biological fouling and control at RAS Abu Jarjur RO plant—a new approach. *Desalination* 74:69–84.

Allison, D.G., and J.W. Sutherland. 1987. The role of exopolysaccharides in adhesion of freshwater bacteria. *Journal of General Microbiology* 133:1319–1327.

Applegate, L.E. 1986. Posttreatment of reverse osmosis product waters. *Journal of the American Water Works Association* 76:59–65.

Applegate, L.E., and C.W. Erkenbrecher. 1987. Monitoring and control of biological activity in Permasep seawater RO plants. *Desalination* 65:331–359.

Applegate, L.E., C.W. Erkenbrecher, and H. Winters. 1989. New chloramine process to control aftergrowth and biofouling in Permasep B-10 RO surface seawater plants. *Desalination* 74:51–67.

Argo, D., and H.F. Ridgway. 1982. Biological fouling of reverse osmosis membranes. *Aqua* 6:481–491.

Baier, R.E. 1980. Substrata influences on adhesion of microorganisms and their resultant new surface properties. In *Microbial adhesion to surfaces,* edited by G. Bitton and K.C. Marshall. New York: John Wiley, pp. 60–104.

Belas, M.R., and R.R. Colwell. 1982. Adsorption kinetics of laterally and polarly flagellated vibrio. *Journal of Bacteriology* 151:1568–1580.

Belfort, G. 1977. Pretreatment and cleaning of hyperfiltration (reverse osmosis) membranes in municipal wastewater renovation. *Desalination* 21:285–300.

Brown, C.M., D.C. Ellwood, and J.R. Hunter. 1977. Growth of bacteria at surfaces: Influence of nutrient limitation. *FEMS Microbiology Letters* 1:163–166.

Bryers, J.D., and W.G. Characklis. 1981. Kinetics of biofilm formation within a turbulent flow system. In *Fouling of heat exchanger equipment,* edited by E.F.C. Somerscales and J.G. Knudsen. Washington, D.C.: Hemisphere, pp. 313–334.

Celesk, R.A., and J. London. 1980. Attachment of oral cytophaga species to hydroxyl-apatite-containing surfaces. *Infection and Immunity* 29:68–77.

Characklis, W.G. 1981. Microbial fouling: a process analysis. In *Fouling of heat transfer equipment*, edited by E.F.C. Somerscales and J.G. Knudsen. Washington, D.C.: Hemisphere, pp. 251–291.

Characklis, W.G. 1982. Processes governing primary biofilm formation. *Biotechnology and Bioengineering* 24:2451–2476.

Characklis, W.G. 1990. Microbial fouling. In *Biofilms*, edited by W.G. Characklis and K.C. Marshall. New York: John Wiley, pp. 523–584.

Characklis, W.G., and K.E. Cooksey. 1983. Biofilms and microbial fouling. *Advances in Applied Microbiology* 29:93–138.

Characklis, W.G., and K.C. Marshall. 1990. *Biofilms*. New York: John Wiley.

Characklis, W.G., N. Zelver, and F.L. Roe. 1986. Continuous on-line monitoring of microbial deposition on surfaces. *Biodeterioration* 6:427–433.

Christian, D.A., and T.H. Meltzer. 1986. The penetration of membranes by organism grow-through and its related problems. *Ultrapure Water* (May/June 1986):39–44.

Corpe, W.A. 1980. Microbial surface components involved in adsorption of microorganisms onto surfaces. In *Adsorption of microorganisms to solid surfaces*, edited by G. Bitton and K.C. Marshall. New York: John Wiley, pp. 106–144.

Costerton, J.W. 1984. The formation of biocide-resistant biofilms in industrial, natural and medical systems. *Developments in Industrial Microbiology* 25:363–372.

Costerton, J.W., and E.S. Lashen. 1984. Influence of biofilm on efficacy of biodices on corrosion-causing bacteria. *Mater. Perf.* 23:13–17.

Costerton, J.W., K.-J. Cheng, G.G. Geesey, T.I. Ladd, J.C. Nickel, M. Dasgupta, and T.J. Marrie. 1987. Bacterial biofilms in nature and disease. *Annual Reviews in Microbiology* 41:435–464.

Daniels, S.L. 1980. Mechanisms involved in sorption of microorganisms to solid surfaces. In *Adsorption of microorganisms to surfaces*, edited by G. Ditton and K.C. Marshall. New York: John Wiley, pp. 7–58.

Dawson, M.P., B.A. Huphrey, and K.C. Marshall. 1981. Adhesion: A tactic in the survival strategy of a marine vibrio during starvation. *Curr. Microbiol* 6:195–199.

Defrise, D., and V. Gekas. 1988. Microfiltration membranes and the problem of microbial adhesion. *Process Biochemistry* (August):105–116.

Dempsey, M.J. 1981. Colonization of antifouling paints by marine bacteria. *Bot. Mar.* 24:185–191.

Dlouhy, G., and K. Marquardt. 1984. Moderne Techniken zur Wasseraufbereitung für pharmazeutische, kosmetische und medizintechnische Zwecke. In *Seminar of Hager and Elsässer*, Stuttgart, Germany.

Donlan, R.M., and W.O. Pipes. 1988. Selected drinking water characteristics and attached microbial population density. *Journal of the American Water Works Association* 80:70–76.

Dott, W., and D. Schoenen. 1985. Qualitative and quantitative examination of bacteria found in aquatic habitats. 7th communication, Development of bacterial Aufwuchs on different working materials exposed to potable water. *Zentralblatt für Bakteriology, Hygiene und Parasitenkunde* 180:436–447.

Du Moulin, G.C., K.D. Stottmeier, P.A. Pelletier, A.Y. Tsang, and J. Hedley-Whyte.

1988. Concentration of mycobacterium avium by hospital hot water systems. *Journal of the American Medical Association* 260:1599–1601.

Ebrahim, S., and A. Malik. 1987. Membrane fouling and cleaning at DROP. *Desalination* 66:201–221.

Eisenberg, T.N., and E.J. Middlebrooks. 1984. A survey of problems with reverse osmosis water treatment. *Journal of the American Water Works Association* 76:44–49.

Exner, M., G.J. Tuschewitzki, and E. Thofern. 1983. Untersuchungen zur Wandbesiedlung der Kupferrohrleitung einer zentralen Desinfektionsmitteldosieranlage. *Zentralblatt für Bakteriology, Hygiene and Parasitenkunde* 177:170–181.

Exner, M., G.J. Tuschewitzki, and J. Scharnagel. 1987. Influence of biofilms by chemical disinfectants and mechanical cleaning. *Zentralblatt für Bakteriology, Hygiene and Parasitenkunde* 183:549–563.

Flemming, H.C. 1984. Peracetic acid as a disinfectant: A review. *Zentralblatt für Bakteriology, Hygiene and Parasitenkunde* 179:97–111 (English translation available by the author).

Flemming, H.C. 1987. Microbial growth on ion exchangers. *Water Research* 21:745–756.

Flemming, H.C., and G. Schaule. 1988a. Investigations on biofouling of reverse osmosis and ultrafiltration membranes. Part 1, Initial phase of biofouling. *Vom Wasser* 71:207–223.

Flemming, H.C., and G. Schaule. 1988b. Biofouling on membranes: A microbiological approach. *Desalination* 70:95–119.

Flemming, H.C., and G. Schaule. 1989. Investigations on biofouling of reverse osmosis and ultrafiltration membranes. Part 2, Analysis and removal of surface films. *Vom Wasser* 73:287–301.

Flemming, H.-C., and G. Schaule. 1991. Biofouling on membranes—role of hydrophobicity. In *Microbially Induced Corrosion*. edited by N. Dowling, M.W. Mittelman, and J.C. Danco. Knoxville: 5.101–5.109

Flemming, H.-C., G. Schaule, and E. Gaveras u. M. Beck. 1992. Biofouling von Umkehrosmose-Membranen bei der Reinwasser-Herstellung. In *Herstellung von Reinwasser,* edited by K. Marquardt. Ostfildern: Expert Verlag, in press.

Flemming, H.-C.,G. Schaule, and R. McDonogh. 1992. Biofouling on membranes—a short review. In *Biofilms: Science and Technology.* edited by L. Melo, M.M. Fletcher, T.R. Bott, and B. Capdeville. Amsterdam: Kluwer Academic Publ., in press.

Fletcher, M. 1976. The effects of proteins on bacterial attachment to polystyrene. *Journal of General Microbiology* 94:400–404.

Fletcher, M. 1977. The effects of culture concentration and age, time and temperature on bacterial attachment to polystyrene. *Canadian Journal of Microbiology* 23:1–6.

Fletcher, M. 1980. The question of passive versus active attachment mechanisms in non-specific bacterial adhesion. In *Microbial adhesion,* edited by R.C.W. Berkeley et al. Chichester: Horwood, pp. 197–210.

Fletcher, M.M. 1980. The question of passive versus active attachment mechanisms in nonspecific bacterial adhesion. In *Microbial adhesion to surfaces.* edited by R.C.W. Berkeley, J.M. Lynch, J. Melling, P.P. Rutter, and B. Vincent. Chichester: Ellis Horwood, 197–210.

Fletcher, M., and K.C., Marshall. 1982. Are solid surfaces of ecological significance to aquatic bacteria? In *Advances in microbial ecology,* vol. 6. edited by K.C. Marshall. New York: Plenum Press, pp. 199–236.

Geesey, G.G. 1982. Microbial exopolymers: Ecological and economic considerations. *ASM News* 48:9–14.

Geesey, G.G. 1987. Survival of microorganisms in low nutrient waters. In *Biological fouling of industrial water systems,* edited by G. Geesey and M.W. Mittleman. San Diego, Calif: Water Micro Associates, pp. 1–23.

Geesey, G.G., and M.W. Mittelman, 1987. Biological fouling of industrial water systems: A problem solving approach. San Diego, Calif: Water Micro Associates.

Geldreich, E.E., R.H. Taylor, J.C. Blannon, and D.J. Reasoner. 1985. Bacterial colonization of point-of-use water treatment devices. *Journal of the American Water Works Association* 77:72–80.

George, D.M., G. DuMoulin, and E.M. Carney. 1988. Comparative evaluation of a renal dialysis ultrapure water system using epifluorescence microscopy, R2A media, millipore SPC samplers and LAL assay. *Abstracts of the Annual Meeting of the American Society of Microbiology* L-15.

Gilbert, E. 1988. Biodegradability of ozonation products as a function of COD and DOC elimination by the example of humic acids. *Water Research* 22:123–126.

Gordon, A.S., and F.J. Millero. 1984. Electrolyte effects on attachment of an estuarine bacterium. *Applied Environmental Microbiology* 47:495–499.

Graham, S.I., R.L. Reitz, and C.E. Hickman. 1989. Improving reverse osmosis performance through periodic cleaning. *Desalination* 74:113–124.

Große-Böwing, W. 1982. Reinigung und Desinfektion von Membrananlagen. *Deutsche Molkereizeitschrift* 103:83–84.

Hamilton, W.A. 1987. Biofilms: Microbial interactions and metabolic activities. In *Ecology of microbial environments,* edited by M. Fletcher, T.R.G. Gray, and J.G. Jones. Cambridge: Cambridge University Press, pp. 361–385.

Himelstein, W.D., and Z. Amjad. 1985. The role of water analysis, scale control and cleaning agents in reverse osmosis. *Ultrapure Water* (March/April): 32–36.

Ho, L.C.W., D.D. Martin, and W.C. Lindemann. 1983. Inability of microorganisms to degrade cellulose acetate reverse-osmosis membranes. *Applied Environmental Microbiology* 45:418–427.

Hogt, A.H., J. Dankert, and J. Feijen. 1985. Adhesion of a *Staphylococcus epidermis* and *Staphylococcus saprophyticus* to a hydrophobic biomaterial. *Journal of General Microbiology* 131:2485–2491.

Humphries, M., J.F. Jaworzyn, and J.B. Cantwell. 1986. The effect of a range of biological polymers and synthetic surfactants on the adhesion of a marine Pseudomonas sp. strain NCMB 2021 to hydrophilic and hydrophobic surfaces. *FEMS Microbiology and Ecology* 38:299–308.

Humphries, M., J.F. Jaworzyn, J.B. Cantwell, and A. Eakin. 1987. The use of non-ionic ethoxylated and propoxylated surfactants to prevent the adhesion of bacteria to solid surfaces. *FEMS Microbiology Letters* 42:91–101.

Jacobs, P. 1981. The use of reverse osmosis water for the production of parenterals in the hospital pharmacy. *Pharmaceutisk Weekblad* 116:342–350.

Johnson, C., and M. Howells. 1981. Biofouling: New insights, new weapon. *Power,* (April):90–91.

Kaakinen, J.W., and C.D. Moody. 1985. Characteristics of reverse-osmosis membrane fouling at the Yuma desalting test facility. *Symposium on Reverse Osmosis and Ultrafiltration* 359–382.

Kissinger, J.C. 1970. Sanitation studies of a reverse osmosis unit used for concentration of maple sap. *Journal of Milk Food Technology* 33:326–329.

Kissinger, J.C., and C.O. Willits. 1970. Preservation of reverse osmosis membranes from microbial attack. *Food Technology* 24:177–180.

Kjelleberg, S., and M. Hermansson. 1984. Starvation-induced effects on bacterial surface characteristics. *Applied Environmental Microbiology* 48:497–503.

Kozelski, K.J. 1983. Field experience with a simple cooling water biofilm monitoring device. In *Proceedings of the International Water Conference, Engineering Society*, West Pa. 44th, pp. 447–454.

Kutz, S.M., D.L. Bentley, and N.A. Sinclair. 1986. Morphology of Seliberia-like organisms isolated from reverse osmosis membranes. In *Proceedings of the Fourth International Symposium on Microbial Ecology*, edited by F. Megusar and M. Gantar. Slov. Soc. Microbiol., Ljubljana, pp. 584–587.

LeChevallier, M.W., and G.A. McFeters. 1990. Microbiology of activated carbon. In *Drinking water microbiology*, edited by G.A. McFeters. New York: Springer Verlag, pp. 104–119.

LeChevallier, M.W., C.D. Cawthon, and R.G. Lee. 1988a. Mechanisms of bacterial survival in chlorinated drinking water. Proceedings of the International Conference on Water Wastewater Microbiology, vol. 1, 8–11, at Newport Beach, California, Feb. 1988, 24.1–24.7.

LeChavallier, M.W., C.D. Cawthon, and R.G. Lee. 1988b. Inactivation of biofilm bacteria. *Applied Environmental Microbiology*. 54:2492–2499.

Light, W.G., J.L. Perlman, A.B. Riedinger, and D.F. Needham. 1988. Desalination of non-chlorinated surface seawater using TFC-membrane elements. *Desalination* 70:47–65.

Lintner, K., and S. Bragulla. 1988. Reinigung und Desinfektion von Membrananlagen. *Henkel Referate* 24:42–45.

Little, B.J., P. Wagner, J.S. Maki, M. Walch, and R. Mitchell. 1986. Factors influencing the adhesion of microorganisms to surfaces. *Journal of Adhesion* 20:187–210.

Loeb, G.I., and R.A. Neihof. 1975. Marine conditioning films. In *Applied chemistry at protein interfaces*. edited by R.E. Baier. Advances in Chemistry Series 145, Washington, D.C.: Applied Chemistry Society, pp. 319–335.

Loosdrecht, M.C.M., J. Lyklema, W. Norde, and A.J.B. Zehnder. 1989. Bacterial adhesion: a physicochemical approach. *Microbial Ecology* 17:1–15.

Loosdrecht, M.C.M., J. Lyklema, W. Norde, and A.J.B. Zehnder. 1990. Influence of interfaces on microbial activity. *Microbiol. Rev.* 54:75–87.

Loosdrecht, M.C.M., J. Lyklema, W. Norde, G. Schraa, and A.J.B. Zehnder. 1987. The role of bacterial cell wall hydrophobicity in adhesion. *Applied Environmental Microbiology* 53:1893–1897.

Marshall, K.C. 1985. Mechanisms of bacterial adhesion at solid-water interfaces. In *Bacterial adhesion*, edited by D.C. Savage and M. Fletcher. New York: Plenum Press, pp. 133–161.

Marshall, K.C. 1986. Theory and practice in bacterial adhesion processes. In *Proceedings of the Fourth International Symposium on Microbial Ecology,* edited by F. Megusar and M. Gantar, 24–29 Aug. at Slovic Society of Microbiology, Ljubljana, pp. 112–116.

Marshall, K.C. 1988. Adhesion and growth of bacteria at surfaces in oligotrophic habitats. *Canadian Journal of Microbiology* 34:503–506.

Marshall, K.C., and B. Blainey. 1988. The effect of some members of the synperonic PE series of compounds on microbial adhesion to surfaces. Australian Society of Microbiology, Annual Scientific Meeting, Canberra, May 1988.

Marshall, K.C., R. Stout, and R. Mitchell. 1971. Selective sorption of bacteria from seawater. *Canadian Journal of Microbiology* 17:1413–1416.

Marshall, P.A., G.I. Loeb, M.M. Cowan, and M. Fletcher. 1989. Response of microbial adhesives and biofilm matrix polymers to chemical treatments as determined by interference reflection microscopy and light section microscopy. *Applied Environmental Microbiology* 55:2827–2831.

Martyak, J.E. 1988. Reverse osmosis/deionized water bacterial control at the central production facility. *Microcontamination* 1:34–55.

März, F., R. Scheer, and E. Graf. 1990. Microbiological aspects in the production of water for injection by reverse osmosis. *International Journal of Pharmacology* In press.

Matin, A., and S. Harakeh. 1990. Effect of starvation on bacterial resistance to disinfectants. In *Drinking water microbiology,* edited by G.A. McFeters. New York: Springer, pp. 88–103.

McDonough, F.E., and R.E. Hargrove. 1972. Sanitation of reverse osmosis/ultrafiltration equipment. *Journal of Milk Food Technology* 35:102–106.

McEldowney, S., and M. Fletcher. 1986a. Effect of growth conditions and surface characteristics of aquatic bacteria on their attachment to solid surfaces. *Journal of General Microbiology* 132:513–523.

McEldowney, S., and M. Fletcher. 1986b. Variability of the influence of physicochemical factors affecting bacterial adhesion to polystyrene substrata. *Applied Environmental Microbiology* 52:460–465.

McEldowney, S., and M. Fletcher. 1987. Adhesion of bacteria from mixed cell suspension to solid surfaces. *Archives of Microbiology* 148:57–62.

McFeters, G.A., M.W. LeChevallier, A. Singh, and J.S. Kippin. 1986. Health significance and occurrence of injured bacteria in drinking water. *Water Science Technology* 18:227–231.

Miller, P.C., and T.R. Bott. 1981. The removal of biological films using sodium hypochlorite solutions. *Prog. Prev. Fouling Ind. Plant.,* (Pap. Conf.), Harwell, U.K., pp. 121–136.

Mindler, A.B., and A.C. Epstein. 1986. Measurements and control in reverse osmosis desalination. *Desalination:* 343–379.

Mittelman, M.W. 1987. Biological fouling of purified water systems. In *Biological fouling of industrial waters: A problem solving approach,* edited by G.G. Geesey and H.W. Mittleman. San Diego, Calif.: Water Micro Associates, pp. 194–233.

Morita, R.Y. 1988. Bioavailability of energy and its relationship to growth and starvation survival in nature. *Canadian Journal of Microbiology* 34:436–441.

Nagel, R. 1990. Letter to Hager + Elsässer GmbH, Ruppmannstr. 12, D-7000 Stuttgart, Germany.

Nagy, L.A., A.J. Kelly, M.A. Thun, and B.H. Olson. 1982. Biofilm composition, formation and control in the Los Angeles aquaeduct system. In *Proceedings of the American Water Works Association Water Quality Technology Conference,* at Denver, pp. 213–238.

Nichols, W.W. 1989. Susceptibility of biofilms to toxic compounds. In *Structure and function of biofilms.* edited by W.G. Characklis and P.A. Wilderer. New York: John Wiley, pp. 321–331.

Nusbaum, I., and D.G. Argo. 1984. Design, operation and maintenance of a 5-mgd wastewater reclamation reverse osmosis plant. In *Synthetic membrane processes,* edited by G. Belfort. Academic Press, pp. 378–436.

Osta, T.K. 1987. Pretreatment system in reverse osmosis plants. *Desalination* 63:71–80.

Paul, J.H., and W.H. Jeffrey. 1985. Evidence for separate adhesion mechanisms for hydrophilic and hydrophobic surfaces in vibrio proteolytica. *Applied Environmental Microbiology* 50:431–437.

Payment, P. 1989. Bacterial colonization of domestic reverse-osmosis water filtration units. *Canadian Journal of Microbiology* 35:1065–1067.

Pedersen, K. 1982. Factors regulating microbial biofilm development in a system with slowly flowing seawater. *Applied Environmental Microbiology* 44:1196–1204.

Pedersen, K. 1990. Biofilm development on stainless steel and PVC surfaces in drinking water. *Water Research* 24:239–243.

Potts, D.E., R.C. Ahlert, and S.S. Wang. 1981. A critical review of fouling of reverse osmosis membranes. *Desalination* 36:235–264.

Pringle, J.H., and M. Fletcher. 1986. Influence of substratum hydration and adsorbed macromolecules on bacterial attachment to surfaces. *Applied Environmental Microbiology* 51:1321–1325.

Probstein, R.F., K.K. Chan, R. Cohen, and I. Rubinstein. 1981. Model and preliminary experiments on membrane fouling in reverse osmosis. In *Synthetic membranes,* vol. 1. *Desalination,* edited by A.F. Turbak. *American Chemical Society Symposium Series* 153, pp. 131–162.

Rechen, H.C. 1985. Microorganisms and particulate control in microelectronic process water systems—pharmaceutical manufacturing technology. *Microcontamination* (July).

Ridgway, H.F. 1987. Microbial fouling of reverse osmosis membranes: genesis and control. In: *Biological fouling of industrial water systems.* edited by G. G. Geesey and M. W. Mittelman. San Diego, Calif.: Water Micro Systems, pp. 138–193.

Ridgway, H.F. 1988. Microbial adhesion and biofouling of reverse osmosis membranes. In *Reverse osmosis technology,* edited by B.S. Parekh. New York: Marcel Dekker, pp. 429–481.

Ridgway, H.F., and B.H. Olson. 1982. Chlorine resistance patterns of bacteria from two drinking water distribution systems. *Applied Environmental Microbiology* 44:972–987.

Ridgway, H.F., A. Kelly, C. Justice, and B.H. Olson. 1983a. Microbial fouling of reverse-osmosis membranes used in advanced wastewater treatment technology: Chemical, bacteriological and ultrastructural analyses. *Applied Environmental Microbiology* 45:1066–1084.

Ridgway, H.F., M.G. Rigby, and D.G. Argo. 1983b. Adhesion of a mycobacterium sp. to cellulose diacetate membranes used in reverse osmosis. *Applied Environmental Microbiology* 47:61–67.

Ridgway, H.F., M.G. Rigby, and D.G. Argo. 1984b. Biological fouling of reverse osmosis membranes: the mechanism of bacterial adhesion. *Proceedings of the Water Reuse Symposium II, The future of water reuse*, at San Diego, California, pp. 1314–1350.

Ridgway, H.F., M.G. Rigby, and D.G. Argo. 1985. Bacterial adhesion and fouling of reverse-osmosis membranes. *Journal of the American Water Works Association* 77:97–106.

Ridgway, H.F., D.M. Rodgers, and D.G. Argo. 1986. *Effect of surfactants on the adhesion of mycobacteria to reverse osmosis membranes*. Paper presented at Semiconductor Pure Water Conference, 16–17 Jan. San Francisco: pp. 133–164.

Ruseska, I., J. Robbins, J.W. Costerton, and E.S. Lashen. 1982. Biocide testing against corrosion-causing oilfield bacteria helps control plugging. *Oil Gas Journal* 80:253–264.

Rutter, P., and B. Vincent. 1988. Attachment mechanisms in the surfce growth of microorganisms. In *Physiological models in microbiology*, edited by M.J. Bazin and J.I. Prosser. Boca Raton: CRC Press, pp. 88–107.

Schaule, G., H.-C. Flemming, and K. Poralla. 1990. Biofouling on membranes: Influence of the membrane material. DECHEMA Biotechnology Conference, 4, 1010–1013.

Schaule, G., H.C. Flemming, and K. Poralla. 1992. Forces involved in primary adhesion of *Pseudomonas diminuta* to filtration membranes. 5th International Conference on Microb. Ecol., Barcelona, September 8–12.

Schoenen, D., R. Schulze-Röbbecke, and N. Schidewahn. 1988. Microbial contamination of water by materials of pipes and hoses. 2d communication, Growth of Legionella pneumophila. *Zentralblatt für Bakteriology, Hygiene und Parasitenkunde* 186:326–332.

Schulze-Röbbecke, R., and R. Fischeder. 1989. Growth and die-off kinetics of mycobacteria in biofilms. In *42d Conference Deutsche Gesekschaft für Mikrobiologie* 4.–6.10 (abstract).

Stanley, P.M. 1983. Factors affecting the irreversible attachment of *Pseudomonas aeruginosa* to stainless steel. *Canadian Journal of Microbiology* 29:1493–1499.

Stenström, T.A. 1989. Bacterial hydrophobicity, an overall parameter for the measurement of adhesion potential to soil particles. *Applied Environmental Microbiology* 55:142–147.

Stenström, T.A., and S. Kjelleberg. 1985. Fimbriae mediated nonspecific adhesion of *Salmonella typhimurium* to mineral particles. *Archives of Microbiology* 143:6–10.

Sutherland, I.W. 1983. Microbial exopolysaccharides: Their role in microbial adhesion in aqueous systems. *CRC Critical Reviews in Microbiology* 10:173–201.

Tadros, T.F. 1980. Particle-surface adhesion. In *Microbial adhesion to surfaces*, edited by R.C.W. Berkeley et al. Chichester: Harwood, pp. 329–338.

Trägardh, G. 1989. Membrane cleaning. *Desalination* 71:325–335.

van der Koij, D., and W.A.M. Hijnen. 1988. Multiplication of a *Klebsiella pneumoniae* strain in water at low concentrations of substrate., vol. 1 In *Proceedings of the*

International Conference on Water Wastewater Microbiology, Newport Beach, California, 20.1–20.7.

Vanhaecke, E., J.P. Remon, M. Moors, F. Raes, D. de Rudder, and A. van Petegheim. 1990. Kinetics of *Pseudomonas aeruginosa* adhesion to 304 and 316-L stainless steel: Role of cell surface hydrophobicity. *Applied Environmental Microbiology* 56:788–795.

Walker, S.J. 1986. The growth of biological films on reverse osmosis membranes. Ph.D. diss., John Hopkins University.

Wallhäusser, K.H. 1983. Durchwachs- und Durchblaseffekte bei Langzeit-Sterilfiltrationsprozessen. *Pharm. Ind.* 45:527–531.

Whitfield, C. 1988. Bacterial extracellular polysaccharides. *Canadian Journal of Microbiology* 34:415–520.

Whittaker, C., H. Ridgway, and B.H. Olson. 1984. Evaluation of cleaning strategies for removal of biofilms from reverse-osmosis membranes. *Applied Environmental Microbiology* 48:395–403.

Winfield, B.A. 1979a. The treatment of sewage effluents by reverse osmosis: pH based studies of the fouling layer and its removal. *Water Research* 13:561–564.

Winfield, B.A. 1979b. A study of the factors affecting the rate of fouling of reverse osmosis membranes treating secondary sewage effluents. *Water Research* 13:565–569.

Winters, H. 1987. Control of organicd fouling at two seawater reverse-osmosis plants. *Desalination* 66:319–325.

Winters, H., and I.R. Isquith. 1979. In-plant microfouling in desalination. *Desalination* 30:337–399.

Winters, H., I.R. Isquith, W.A. Arthur, and A. Mindler 1983: Control of biological fouling in seawater reverse osmosis. *Desalination* 47:233–238.

Yanagi, C., and K. Mori. 1980. Advanced reverse osmosis process with automatic sponge ball cleaning for the reclamation of municipal sewage. *Desalination* 32:391–398.

Zips, A., G. Schaule, H.C. Flemming, and U. Faust. 1990. *Ultrasonic energy as a tool for removal of biofilms and measurement of adhesion strength.*

Zips, A., G. Schaule, and H.-C. Flemming. 1990. Ultrasound as a mean for detachment of biofilms. *Biofouling* 2:323–333.

ZoBell, C.E., and C.W. Grant. 1943. Bacterial utilization of low concentrations of organic matter. *Journal of Bacteriology* 45:555–564.

7

Considerations in Membrane Cleaning

Zahid Amjad

The BFGoodrich Company
Brecksville Research Center

Kenneth R. Workman

Arrowhead Industrial Water, Inc.
A BFGoodrich Company

Donald R. Castete

Arrowhead Industrial Water, Inc.
A BFGoodrich Company

INTRODUCTION

Reverse osmosis (RO) is an accepted unit of operation in a variety of industries. RO plants are used throughout the world as self-contained systems or as pretreatment to ion exchange (IX) equipment. RO systems are used in a large variety of industries: semiconductor, pharmaceutical, cosmetic, power, food and beverage, waste water reclamation, and potable water production from seawater and brackish waters (Koseoglu 1990; Parise et al. 1988; Pepper 1990; Pittner 1986; Sackinger 1985; Simpson 1983; Treffry-Goatley et al. 1983). The removal of dissolved ions, suspended matter, and microorganisms is of paramount importance in these applications.

The economic feasibility of RO systems largely depends on maximizing the permeation rate and the membrane life. Studies have indicated that RO system efficiency depends on the maintenance of the membrane in an unfouled condition. Probably the greatest problem experienced in the use of these systems is fouling by deposits (Arora and Trompeter 1983; Flemming and Schaule 1988; Himelstein and Amjad 1985; Paul and Rahman 1990; Paulson 1987; Ridgway et al. 1985). In many cases, fouled membranes cause reduction in flux rate and operating efficiency, leading to unscheduled shutdown, lost production time, replacement of membrane, and the like, thereby increasing operating costs. Prevention of membrane fouling is one of the most important aspects of an RO system application.

The successful and reliable operation of an RO system requires the application of the best possible methods to prevent membrane fouling. Efforts to control membrane fouling in RO systems have focused on feed-water pretreatments, including acid feed and the addition of antiscalants (Amjad 1987, 1989a; Amjad et al. 1988; Logan 1987; Reitz 1984). Despite the success of these approaches, membrane fouling remains a problem. Experience suggests that feed-water pretreatment alone is inadequate. Therefore, regular cleaning should be an integral part of system operation (Amjad 1989b; Graham et al. 1989; Luss 1986).

Fouling problems are often not apparent because system performance appears to be adequate. However, several common foulants may be present in minute quantities. If these foulants are allowed to accumulate on the membrane, a gradual, but serious, deterioration of performance will occur. Eventually, the membrane will become irreversibly fouled, necessitating replacement.

Membrane failure problems are caused by a number of conditions, both mechanical and chemical. When membrane failures occur, it is important to identify correctly the root cause and take corrective actions to prevent recurrences. In any instance of membrane failure, deposit analysis is a key factor in solving the problem.

Figures 7.1 to 7.4 illustrate examples of fouled membranes. Figure 7.1 demonstrates heavy scaling resulting from inadequate pretreatment. Note the scaling blinding the brine flow channels and destroying the permeate bundle. Figure 7.2 shows extensive fouling of a low-pressure polyamide composite membrane element by iron-reducing bacteria. An organically fouled membrane is shown in Figure 7.3. In this case, the element was fed with surface water. A breakdown in the pretreatment system allowed higher than usual concentrations of tannins and other organics that caused irreversible fouling. Note that the element was fed from the right, as evidenced by the increased density of the foulant layer. Figure 7.4 shows a low-pressure composite membrane coated with a layer of silt, drawn from a river bottom by a broken diffuser manifold.

Types of membrane deposits and various analytical techniques used to identify these deposits are covered in Chapters 6 and 8, respectively. This chapter reviews membrane cleaning, rejuvenation, and sanitation techniques. It also discusses troubleshooting RO systems, selection criteria in membrane-cleaning programs, and case histories.

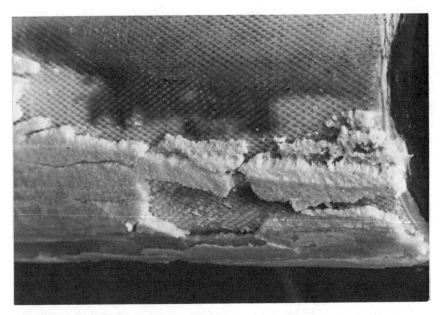

FIGURE 7.1.   A heavily scaled RO membrane.

FIGURE 7.2.   A low-pressure polyamide composite membrane fouled by iron-reducing bacteria.

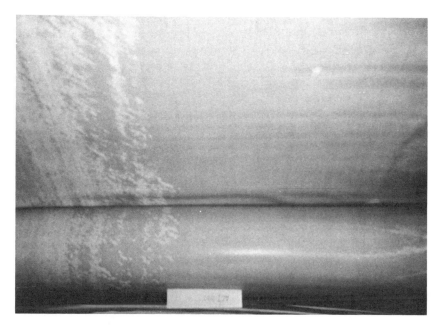

FIGURE 7.3.   An organically fouled membrane.

FIGURE 7.4.   A low-pressure composite membrane coated with a layer of silt.

DEPOSIT REMOVAL TECHNIQUES

Techniques for removing deposits from fouled membranes include mechanical cleaning (i.e., direct osmosis, flushing with high-velocity water, spongeball or brush cleaning, air sparging, etc.), chemical cleaning (use of chemical agents), and a combination of mechanical, ultrasonic, and chemical cleanings (Amjad 1989b; Belfort 1975; Lipinski and Chang 1989). The most prevalent technique is chemical cleaning using specially formulated membrane cleaners.

A large number of chemical agents are available for removing scale and other deposits. Chemical cleaning essentially involves the use of chemicals to react with deposits, scales, corrosion products, and other foulants that affect flux rate and product water quality. These chemical agents can be classified into four categories:

- Acids
- Alkalies
- Chelants
- Formulated products

These chemical cleaning agents are discussed in the following paragraphs.

Acids

Various inorganic, organic, and mixed organic acids have been used for the removal of deposits in general cleaning operations in industry (for heat exchangers, boilers, RO units, and ultrafiltration (UF) membranes), in homes (laundry machines, dishwashers), and in laboratories. The most commonly used acids and their salts have included hydrochloric, phosphoric, and sulfuric acids, organic acids (i.e., oxalic, citric, lactic, malic, sulfamic), and acid salts such as sodium and ammonium acid citrates.

Calcium carbonate, a commonly encountered scale in RO systems, is the easiest deposit to remove. The two methods of removal are (1) decreasing the range for feed pH to 3 to 4 for a sufficient period of time, usually about 30 to 60 minutes, and (2) the use of acid as a batch treatment when the system is off-line. Although hydrochloric acid or sulfuric acid can be used, careful monitoring is required to ensure that the feed does not fall below pH 1.5, preferably not below pH 2. A phosphoric acid solution adjusted to a pH of 1.5 to 2 can also be used. Although any acid can be used, provided it does not attack the membrane or produce an insoluble precipitate, care should be exercised with the use of acid cleaning agents, because prolonged exposure membranes with cleaning solutions of to pH 2 and below, especially at elevated temperatures, may seriously damage cellulose acetate (CA) membranes.

Citric acid (2-hydroxy-1,2,3-propane tricarboxylic acid) has been used in specialty cleaning applications for more than 30 years. The first reported large-scale use of citric acid for chemical cleaning was at the Philo Station of the American Electric Power Corporation (Frankenberg et al. 1958). This was the preoperational chemical cleaning of a large new steam generator. Since then, several new citric-acid–based formulations have been developed and applied to remove rust and water-formed deposits from equipment surfaces, including those of heat exchangers and RO membranes.

Citric acid, although an effective deposit-removing agent for carbonate and iron-based scales, suffers from the disadvantage that it forms a complex with ferrous iron, which has a limited solubility. This product can form when the majority of the citric acid has been reacted. The problem of ferrous citrate precipitation during cleaning has been overcome by the development of ammoni-ated citric acid (i.e., citric acid neutralized to about pH 4), which forms a more soluble ferrous ammonium citrate salt. It should be noted that, like citric acid, other organic acids, such as tartaric acid, hydroxyacetic acid, and formic acid, when used as a sole acid cleaning agent, also form salts with ferrous ions that have limited solubilities.

Sulfamic acid, an amidosulfonic acid, has also been widely used for removing water-formed deposits from equipment surfaces. Sulfamic acid offers several advantages over hydrochloric acid in terms of less corrosiveness, easier trans-port, and absence of acid fumes. However, sulfamic acid's two major dis-advantages are that it reacts slowly and is more expensive.

Oxalic acid, the simplest of the dibasic acids, is very effective in removing deposits containing iron. Oxalic acid also forms salts with calcium ion, which exhibit extremely poor solubilities. The problem of postprecipitation of calcium oxalate salt, however, may be overcome by incorporating a scale inhibitor in the cleaning formulations.

Acids are effective in removing calcium-based scales such as calcium carbon-ate and calcium phosphate, iron oxides, and metal sulfides. However, acids have limited utility in removing scales formed by silica and metal silicates. Acids are also relatively ineffective in removing biological suspended solids and other related organic foulants. In addition to inorganic and organic acids, homopoly-mers and copolymers of acrylic acid have been tried in the removal of deposits from fouled heat exchanger and membrane surfaces (Lipinski and Chang 1989).

Alkalies

Alkaline cleaning solutions include phosphates, carbonates, and hydroxides. These materials function like detergents to loosen, emulsify, and disperse depos-its. The detergency of alkaline cleaners is usually increased by the addition of surfactants in order to wet oil, grease, dirt, and biomatter. Alkaline cleanings are often alternated with acid cleanings to remove particularly difficult deposits such as silicates.

Chelants

In addition to strong acids and alkalies, chelants have also been used to remove deposit from fouled membranes. Commonly used chelants include ethylenediamine tetracetic acid (EDTA), phosphonocarboxylic acid, gluconic acid, citric acid, and polymer-based chelants (Nagarajan 1985; Schaffer and Woodhams 1977). Gluconic acid is usually effective in chelating ferric ion (Fe^{3+}) in strong alkaline solution.

EDTA is a more powerful chelating agent than citric acid because it offers more binding sites. The complexing constants of alkaline earth metal—EDTA and—citrate complexes are shown in Table 7.1. The complexing constant data (Table 7.1) clearly show that EDTA is more effective than citric acid in forming soluble complexes with calcium, magnesium, iron, and barium, and is therefore generally used to dissolve alkaline earth metal sulfates.

Formulated Products

The limitations of commodity chemicals have led membrane manufacturers and others to publish nonproprietary formulations for use as membrane cleaning agents. Table 7.2 presents a listing of several nonproprietary formulations. In addition, several companies, including Argo Scientific, The BFGoodrich Company (parent company of Arrowhead Industrial Water, Inc.), and HB Fuller

TABLE 7.1   Stability
Constants for Metal
Chelates of EDTA and
Citric Acid

Metal Ion	Log K	
	EDTA	Citric Acid
Ba^{2+}	7.86	2.98
Ca^{2+}	10.69	4.68
Cu^{2+}	18.80	4.35
Fe^{2+}	14.32	3.08
Fe^{3+}	25.10	12.50
Mg^{2+}	8.79	3.29
Mn^{2+}	13.87	3.67
Ni^{2+}	18.62	5.11
Pb^{2+}	18.04	6.50
Zn^{2+}	16.50	4.71

From J. Roger Hart (W.R. Grace Co.). (Reprinted by permission.)

TABLE 7.2   Generic Membrane Cleaning Formulations

Cleaner	Scale/Metal Oxides	Colloidal/ Particulate	Biological	Organics
Hydrochloric acid 0.5% (wt)	X			
Citric acid 2% (wt) and ammonium hydroxide (pH 4.0)	X			
Phosphoric acid 0.5% (wt)	X			
Sodium hydroxide pH 11–11.9		X	X	
Trisodium phosphate or sodium tripoly phosphate 1% (wt), sodium salt of ethylenediaminetetraacetic acid 1% (wt), and sodium hydroxide—pH 11.5–11.9		X	X	
Sodium hydrosulfite 1% (wt)	X			
BIZ* 0.5% (wt)			X	
Citric acid 2.5% (wt) and ammonium bifluoride 2.5% (wt)	X	X		

*Trademark of Proctor and Gamble, USA, for a detergent sold in the United States.

Company (Monarch Division), recognized the need for specialized membrane cleaning chemicals. There followed the development of a number of proprietary formulated cleaners for removing deposits, as summarized in Table 7.3. The nonproprietary and formulated membrane cleaning products are superior to the more traditional, single-function commodity-type cleaners such as citric acid (used for removing calcium carbonate scaling) and laundry detergents with enzyme additives (used for removing biofoulants and certain organics).

MEMBRANE CLEANER SELECTION CRITERIA

Clean membranes are critical for maintaining the efficient operation of RO systems. Because of the many possible foulants, membrane cleaning is a complex subject. Characterizing the deposits on fouled membranes is essential to the selection of the most economical and effective cleaner. Analysis of feed waters and spent cartridge filters, as well as evaluation of chemical changes (if any) resulting from pretreatment provides insight into foulant characteristics.

Once the magnitude and types of deposits are identified, membrane cleaning is required to restore design performance. If a deposit analysis reveals a variety of foulants, a traditional chemical cleaning may not suffice. Acid cleaning has a limited effect on sand, clay, and biological matter. Alkaline cleaners can do little to dissolve and disperse hardness scales. Figures 7.5a and b, respectively, show a scanning electron micrograph and EDX spectrum of a fouled membrane. The

TABLE 7.3   Customized Membrane Cleaning Compounds

Cleaner	Scale/Metal Oxides	Colloidal/ Particulate	Biological	Organics	Source[1]	Membrane Type(s)[2]
Arro-Clean™ 2100	X	X			BFG	B
Arro-Clean™ 2200		X	X		BFG	B
Arro-Clean™ 2300	X	X	X		BFG	B
Arro-Clean™ 2400				X	BFG	B
Bioclean 103A	X				AS	CA
Bioclean 107A		X	X		AS	CA
HPC 303	X				AS	CA
HPC 307		X			AS	CA
Bioclean 511			X		AS	PA
IPA 403	X				AS	PA
IPA 411		X				
Poly Blue 202				X	AS	B
Filtra Pure™ Iron Remover	X				HBF	B
Filtra Pure™ CA		X	X		HBF	CA
Filtra Pure™		X	X		HBF	PA
Filtra Pure™ HF		X	X		HBF	PA
Monarch Membrane Acid 23	X				HBF	B
Monarch Enzyme 95, DDS		X	X		HBF	B
Monarch Enzyme 96		X	X		HBF	PA
Monarch RO Cleaner 115		X				
Monarch HPH-RO		X				
Monarch Enzyme		X	X		HBF	CA
Monarch Chelate	X				HBF	B

[1]Information source and provider:
   BFG = Arrowhead Industrial Water, Inc. A subsidiary of The BFGoodrich Company, Lincolnshire, Illinois
   AS = Argo Scientific, San Marcos, California
   HBF = HB Fuller Co., Monarch Division, Minneapolis, Minnesota
[2]Membrane types:
   CA = Cellulose acetate
   PA = Polyamide/thin film composite
   B = Both CA and PA

deposit analysis identified a mixture of foulants, thereby suggesting that in such cases traditional cleaning agents such as citric acid or alkaline cleaners would not be effective.

In selecting a cleaning program, the following factors should be considered:

- Cleaning equipment requirements
- Membrane type and cleaner compatibility
- System's materials of construction
- Foulant identification
- Requirements for, and impact resulting from, discharging spent cleaning solutions

SYSTEM TROUBLESHOOTING

Troubleshooting is required when RO system performance changes at unanticipated levels (Bukay and Berne 1986; Graham et al. 1989). RO system troubleshooting is required when one or more of the following occurs:

- Normalized production changes by 15 percent.
- Normalized salt passage changes by 50 percent.
- Corrected differential pressure (d/p) changes by 15 percent.

Troubleshooting an RO system that is not meeting performance expectations consists of the following steps:

1. Verify instrument operation.
2. Review operating data.
3. Evaluate potential mechanical and chemical problems.
4. Compare current feed-water analysis with design criteria.
5. Identify foulants.

The following paragraphs discuss each of these five steps.

The first step in troubleshooting an RO system is to verify instrument accuracy (flow meters, pressure gauges, temperature gauge, total dissolved solids (TDS)/conductivity, and pH meters) and recalibrate as necessary (Bukay and Berne 1986).

The second step in the RO system troubleshooting process is to examine operating logs and review trend plots of performance data, including normalized salt rejection, normalized production, and corrected d/p. Procedures for normalizing salt rejection and production are available (Bukay 1986; McPherson et al. 1982). Essentially, normalization is the mathematical procedure for correcting actual production and salt rejection values to standard conditions (usually start-up conditions). The effects of temperature, pressure, and concentration are incorporated into the normalization process. Assume, for example, that a system has experienced a 25 percent drop in production, but with little effect (or with a slight improvement) in salt rejection. The feed-water (a surface supply) temperature has decreased 8°C since start-up. No other conditions have changed significantly during this time. Normalizing the production data indicates that the apparent loss of production is due almost entirely to the lower feed-water temperature. The feed pressure is raised by 25 percent to maintain the required production. In some cases, it may be necessary to factor the effects of membrane age on production and salt rejection. Consult the membrane manufacturer to determine whether this correction is needed. Differential pressure is corrected by relating the ratio of actual d/p and average feed-brine flow multiplied by the start-up average feed-brine flow to the start-up d/p.

a

FIGURE 7.5.   Scanning electron micrograph (a) and EDX spectrum (b) of a fouled membrane.

The third step in troubleshooting RO systems is to determine whether mechanical problems or chemical upsets are the culprit. Examples of mechanical problems are damaged O-rings, damaged or missing feed tubes in hollow fiber permeators, damaged brine seals, pump failures, inaccurate instrumentation, piping and valving failures, defective elements, missing or improperly installed cartridge filters, and filter malfunctions. Chemical upsets include the following:

1. *Improper acid addition.* High dosages may damage membranes or cause sulfate-based scaling (if H_2SO_4 is used), and low dosages may lead to carbonate or metal hydroxide-based scaling and/or fouling.

2. *Improper scale-inhibitor addition.* High dosages may lead to fouling and low dosages can lead to scaling.

The fourth step in the troubleshooting process is to compare a recent feed-water analysis with the design-basis analysis. Changes in the feed-water chemistry may result in the need for additional pretreatment or modifications to the existing pretreatment program. Fouling or scaling conditions may occur with the new water chemistry.

If mechanical problems, chemical upsets, and water chemistry changes have

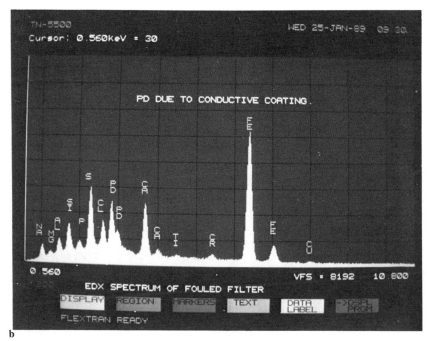

FIGURE 7.5.   *(continued)*

been ruled out, the fifth step in the RO system troubleshooting process should be taken. Specifically, efforts should be focused on the identification of foulants. Two types of analysis have been effective in identifying RO system foulants. Their procedures are outlined in the following paragraphs.

1. Analyze the feed, brine, and product streams for inorganic constituents, total organic carbons (TOC), turbidity, pH, TDS, total suspended solids (TSS) (feed only), Silt Density Index (SDI) (feed only), and temperature. This step proves invaluable for mass balance calculations and the identification of potential scaling components. The SDI, TSS, and turbidity measurements can provide a good indication of particulate-matter fouling. TOC measurements give an indication of organic fouling.

2. Digest and analyze a feed-water cartridge filter (preferred method) or an SDI filter pad. This technique is quite effective in determining the composition of likely suspended-matter foulants. In addition, examination of the operating history may provide insight as to the identity of the foulant(s). Table 7.4 lists various fouling regimes and the resulting effects on RO performance. If available data still do not support the presence of a particular foulant, then RO membrane element autopsies should be considered. Such autopsies involve two phases:

TABLE 7.4   Summary of Foulants and Likely Performance Effects

Foulant	Description	Effects on RO Performance	Method of Control/Cleaning
		(General order of appearance)	
Scale	Precipitate of sparingly soluble salts (minerals) caused by the concentration of salts in the feed/brine solution during passage across the membrane surface. Example: $CaCO_3$, $CaSO_4 \cdot 2H_2O$, $BaSO_4$, $SrSO_4$, SiO_2	• Major loss of salt rejection • Moderate increase in d/p • Slight loss of production • The effects usually occur in the final stage	• Lower recovery • Adjust pH • Use scale inhibitor • Clean w/citric acid or EDTA-based solution • Clean silicate-based foulants with ammonium bifluoride-based solutions
Colloidal Clays/Silt	Agglomeration of suspended matter on the membrane surface. Example: SiO_2, $Fe(OH)_3$, $Al(OH)_3$, $FeSiO_3$	• Rapid increase in d/p • Moderate loss of production • Moderate loss of rejection • The effects usually occur in the first stage	• Filtration • Charge stabilization • Higher feed-brine flows • Clean w/EDTA, STP, or BIZ-type detergents at high pH • Clean silicate-based foulants with ammonium bifluoride-based solutions • Lower recovery
Biological	Formation of bio-growth upon membrane surface. Example: Iron reducing bacteria, sulfur reducing bacteria, mycobacteria, *Pseudomonas*	• Major loss of production • Moderate loss of salt rejection • Possible moderate increase in d/p • The above effects occur slowly, steadily	• Sodium bisulfite addition • Chlorination with or without activated carbon filtration • Clean with BIZ-type detergents or EDTA-based solutions at high pH • Shock disinfection program with formaldehyde, hydrogen peroxide, peracetic acid, etc.
Organic	Attachment of organic species to the membrane surface. Example: polyelectrolytes, oil, grease	• Rapid and major loss of production • Stable or moderate increase in salt rejection • Stable or moderate increase in d/p	• Filtration with GAC • Cleaning is rarely successful but isopropanol or proprietary solutions have been effective

a. *Nondestructive*. The nondestructive method involves isolating an element for study. The element may be tested on-site or sent to a company that specializes in this technique. Generally, the element is physically examined for damage, buildup of foulant, color, smell, and weight. If possible, foulant samples are collected for analysis. The element is then performance tested at the manufacturer's standard test conditions. Based on the available data, a cleaner is selected from those included in Tables 7.2 and 7.3, which list several generally acceptable membrane-cleaner formulations (DuPont, 1982; FilmTec 1988a; Luss 1988; UOP 1989a,b,c). Table 7.5 lists the steps involved in a standard membrane cleaning, and Figure 7.6 is a schematic of a cleaning system (DuPont 1982; FilmTec 1988a). After the cleaning procedure is carried out, the element is performance tested. If performance is improved, the procedure(s) should then be

TABLE 7.5   General Membrane Cleaning Procedure

1. Flush unit to be cleaned with RO product water for 15 minutes at 50 psig and at 75 percent of the maximum flow for that size (diameter) of element. Clean one stage at a time.

2. Prepare cleaning solution according to supplier's instructions. Be sure cleaner is thoroughly mixed. Adjust pH as necessary to be within membrane manufacturer's limits. Adjust temperature to maximum permissible by the membrane manufacturer.

3. Pump cleaning solution through the unit and to the drain at a high flow rate (75 percent maximum). When the solution is observed at the drain (after 1 to 2 minutes), turn off the pump and close all valves. Allow unit to soak for at least 15 minutes.

4. Repeat Step 3 two times. Monitor the spent cleaning solution for color changes. If color pickup remains significant, continue Step 3 until there is little color change.

5. Recirculate the brine to the cleaning tank for 45 minutes after Step 4 is finished.

6. Direct brine and product to drain. This will help empty the cleaning tank faster.

7. Flush the unit with RO product water for 15 minutes at moderate flow rates (50 percent of maximum) and at 50 psig. Flush water temperature should be kept above 20°C to prevent certain detergents from precipitating (e.g., sodium lauryl sulfate).

8. Continue flush cycle with RO product water (preferred method) or RO feed water for at least 30 minutes at high flow rates.

9. Before stopping the flush cycle, check the brine stream for the following conditions:
    • Brine pH is within 1 unit of the feed pH.
    • Brine conductivity is within 100 micromhos/cm of the feed conductivity.
    • Brine is not foamy.
    If all three conditions are met, then this cleaning treatment is completed. Additional cleanings may be performed or the unit may be put into operation.

10. Consult your system vendor, membrane manufacturer, and cleaner supplier for additional information. This procedure is one of many used in the industry. Your specific conditions will determine appropriate cleaner type(s) and procedures.

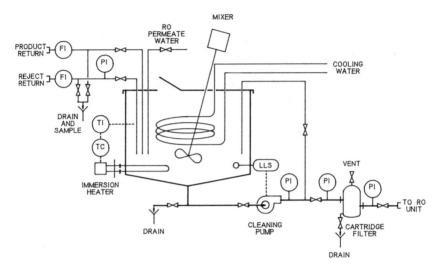

FIGURE 7.6.   Schematic of a typical membrane cleaning system.

tried in the field. If performance is not significantly improved, then additional cleanings, different cleaners, or a destructive autopsy will be needed.

b. *Destructive.* The destructive autopsy is a procedure whereby the element is sacrificed in order to obtain more extensive data. Samples of the membrane are then examined by techniques such as infrared (IR) spectroscopy, scanning electron microscopy (SEM), and other analytical methods to identify the foulant on the membrane or to determine the extent of damage experienced by the membrane.

Information obtained through these techniques usually identifies the system failure. Steps can be taken to clean the unit, modify the pretreatment system, replace elements, change operating procedures, or rejuvenate the elements. The last section of this chapter presents two case studies that apply the techniques discussed.

REJUVENATION AND STERILIZATION

Rejuvenation

Rejuvenation is the enhancement of salt-rejection characteristics by the application of chemical treatments to the surface of semipermeable membranes. During the life of RO membranes, degradation of performance will be experienced. Cleaning treatments generally have varying degrees of success. In some cases, a noticeable loss of salt rejection will have occurred, which may be due to surface defects, abrasion damage, chemical attacks (cleaning compounds, oxidants), and/or hydrolysis. To restore the lost salt-rejection characteristics, rejuvenation chemicals should be tried.

Rejuvenation techniques have been studied by membrane manufacturers and the government during the 1970s and early 1980s (Cadotte et al., 1981; Dalton et al. 1978; Du Pont 1982; Guy and Singh 1982; Kremen et al. 1975; UOP undated). Two mechanisms have been proposed to explain the rejuvenation process: "surface treatment," which is, in effect, a surface coating on the membrane, and "hole plugging." Table 7.6 lists three membrane rejuvenation agents in use today. Table 7.7 outlines a general procedure for membrane rejuvenation.

Membrane rejuvenation typically increases salt rejection to at least 94 percent. The keys to successful application of these products are (1) to thoroughly clean the RO membranes and (2) to follow carefully the instructions supplied by both membrane manufacturers and chemical suppliers. These procedures will not work when either physical damage of the membrane is great or the O-rings are damaged. When salt rejection is below 75 percent, successful rejuvenation is unlikely; if below 45 percent, rejuvenation treatments are completely ineffective.

Sterilization (Sanitization) Techniques

Several procedures are available to sterilize RO membranes and systems. Hydrogen peroxide (0.25 percent wt), sodium bisulfite/glycerine (0.2 to 1 percent/16 to 20 percent wt), sodium bisulfite (0.5 to 1 percent), formaldehyde (0.25 to 1 percent), and copper sulfate (0.1 to 0.5 ppm) have all been successfully used as disinfectants (DuPont 1982; FilmTec 1988a,c; Katz et al. 1976). For storage purposes, formaldehyde, sodium bisulfite, and the sodium bisulfite/glycerine solutions have been proven effective and membrane compatible. Table 7.8 details a generic sterilizing procedure with formaldehyde. Sterilization is undertaken to kill microorganisms or to prevent microbial growth during periods of storage.

It is extremely important to ensure bio-free conditions with CA-based membrane systems. Some bacterial strains degrade CA materials (Canter and Mechalas 1969). During normal operation, feed water to CA RO systems is chlorinated. This usually proves effective for keeping the system disinfected. The difficulty arises when the RO unit is shut down for extended periods of time (>2 to 5 days). It is then necessary to sterilize the system.

Because synthetic RO membrane systems can tolerate little (<1,000 ppm hrs Cl_2) to no chlorine, microbiological activity is almost always present. Although these membranes are quite resistant to microbial degradation, fouling as a result of bioactivity can be detrimental to RO system performance (Argo and Ridgway

TABLE 7.6   Rejuvenation Agents

Trade Name	Chemical Name	Comments
Colloid 189*/zinc chloride	Polyvinyl acetate copolymer in an ammonical solution	• Hole plugging • Proven effective on most CA-type spiral wound membranes that have been well cleaned • Applied in situ • Temporary improvement (1 month to 6 months) • Erratic improvements on poly(ether) urea and poly amide membranes • Flux loss generally <15 percent • Differential pressure increase may be unacceptable • Cleaning treatments will remove the coating • Usually very effective when salt rejection is above 75 percent
PTA/PTB (Lutonol M40 from BASF)	Polyvinyl methyl ether/ tannic acid	• Proven effective on DuPont B-9 and B-10 fiber permeators • Applied in situ • Either has been applied individually with some success • Surface coating • PTB wears away with time • PTA is generally durable • Cleaning treatments may remove PTA and will readily remove PTB • Usually very effective when salt rejection is above 80 percent • Flux loss below 5 percent • Little, if any, increase in d/p • Follow membrane manufacturer's instructions precisely for each product
Vinac/Gelva	Polyvinyl acetates	• Surface treatment • Proven effective on CTA hollow-fiber membranes • Applied in situ • Temporary (1- to 12-month improvement) • Flux loss generally above 5 percent • Cleaning treatments will remove the coating • An increase in d/p may be encountered • Usually very effective when salt rejection is above 75 percent

*Trade name

TABLE 7.7   General Membrane Rejuvenation Procedure

1. Thoroughly clean the system.

2. Restart unit. Measure performance characteristics.

3. Verify no mechanical problems—bad O-rings, damaged brine seals. Probe system if necessary. Elements with high productivity and low salt rejection (<45%) should be replaced. Rejuvenation chemical cannot coat/plug gross defects.

4. Set feed pH to 7.0 to 8.0 (or turn off acid pump).

5. Prepare rejuvenation solution (Colloid 189) for chemical addition. Use acid pump if practical. Add rejuvenation agent (12% solution) to feed tank. Add NH_4OH to solution to obtain a target pH of 8 to 9. Dilute with RO product water (pH >7.0) if necessary to ensure metering pump can inject 10 to 20 ppm (active) of rejuvenator.

6. Start addition. Closely monitor product TDS, production and d/p.

7. When performance stabilizes, or production losses or d/p increases exceed requirements, then discontinue addition (generally within 1 hour).

8. Flush out chemical feed tank and tubing with product water (pH adjusted >7.0 with NH_4OH) for 15 minutes.

9. Start acid addition if part of normal operation.

10. Prepare second solution (zinc chloride) for addition. Mix powder with product water to make a 5 percent (wt) solution. Adjust to pH 4.0 with HCl.

11. Add to RO feed stream at 10 ppm for 60 minutes, then continue normal operation.

12. Performance should stabilize within 72 hours.

Note: Consult membrane manufacturer for its specific procedure.

TABLE 7.8   General Procedure for Membrane Sanitization

1. Flush unit with RO product water for 10 minutes.

2. Prepare a 1 percent (wt) formaldehyde solution in the cleaning tank if possible. Use RO product water for the dilution of formaldehyde.

3. Flush the solution through the unit with the cleaning pump at 50 psig and normal brine flows. Direct the first 20 percent of solution to waste. Then recirculate both product and brine lines to cleaning tank.

4. Continue recirculation for 1 hour or until temperature reaches 85°F for CA membranes and 95°F for synthetic membranes, whichever occurs first.

5. Drain cleaning tank and piping to waste. Close up the unit to make sure RO membranes remain moist during the storage/disinfection period. Certain manufacturers request that the vessels be kept completely full during the storage period.

6. When placed back into operation, direct product and brine streams to waste. The unit must be flushed at higher brine flow than normal (lower recoveries) for at least 2 hours.

Note: Refer to membrane manufacturer's specific procedures and warnings. Consult with local, state, and federal regulatory agencies before disposal of these solutions.

1982; Arora and Trompeter 1983; Belfort 1975; Ridgway et al. 1983). Sterilizing the system to kill microorganisms before applying a bio-cleaner has, in many cases, improved the efficiency of the cleaner—in addition to the benefit of sterilizing the system. As for CA membranes, the synthetic membranes should be sterilized before extended periods of downtime (>2 days). If feed water or product water flushes are routinely practiced during shutdowns, 5 days may elapse before scheduling a sterilization.

Sterilization is a necessary process in the maintenance of RO systems. Sterilizing agents, however, pose a health hazard and a serious disposal concern. Consult with suppliers to determine whether the sterilizing agents can be neutralized before disposal.

CASE HISTORIES

Case 1

Background
The RO system at a southwestern U.S. chemical manufacturing plant had been in service for 3 months. This system was designed to produce 100 gallons per minute (gpm) from a surface water supply. The permeate from the RO system was used as feed to a deionization (DI) system in a boiler feed application. Pretreatment consisted of multimedia filtration, followed by 5 μm cartridge filtration. No chemical pretreatment was used. Polyamide thin film composite membranes were used in this system. The system was cleaned twice during the first 2 months of operation. Two commercial RO cleaners designed to remove colloidal foulants and biological foulants were used for each of the cleanings. During this period, normalized permeate flow decreased by 38 percent, corrected differential pressure increased by 43 percent, and normalized salt passage increased by 75 percent.

Problem Evaluation and Recommendations
The rapid flux decline of the system indicated the presence of foulants rather than mineral scale. Two cartridge filters were digested from successive changeouts. The results (see Table 7.9) indicated a large variation in the feed-water fouling characteristics. The first filter digestion identified the major potential foulants as aluminum, iron, and silica. The second filter digestion indicated the presence of calcium as a precipitate (carryover from the plant clarifier). A product designed to remove metal oxides and metal silicates (BFGoodrich's MT® 3100 Cleaner) was recommended. This product is also capable of removing calcium carbonate. The system was cleaned in the field using the customer's cleaning skid. The first step was to flush the bank to be cleaned with permeate water. The next step was displacement of the water in the stage with cleaning solution. MT 3100 Cleaner

TABLE 7.9   Filter Digestion Analyses

Component (ppm)	Sample Number 1		Sample Number 2	
	Outer	Inner	Outer	Inner
Alkalinity (as CaCO₃)	84.30	4.56	552.69	91.08
Al	6.53	6.70	16.73	6.60
Ba	0.13	0.16	2.94	1.29
B	0.08	0.10	5.13	5.17
Ca	16.46	22.64	3482.00	1996.00
Cr	0.04	0.05	<DL	<DL
Co	<DL	<DL	<DL	<DL
Cu	0.04	0.04	0.22	0.08
Fe	36.90	30.90	93.74	65.79
Pb	<DL	<DL	<DL	<DL
Mg	5.09	6.20	163.43	65.79
Mn	0.16	0.11	1.52	0.63
Ni	<DL	<DL	<DL	<DL
P	9.40	10.85	5.80	2.90
K	<DL	<DL	2.82	<DL
Si	11.53	8.92	58.04	57.97
Na	NA	NA	NA	NA
Sr	0.04	0.18	8.71	3.83
SO₄	<DL	<DL	<DL	<DL
Zn	2.18	0.53	2.18	0.53
pH	7.14/6.81*	7.68/6.79*	7.00/8.60*	7.00/8.50*

All concentrations are ppm's unless otherwise noted.

<DL = less than detection limit.

NA = not analyzed (owing to composition of solution).

Values are normalized to 1 gram filter/40 grams solution.

*Initial pH water/final pH water after 24 hour soak of filter.

was used to remove metal oxides, metal silicates, and hardness scales. The stage was soaked for 30 minutes. The spent cleaner was displaced with fresh cleaner and soaking was continued. Owing to the large quantity of foulants present in the membranes, six soak-flush cycles were used before the spent cleaning solution color appeared clear. After the soaks, a 30-minute recirculation was used for final polishing. The stage was then thoroughly flushed with permeate. The same steps were subsequently applied to the second stage. Upon completion, the RO system was put back in service.

Cleaning Program Performance

Figure 7.7 shows the normalized permeate flow, corrected differential pressure, and normalized salt rejection before and after cleaning. Referring to Figure 7.7, "Initial" indicates the performance of the system at start-up. "BC1," "BC2," and

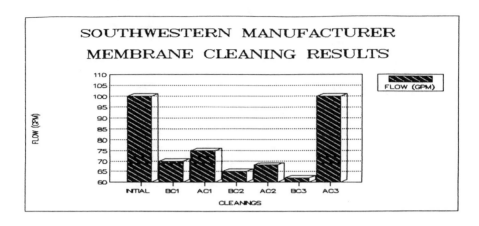

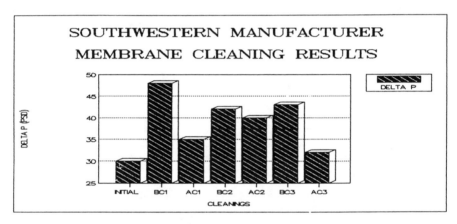

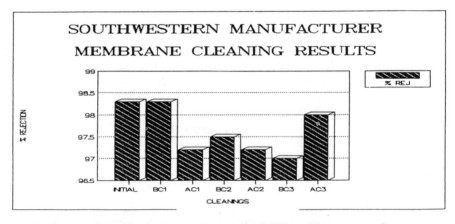

FIGURE 7.7. **Normalized premeate flow, corrected differential pressure, and normalized salt rejection before and after cleaning of RO system at a southwestern U.S. chemical manufacturing plant.**

"BC3" indicate performance before each of the cleanings. "AC1," "AC2," and "AC3" indicate performance after each of the cleanings. The data shown in Figure 7.7 normalized permeate flow, corrected differential pressure, and normalized salt rejection. The normalized product flow was restored to within manufacturer's specifications. The differential pressure was reduced to within 2 psig of the original value. The salt rejection was restored to manufacturer's specifications. These data indicate that the cleaning program was very effective.

The increase in permeate flow and salt rejection reduced ion exchange regeneration costs by $2,000 a month. The membrane cleaning cycle decreased from once every 3 weeks to once every 2 months. The life of the membrane was extended by 2 years, resulting in a savings of $9,000 a year, and the overall program resulted in a savings of $36,000 a year.

Case 2

Background
The RO system at a southern chemical manufacturing plant had been in service for 4 months. This system was designed to produce 15 gpm from a surface water supply. The supply water was treated at an off-site facility using clarifiers with polymer addition. The permeate from the RO system was used as process water for the plant. Pretreatment consisted of multimedia filters with polymer addition, acid feed (HCl), sodium bisulfite addition for dechlorination, and 5 μm cartridge filters. Polyamide thin film composite membranes were used in the system. During this initial 4-month operating period, the RO system's normalized permeate flow decreased by 35 percent, corrected differential pressure increased by 72 percent, and normalized salt passage increased by 15 percent. Periodic flushes with HCl had been used to partially restore membrane flux.

Problem Evaluation and Recommendation
System operation initially involved several extended periods of shutdown. The periods of downtime, in conjunction with flux decline throughout the system, indicated potential biological fouling. In addition, several instances of flux decline in the first stage were noted. An SDI pad was allowed to run until plugged. The pad was digested (see Table 7.10), and results indicated the presence of aluminum, iron, magnesium, and silica. The SDI pad digestion indicated the potential for metal silicates fouling. As a result, a product designed to remove biological foulants (BFGoodrich's MT 5000 Cleaner) was recommended. In addition, it was recommended that the first stage should be cleaned with a product (BFGoodrich's MT 3100 Cleaner) designed to remove metal silicates.

The system was cleaned in the field using the customer's cleaning skid. The first step was to flush the bank to be cleaned with permeate water. The next step was displacement of the water in the bank with cleaning solution. MT 5000

TABLE 7.10   SDI Pad Digestion Analysis

Component (ppm)	As ppm Lab Solution	As ugm Pad	Component % of Total
Al	2.89	57.8	2
Ba	<DL	0	0
B	4.69	93.8	4
Ca	<DL	0	0
Co	<DL	0	0
Cu	<DL	0	0
Fe	8.48	169.6	7
Pb	<DL	0	0
Mg	17.64	352.8	15
Mn	0.12	2.4	0
P	<DL	0	0
K	<DL	0	0
Si	53.33	1066.6	46
Na	4.96	99.2	4
Sr	0.096	1.92	0
Zn	0.85	17.0	1
Alkalinity	10.05	201.0	9
SO_4	<DL	0	0
TOC	13.03	260.6	11
pH*	5.10/5.90	—	—
Total		2323.0	100

All concentrations are ppm's unless otherwise noted.

<DL = less than detection limit.

*Initial pH water/final pH water after 24 hour soak.

Cleaner was used to remove biological fouling. The bank was soaked for 30 minutes. The spent cleaner was displaced with fresh cleaner and soaking contiued. Three soak-flush cycles were used before the color appeared clear. After the soaks, a 30-minute recirculation was used for final polishing. The bank was then thoroughly flushed with permeate. The same steps were subsequently applied to the second bank. Upon completion, the system was put back in service and allowed to stabilize before readings were taken for normalization. After normalization data were examined, it was evident that the first bank performance was still below normal. A second cleaning was performed using MT 3100 Cleaner on the first bank only. The previously described procedures were used for this cleaning.

Cleaning Program Results

Referring to Figure 7.8, "Initial" indicates the performance of the system at start-up. "BC1" and "AC1" indicate the performance before and after cleaning

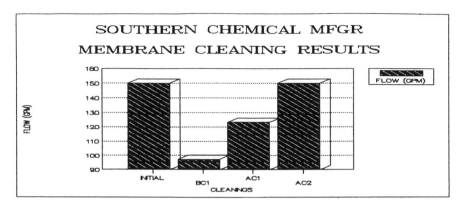

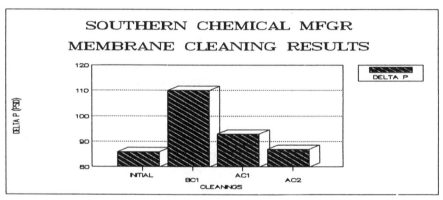

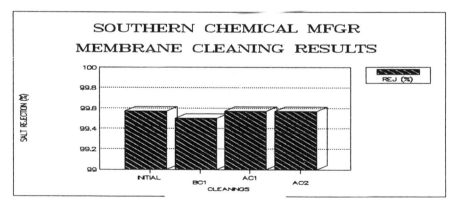

FIGURE 7.8.   Normalized permeate flow, corrected differential pressure, and
normalized salt rejection before and after cleaning of RO system at a southern chemical
manufacturing plant.

with MT 5000 Cleaner, respectively. "AC2" indicates the performance after cleaning with MT 3100 Cleaner. Figure 7.8 shows the normalized permeate flow, corrected differential pressure, and normalized salt rejection before and after cleaning. The normalized product flow was restored to within manufacturer's specifications. The differential pressure was reduced to the original value. The salt rejection was restored to manufacturer's specifications.

The cleaning procedure resulted in restoration of system performance. Membrane replacement was projected to be at a minimum 3-year life cycle. Cleaning frequency and costs will be reduced as a result of using the correct products and procedures. Less system downtime is expected.

References

Amjad, Z. 1987. Advances in scaling and deposit control agents for reverse osmosis systems. *Ultrapure Water* 4(6):34–37.

Amjad, Z. 1989a. Deposit control agents for reverse osmosis applications. *Ultrapure Water* 6(5):57–60.

Amjad, Z. 1989b. Advances in membrane cleaners for reverse osmosis systems. *Ultrapure Water* 6(6):38–42.

Amjad, Z., J.P. Hooley, and K.W. Workman. 1988. Copolymer based reverse osmosis water treatment programs: New developments which expand their applications. Presented at the National Water Supply Improvement Association, Biannual Conferences, 31 July–4 Aug., at San Diego.

Argo, D.G., and H.F. Ridgway. 1982. Biological fouling of reverse osmosis membranes at water factory 21. In *Proceedings of the Water Supply Improvement Association,* Honolulu.

Aiuia, M.L., and K.M. Trompeter. 1983. Fouling of RO membranes in wastewater applications. *Desalination* 48:299–319.

Belfort, G. 1975. Cleaning of reverse osmosis membranes in wastewater renovation. *AIChE Symposium Series No. 151.* 71:76–80.

Bukay, M., and W. Berne. 1986. RO troubleshooting, Parts 1–5. *Ultrapure Water* 3(2,3,4):56–59.

Cadotte, J.E., R.S. King, R.J. Majerle, C. Hultman, and R.J. Petersen. 1981. Posttreatment process for reverse osmosis membranes. Final report. OWRT, U.S. Department of Interior.

Canter, P.A., and B.J. Mechalas. 1969. Biological degradataion of cellulose acetate reverse osmosis membranes. *Journal of Polymer Science* 28(C):225–241.

Dalton, G.L., H.S. Pienaar, and R.D. Sandersen. 1978. Supplemental polymer coatings in reverse osmosis membrane improvement and regeneration. *Desalination* 24:235–248.

DuPont 1982. Bulletins 507, 508, 509, 510. *Permasep® engineering manual.* Wilmington, Del. Polymer Products Department, Permasep Products.

FilmTec. January 1988a. Technical bulletin FT30. *Reverse osmosis membrane biological protection and disinfection.* Minneapolis.

FilmTec. February 1988b. Technical bulletin. *Cleaning procedures for FilmTec FT30 Elements.* Minneapolis.

FilmTec. February 1988c. Technical bulletin. *Disinfecting RO systems with hydrogen peroxide.* Minneapolis.

Flemming, H.C., and G. Schaule. 1988. Biofouling on membranes: A microbiological approach. *Desalination* 70:95–119.

Frankenberg, T.T., A.G. Lloyd, and E.F. Morris. 1958. Operating experience with the first commercial supercritical pressure steam-electric generating unit at the Philadelphia plant. In Proceedings of American Power Conference.

Graham, S.I., R.L. Reitz, and C.E. Hickman. 1989. Improving reverse osmosis performance through periodic cleaning. *Desalination* 74:113–124.

Guy, D.B., and R. Singh. 1982. On-site regeneration of reverse osmosis membranes. *Water Supply Improvement Association Journal* 9(1):35–44.

Himelstein, W., and Z. Amjad. 1985. The role of water analysis, scale control, and cleaning agents in reverse osmosis. *Ultrapure Water* 2:32–35.

Hydranautics. (Undated.) Hydranautics Water Systems technical service bulletin. *RO membrane foulants and their removal from CAB1, CAB2 and CAB3 elements.*

Katz, D.B.A., H. Laney, J.A. Linquist, and E.C. Perisike. 1976. Formaldehyde disinfection to eliminate bacterial contamination of deionizers. *dialysis and transplantation,* 5(3):42.

Koseoglu, S.S., and D.E. Engleau. 1990. Membrane applications and research in the edible oil industry: An assessment. *Journal American Oil Chemists Society.* 67:239–249.

Kremen, S., G.E. Foreman, and J.M. Chirrick. 1975. U.S. Patent No. 3,877,978.

Kuepper, T.A., and R.S. Chapler. 1980. *Field performance improvement for reverse osmosis units.* Port Hueneme, Calif.: Naval Construction Battalion Center.

Lipinski, R., and K. Chang. 1989. Potential new approach for alkaline on-line scale removal. Paper No. 148 *Corrosion 89.* Paper presented at the National Association of Corrosion Engineers, Houston, Texas.

Logan, D.L. 1987. Deposit control in cellulosic and polyamide RO systems. Paper No. 333, *Corrosion 87.* Paper presented at the National Association of Corrosion Engineers, Houston, Texas.

Luss, G. 1986. Cleaning RO units used for potable water production. *Water World News* March/April, pp. 18–19.

Luss, G. 1988. Consideration in membrane cleaning. Paper presented at the Sixth Annual Membrane Technology/Planning Conference, Cambridge, Massachusetts, Nov. 1–3.

McPherson, H., E. Vilalobos, H. Banados, and E. Krueger. 1982. The control of fouling in reverse osmosis systems: Two case histories. Paper presented at the Tenth Annual Conference Water Supply Improvement Association, July 25–29 at Honolulu.

Moody, C.D., J.W. Kaakinen, J.C. Lozier, and P.E. Laverty. 1983. Yuma desalinating test facility: Foulant component study. Paper presented at First World Congress on Desalination and Water Reuse, May 23–27, at Florence, Italy.

Nagarajan, M.K. 1985. Multi-functional polyacrylate polymers in detergents. *Journal American Oil Chemists Society* 62(5):949–955.

Parekh, B.S., editor. 1988. *Reverse osmosis technology.* New York: Marcel Dekker, Inc.

Parise, P.L., B.S. Parekh, and R.T. Smith. 1988. Reverse osmosis for producing pharmaceutical grade waters in reverse osmosis technology: Application high purity water production, edited by B.S. Parekh. New York: Marcel Dekker, Inc.

Paul, D.H., and A.M. Rahman. 1990. Reverse osmosis membrane fouling: The final frontier. *Ultrapure Water* 7(3):25–36.

Paulson, D.J. 1987. An overview of and definition for membrane fouling. Paper presented at the Fifth Annual Membrane Technology/Planning Conference, Cambridge, Massachusetts, October 21–23.

Pepper, D. 1990. RO for improved products in the food and chemical industries and water treatment. *Desalination* 77:55–77.

Pittner, G.A. 1986. Reverse osmosis beats ion exchange for cogen plant feedwater treatment. Paper presented at 6th International Water Conference on Cogeneration, October 15–17, at Orlando, Florida.

Reitz, R.L. 1984. Development of a broad spectrum antiscalant for reverse osmosis systems. Paper presented at the 12th Annual Conference Sponsored by Water Supply Improvement Association, May 13–18, at Orlando, Florida.

Ridgway, H.F., A. Kelley, C. Justice, and B.H. Olson. 1983. Microbial fouling of reverse osmosis membranes used in advanced wastewater treatment technology: Chemical bacteriological and ultrastructural analyses. *Applied and Environmental Microbiology* 45(3):1066–1084.

Ridgway, H.F., M.G. Rigby, and D.G. Argo. 1985. Bacterial adhesion and fouling of reverse osmosis membranes in wastewater applications. *Desalination* 48:299.

Sachs, S.B., and E. Zisner. 1977. Wastewater renovation by sewage ultrafiltration. *Desalination* 20:203–215.

Sackinger, C.T. 1985. Seawater reverse osmosis at high pressures. *Desalination* 54:119–130.

Schaffer, J.F., and R.T. Woodhams. 1977. Polyelectrolyte builders as detergent phosphate replacements. *Industrial Engineering Chemistry, Product Research Development* 16:3–11.

Simpson, M.J., and M.J. Groves. 1983. Treatment of pump/paper bleach effluents by reverse osmosis. *Desalination* 47:327.

Trace Metal Data Institute. 1980. Bulletin 609. An analysis of circumventing reverse osmosis fouling. El Paso, Texas (April).

Treffry-Goatley, K., C.A. Buckley, and G.R. Groves. 1983. Reverse osmosis treatment and reuse of textile dyehouse effluents. *Desalination* 47:313–320.

UOP Fluid Systems. 1989a. *Cleaning instructions for TFC elements.* San Diego, Calif.

UOP Fluid Systems. 1989b. *Cleaning instructions for TFCL elements.* San Diego, Calif.

UOP Fluid Systems. 1989c. *Cleaning instructions for ROGA elements.* San Diego, Calif.

UOP Fluid Systems. Undated. *ROGA® sizing rejuvenation treatment.* San Diego, Calif.

U.S. Department of Interior. 1980. *A Preliminary review of regenerating agents for sea water membranes.* Office of Water Research and Technology. Contract No. 14-343-0001-9524.

8

Analytical Techniques for Identifying Reverse Osmosis Foulants

James D. Isner
Robert C. Williams

The BFGoodrich Company
Avon Lake Technical Center

INTRODUCTION

Reverse osmosis (RO) technology is now recognized as a valuable and indispensable component of chemical process industries. The applications of RO include desalination of seawater and brackish water, ultrapure water for semiconductor manufacturing, food processing, metal finishing, dewatering of electrode deposition paint baths, and distilling and brewery waste treatment.

The efficient use of RO systems is limited most notably by membrane fouling. The commonly encountered foulants include precipitated salts, colloids, metal oxides, oil droplets, and microorganisms. In many cases the accumulation of these deposits results in reduced system efficiency, poor water quality, equipment or membrane failure, and increased production cost. Identification of the causes of deposit formation, or membrane fouling, ranks among the most important aspects of the application and efficient utilization of RO systems.

Often a fouling problem is not apparent, and the performance of the system may appear to be adequate. Several common foulants, however, can be present in minute quantities. If these foulants are allowed to accumulate on the membrane, a gradual but serious deterioration of performance will occur.

Historically, this problem has been addressed by the application of a variety of general-purpose cleaners and antiscalants, not specifically designed for RO systems. These products are inadequate in cleaning membranes and ineffective in preventing membrane scaling. Moreover, some of the cleaners not specifically designed for RO membranes can actually damage the membranes. Eventually, the membranes may become irreversibly fouled and require replacement. Because membrane replacement is among the highest cost components in RO operation, it is important to protect the system before irreparable damage occurs.

Over the course of years, considerable experience has been gained through the examination of failed membranes; in that process autopsies have been performed on nearly every membrane type from all the major manufacturers, including spiral wound, tubular, and hollow fiber configurations. Membrane surfaces including cellulose acetate (CA) and triacetate, thin film composite polyamide, and others have also been studied (Amjad et al. 1988). Through this experience some understanding of deposition mechanisms has been gained. This information has been used to develop products and methods to remove scale and fouling layers, and to design antiscalants to prevent the precipitation of mineral scales. This chapter focuses on the use of several analytical techniques that can be used to characterize the structure, composition, and nature of foulants.

ANALYTICAL TECHNIQUES

There are many methods that may be used to solve fouling problems. Table 8.1 is a list of eight such methods, including the type of information and advantages and disadvantages associated with the techniques.

TABLE 8.1   Analytical Methods for RO Membrane Analysis

Technique	Information	Advantages	Disadvantages
1. Optical microscopy	M	Cost, time	Limited information
2. Scanning electron microscopy	M	Time, sample size	Cost, sample must be dry
3. X-ray fluorescence	E	LDL	Sample size, time
4. Atomic absorption	E	LDL, sample size	Sample prep, time
5. Infrared spectroscopy	C	LDL, time, cost	Interpretation
a. Transmission	C	LDL, time, sample size	Sample should be homogeneous
b. Reflection	C	Time, sample prep	Surface should be flat, smooth and homogeneous
6. Energy-dispersive X-ray	E	Time, sample size	LDL, sample must be dry
7. ESCA	E & C	LDL, surface sensitivity	Cost, interpretation
8. Auger	E & C	LDL, surface sensitivity	Cost, interpretation

E = elemental
M = morphology
C = composition
LDL = lower detectable limit

The types of analytical techniques successfully used in solving RO fouling problems, which are discussed in this chapter, include the following:

- Optical microscopy
- Scanning electron microscopy
- Energy dispersive X-ray analysis
- Infrared spectroscopy

There is also a brief discussion of other techniques included in Table 8.1. These have special capabilities, but have not yet been used extensively for RO problem solving.

Optical Microscopy

Optical microscopy is a useful tool in identifying many common foulants by their color, size, crystalline structure, refractive index, or other identifying characteristics. Frequently, these determinations can be made using an inexpensive microscope equipped with little more than transmitted and reflected illumination capabilities and a maximum of $300\times$ magnification. An example is shown in Figure 8.1, which is iron oxide at $250\times$ using transmitted light illumination. The color, size, and agglomerate nature of the iron oxide make its identification simple and quick.

The use of polarized light illumination gives the scientist another optical property that can be used for particle identification. A material such as calcium sulfate has a unique morphology, which makes it very easy to differentiate from a substance such as iron oxide (see Figure 8.2). With the use of polarized light,

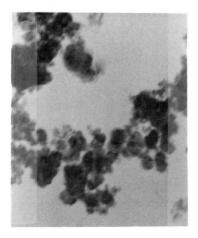

FIGURE 8.1.   Optical micrograph of iron oxide ($250\times$).

FIGURE 8.2.   Optical micrograph of iron oxide and calcium sulfate dihydrate crystals (250×).

calcium sulfate exhibits an optical property known as birefringence, which can occur when a material has more than one refractive index.

Many times, however, the optical microscope is limited in its ability to adequately characterize foulants on a filter. Unless the foulants can be removed from the filter and mounted on a glass slide so that light can be transmitted through the sample, many of the techniques such as polarized illumination cannot be used. Second, limited depth of focus and magnification limitations often offset the cost and analysis time advantages offered by optical microscopy. In Figures 8.3 and 8.4, a mixture of Dixie clay and fumed silica is shown with plane and polarized light illumination. Although some particles show birefringence, it is not clear which particles are clay and which are silica.

Scanning Electron Microscopy

The increased depth of focus and higher magnification capabilities of the scanning electron microscope (SEM) can overcome some of the shortcomings of optical microscopy. SEM micrographs of silica, clay, iron oxide, and calcium sulfate are shown in Figures 8.5, 8.6, 8.7, and 8.8, respectively. Individually,

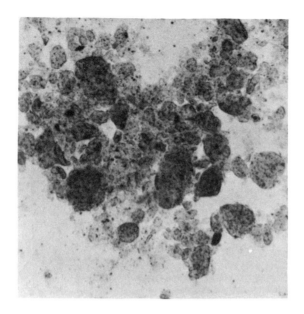

FIGURE 8.3.   Optical micrograph of clay and silica (250×).

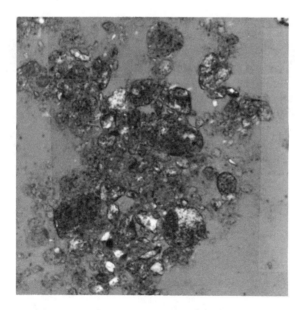

FIGURE 8.4.   Optical micrograph of clay and silica using crossed polarized light (250×).

FIGURE 8.5.   Electron micrograph of silica (5,000×).

FIGURE 8.6.   Electron micrograph of clay (5,000×).

FIGURE 8.7.   Electron micrograph of iron oxide particles (1,000×).

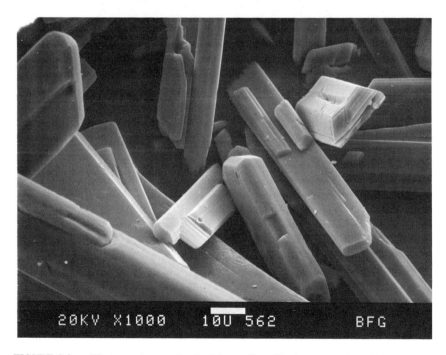

FIGURE 8.8.   Electron micrograph of calcium sulfate dihydrate crystals (1,000×).

the samples show unique packing characteristics; however, these characteristics cannot be distinguished when the samples are mixed into a slurry and filtered onto a 0.1 μm Millipore filter. Figures 8.9 and 8.10 are SEM micrographs of the mixed sample and of a 0.1 μm Millipore filter pad, respectively. As can be seen in Figure 8.9, it is difficult to identify and quantify by SEM alone the various species often present.

There are ancillary techniques that, when used in conjunction with SEM, can help identify mixtures such as those in Figures 8.9 and 8.10. To understand these ancillary techniques, it is first necessary to discuss what happens when an electron beam strikes the surface of a specimen and how the various detectable interactions can lead to a better understanding of the sample.

The interaction of an electron beam with a specimen creates a series of events that can be monitored by various detectors and probes to identify or characterize the sample in a number of ways. Figure 8.11 is a schematic illustration of some of these interactions, with emphasis on those that are most typically associated with the SEM (Hayat 1974, Goldstein and Yakowitz 1975). The techniques explained and illustrated in this text include secondary electron imaging, back-

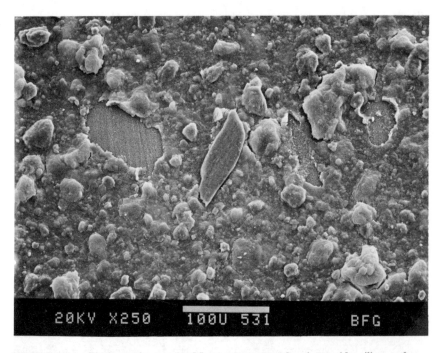

FIGURE 8.9.   **Electron micrograph of four components (clay, iron oxide, silica, and calcium sulfate dihydrate crystals) on Millipore filter (250×).**

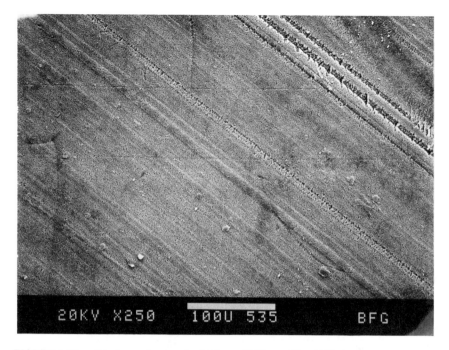

FIGURE 8.10.   Electron micrograph of 0.1 μm Millipore filter (250×).

scattered electron imaging, and a variety of methods associated with the detection of X-rays that are emitted during the bombardment of the specimen by the electron beam.

Secondary Electron Imaging

The secondary electrons, which are responsible for the images most commonly associated with the SEM, such as those seen in Figures 8.9 and 8.10, come from a volume located between 50 and 500 Å below the surface of the sample (Hayat 1974). The exact depth of the volume depends on the accelerating voltage chosen for the primary electron beam and the nature of the samples, especially the atomic numbers of the species present. Samples with low atomic numbers permit deeper penetration of the electron beam than specimens with higher atomic numbers. Moreover, the penetration of the sample is increased by increasing the accelerating voltage (Duncumb and Shields 1963). Therefore, the combination of a higher atomic number sample and low accelerating voltage would yield information from a depth near the surface, whereas a lower atomic number sample combined with higher accelerating voltage would give information from far below the surface as well. For the examination of RO filter membranes, moderate accelerating voltages from 15 to 25 kV are normally used, and the materials imaged are usually low in atomic number. Secondary electrons are ejected from the sample as a result of inelastic collisions between the primary beam and the outer electrons of the specimen (Goldstein and Yakowitz 1975). These secondary electrons have relatively low energy and are collected by a detector positively charged with 40 to 200 V (Hayat 1974). For reasons that are beyond the scope of this chapter, SEM samples must be conductive under most conditions. Therefore, a thin (100 to 200 Å) coating of metal such as gold, palladium, or even carbon, must be placed on the sample by means of a vacuum evaporator or a sputter-coating device.

Backscattered Electron Imaging

From a depth of the sample just below that from which the secondary images are formed (X_B in Figure 8.11), comes another type of emitted electrons known as *backscattered electrons*. These are primary electrons that exit the sample surface after undergoing multiple elastic scattering events with little loss of energy (Goldstein and Yakowitz 1975). The contrast for images produced from backscattered electrons depends on the average atomic number of the specimen, if the sample is relatively flat (Hayat 1974).

Figure 8.12 is a split-screen micrograph of an RO membrane foulant that

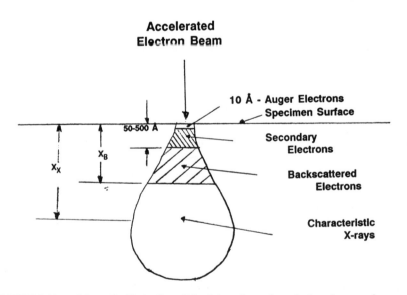

FIGURE 8.11.   Schematic illustration of the interactions of an electron beam and specimen (Hayat 1974, Goldstein and Yakowitz 1975).

FIGURE 8.12.   Micrograph of RO foulant showing secondary image on the left, and the backscattered image on the right (2,000×).

shows the secondary electron image (SE) on the left, and the backscattered electron image (BSE) on the right. The BSE image exhibits a more pronounced difference in the brightness level from particle to particle. This difference was confirmed by elemental analysis to correspond to varying levels of sulfur. To work well, however, BSE requires a relatively flat surface; unfortunately, many of the highly fouled membranes analyzed are not sufficiently flat for BSE imaging, owing to the buildup of foulant material on the membrane surface.

Energy Dispersive X-ray Analysis

It is usually not enough to know the size, shape, and relative atomic number of the species found on a membrane. To make proper recommendations concerning the treatment of the system to eliminate fouling and/or to restore the membranes to their normal efficiency, more accurate information about the composition must be obtained. The most common and most useful analytical technique in this endeavor is energy dispersive X-ray (EDX) analysis, which allows the determination of the elemental composition of the area being imaged by the SEM.

The characteristic X-rays shown in Figure 8.11 are produced by interactions of the primary incident bombarding electrons with the inner shell electrons of the atoms of the specimen. Sometimes, the incident electron will remove an inner shell electron and an outer shell electron will fill the vacancy. In the process of this returning of the atom to its ground state, a photon of X-ray radiation is emitted. The energies of these emitted X-ray photons are specific for, and characteristic of, the atoms of a given atomic number (Goldstein 1975). A spectrum of the intensities of the detected photons versus their energies will then be characteristic of the elements present in the sample. The X-rays are measured by a detector located inside the SEM chamber, and the data are analyzed and plotted using a computer which, until recently, has always been separate from the SEM.

The depth from which these X-rays are emitted is shown as X_x in Figure 8.11 and is dependent on the atomic number of the sample and the accelerating voltage used for the primary electron beam. This is the same dependence described earlier for the formation of the secondary electron images, as shown in Figure 8.11. Although the dependence is the same, the specific analysis depths differ for the three techniques. Using a normal accelerating voltage of 20 kV, the X_x is about 2.5 μm for a light element such as aluminum. Using the same conditions, the depth for X-ray production for copper, which is heavier and has a density almost three times that of aluminum, is only 0.6 μm (Goldstein and Yakowitz 1975).

Most EDX detectors are able to detect elements with atomic numbers greater than 10 (i.e., for sodium and higher), but many newer instruments can now detect elements as light as beryllium (which has an atomic number of 4). More information about these advanced systems and the types of analyses they can perform are presented later in this chapter.

The EDX spectrum of the area shown in Figure 8.9 is seen in Figure 8.13. The heights of the aluminum (AL) and silicon (SI) peaks indicate a high level of both clay and silica. Smaller amounts of iron (FE), calcium (CA), and sulfur (S) are also present. The iron is most likely present as the oxide, and the combination of calcium and sulfur together in the same area probably indicates calcium sulfate. However, EDX can identify only elements present; the definitive identification of calcium sulfate requires IR spectroscopy or some other molecular spectroscopic technique. The tall peak labeled "PD" in Figure 8.13 is palladium, which in this case is the conductive coating sputtered onto the sample to reduce damage caused by the high-energy electron beam.

Low Atomic Number EDX Analysis

As previously mentioned, most EDX systems do not allow the detection of elements with atomic numbers less than 10. In the past several years, however,

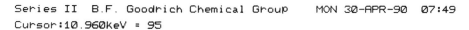

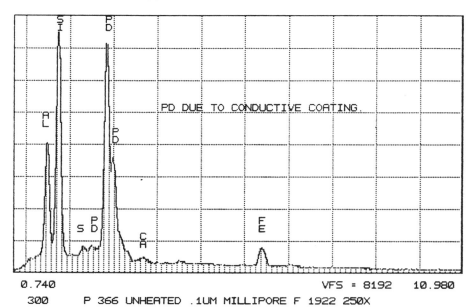

FIGURE 8.13.   EDX spectrum of the area shown in Figure 8.9.

various manufacturers of EDX systems have offered specialized detectors with optional thin-window or no-window modes of operation that allow the detection of beryllium and heavier elements. These detectors are commonly called "low-Z" units, where Z refers to the atomic number. During the early existence of these detectors, there were problems with their durability and the associated maintenance costs. Improvements in both the detector design and in the quality of the vacuum in the SEM sample chamber have made such instruments much more common. The analyses for the very light elements are only qualitative, but their use can often help to avoid other analytical techniques that may require much larger samples and be much costlier to perform. Additional benefits provided by these detectors are improved resolution and detection efficiencies for those elements that are near the detection limit of normal detectors. For instance, low levels of sodium and magnesium, which might well be lost in the background of a normal EDX spectrum, may be detected quite easily with the use of a low accelerating voltage and a low-Z detector.

An example of this type of analysis is shown in Figure 8.14. The EDX spectrum using a very thin window shows the presence of carbon and oxygen, not normally detected, and the detection of sodium and magnesium is much more definitive than that achieved with use of a normal detector. This improvement in

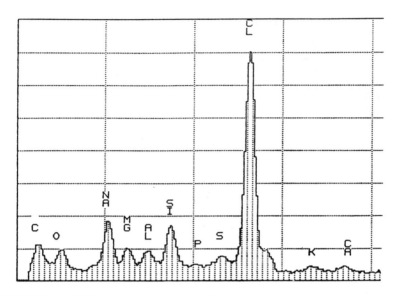

FIGURE 8.14.   EDX spectrum from a low Z-detector.

the detection of elements such as sodium and magnesium is probably more significant to the analysis of RO foulants than is the ability to detect elements with atomic numbers lower than 10.

EDX Mapping

Some very enlightening information can be gained from systems that enable the SEM and EDX components to interact with each other through a digital beam control process that can produce high quality, elemental dot maps. These maps not only show qualitatively which elements are present, but also show which element or group of elements is associated with a particular part of the SEM image. Figure 8.15 indicates the location of iron, sulfur, silicon, and their combinations in the area shown in Figure 8.9.

More advanced EDX systems have the ability to quantify the percentage of area covered by an individual element (often limited to three or four elements), as well as by their combinations. An example of this feature is given in Figure 8.16, which shows the areas covered by sulfur (S), aluminum (AL), silicon (SI), and iron (FE) as well as their combinations. Information provided in Figure 8.16 also indicates that 48 percent of this area is not covered by any of these elements. Figure 8.17 illustrates the ability to show the area covered by any individual element, such as silicon in this example.

Figure 8.18 shows the dot maps for S and CA from an area of an RO membrane shown by SEM, as in Figure 8.19. From the EDX maps it is clear that

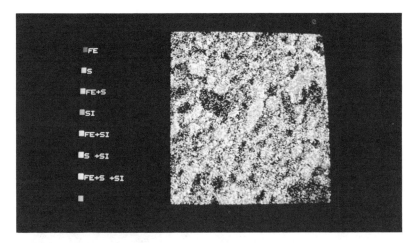

FIGURE 8.15.   EDX map of Figure 8.9 area.

nearly all of the detected sulfur is from the membrane itself. An EDX spectrum of a control membrane is shown in Figure 8.20 and confirms that this particular type of membrane has a sulfur-containing layer somewhere in the top micron of its construction. With the use of the same technique illustrated in the previous paragraph, it was revealed that whereas three-fourths of the foulant contains

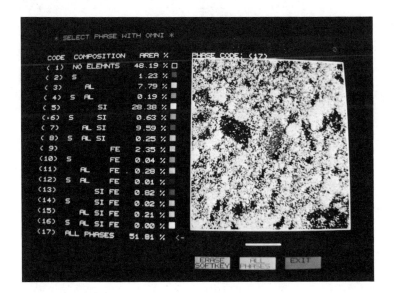

FIGURE 8.16.   EDX map showing area % coverage by S, Al, Si and Fe.

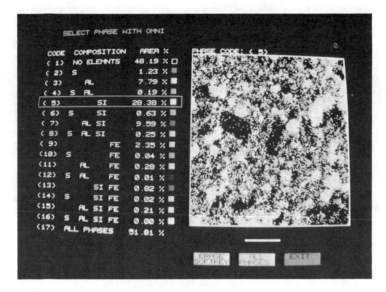

FIGURE 8.17.   EDX map showing area % coverage by Si.

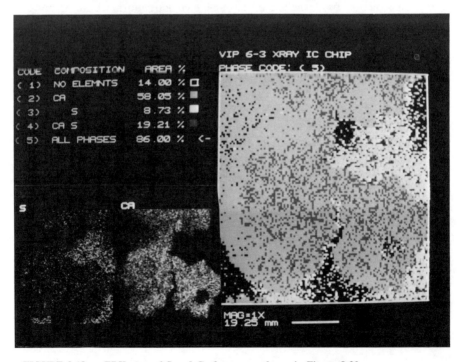

FIGURE 8.18.   EDX map of S and Ca from area shown in Figure 8.20.

FIGURE 8.19.   SEM of fouled RO filter (1,000×).

only calcium (probably present as calcium carbonate), the other one-fourth is a combination of calcium and sulfur, which may explain the unusual morphology.

The area of another membrane, shown in Figure 8.21, was mapped to reveal the locations of calcium and phosphorous, as seen in Figure 8.22. This analysis found that, most often, phosphorous and calcium were present together, although 1.72 percent of the phosphorous present was not associated with calcium. This particular membrane appears to be badly fouled, because 59 percent of the membrane area is covered with calcium and only 11.5 percent of the area examined did not contain some elements indicative of fouling.

Biological Foulant Examples

Two interesting cases of membranes that were coated with biological species are seen in Figures 8.23 to 8.28. In the first example, the foulant seen in Figures 8.23 and 8.24 is a classic example of bacteria. At lower magnifications, such as those possible with an optical microscope, this identification would not be possible. EDX analysis of the membrane identified the other more amorphous-

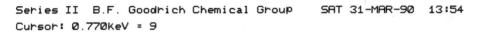

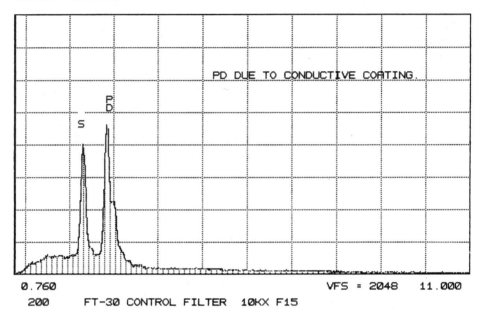

FIGURE 8.20.   EDX of unused control membrane used in Figures 8.18 and 8.19.

looking substance on the membrane as clay. The EDX spectrum shown in Figure 8.25 reveals that in addition to the palladium present due to the conductive coating, the sample contains significant sulfur (S), aluminum (AL), and silicon (Si). An EDX analysis of an unused control membrane showed that it contained about the same level of sulfur as the fouled membrane, which indicates that the sulfur observed arises from the membrane substrate (probably polysulfone), rather than from the foulant. The aluminum and silicon are in the normal ratio for clay, which is a common foulant.

In the second case, the standard SEM-EDX analysis found that the membrane had clay at moderate levels, as seen by the similarity of Figure 8.26 (with the exception of the chlorine and calcium peaks) to the spectrum in Figure 8.25. Careful examination of a few particularly large agglomerates on the membrane revealed blood cells from an unknown source, as seen in Figures 8.27 and 8.28. The blood was not found over a very large portion of the membrane, but its discovery brings to mind some very important considerations in performing microscopic examinations of this type. The analyst must make every attempt to obtain a representative piece of membrane. Because the area imaged and an-alyzed by SEM-EDX is quite small, it is important to examine as much of the

FIGURE 8.21.   SEM of fouled RO membrane (1,050×).

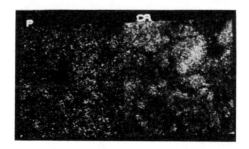

```
NO ELEMENTS: 11.55 %
CA           59.19 %
     P        1.72 %
CA P         27.53 %
ALL PHASES: 100.00 %
```

FIGURE 8.22.   EDX map of Ca and P of area from Figure 8.21.

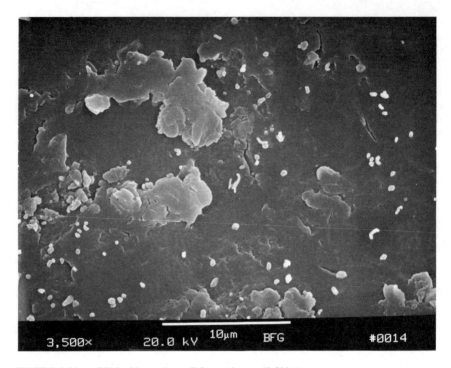

FIGURE 8.23.   SEM of bacteria on RO membrane (3,500×).

sample as possible and to obtain micrographs and spectra from the several different areas. It is very difficult not to dwell on the "unusual" because, as in the case of the blood cells, the unusual is frequently more interesting than the representative areas.

Microscopy Shortcomings

Although SEM-EDX systems have proven to be extremely valuable in analyzing fouled RO systems, there are some obvious shortcomings. Because the samples are placed in a high vacuum, they must be dried before they are coated with palladium (or other conductive elements such as carbon, gold, or gold-palladium alloys). As previously stated, the very small sample size makes it difficult to form conclusions concerning a large membrane. EDX equipment is expensive to purchase and maintain and requires a higher level of training to operate than some other test methods. Finally, most EDX detectors are not sensitive to the lighter elements and cannot detect less than about 0.2 wt. percent of many elements.

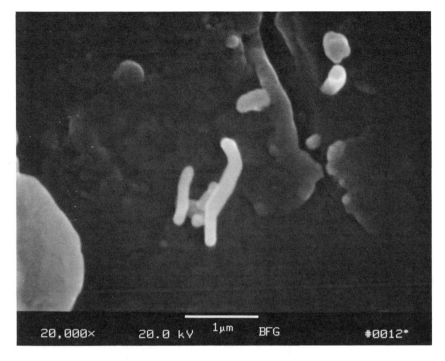

FIGURE 8.24.   SEM of bacteria on RO membrane (20,000×).

Both optical and electron microscopy and their related techniques can do an adequate job of identifying many inorganic foulants, but there are many times when it is necessary to differentiate between elemental sulfur and sulfite or sulfate, for example. Elemental sulfur is usually evidence of bacterial contamination, whereas sulfite or sulfate more likely indicates fouling by an inorganic material. Other kinds of biological fouling or membrane hydrolysis may also have occurred. In these and other instances, different analytical techniques are required to supplement microscopy. One such technique often employed is IR spectroscopy.

Infrared Spectroscopy

The information available from IR spectroscopy is complementary to that available from optical microscopy and SEM-EDX. EDX detects elements, but provides no information about their chemical bonding (although joint element mapping does provide some inferential information). Conversely, in most cases IR detects the bonds between atoms in a molecule and can thus provide information about functional groups and the chemical structure of the analyte. Just as

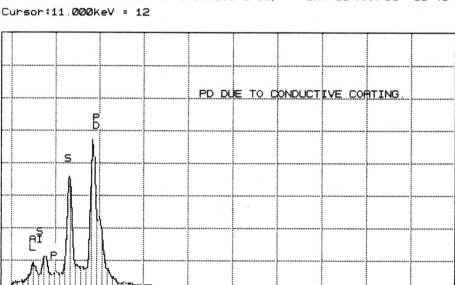

FIGURE 8.25.   EDX spectrum of membrane from Figures 8.23 and 8.24.

EDX has difficulty detecting low-Z elements, however, IR also has some blind spots. Specifically, IR is generally unable to detect nonpolar bonds between like atoms in diatomic molecules such as O_2, N_2, Cl_2, etc. By the same token, IR is very insensitive to molecular sulfur (S_8), which, as previously stated, is sometimes present on fouled membranes as a result of bacterial action. Moreover, IR is generally blind to compounds composed of purely ionic bonds, such as many of the common salts. A detailed discussion of the physics of IR spectroscopy is beyond the scope of this chapter, but is available in other works (see Colthup et al. 1990; Dyer 1965; Silverstein et al. 1991; Williams and Fleming 1980).

In spite of its limitations, IR provides a powerful tool, whose "fingerprint" can unambiguously identify a host of contaminants on RO membranes. The relatively low cost of a modern, bench-top IR spectrometer, when combined with the relative ease of sample preparation and the speed with which results can be obtained, often makes IR the first technique used after the initial optical microscopic screening.

IR spectroscopy can detect and identify organic components to which EDX is totally blind, as well as some inorganic materials that may have components that are not detected by EDX. Consider, for example, calcium carbonate, a com-

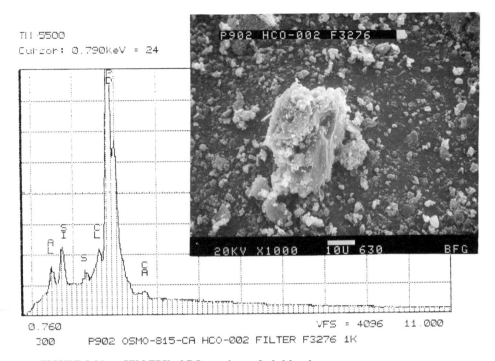

FIGURE 8.26.   SEM-EDX of RO membrane fouled by clay.

monly observed foulant of RO membranes used on well water or spring water. Although EDX can detect the calcium present in calcium carbonate, it is usually blind to the carbonate portion of this material, and to any organic component(s) that may also be present. Conversely, even though the carbonate component can be identified unequivocally by IR spectroscopy, an EDX or other metals analysis is usually needed to identify conclusively the cation present, because the bond between the cation and the carbonate is ionic and therefore not detectable by IR (although the ionic bond does influence the positions of some of the carbonate bands). In addition, IR spectroscopy can often give some indication of chemical changes in the membrane material or substrate, which would not be detected by EDX and which might be only hinted at by microscopic observation of the membrane.

Two IR spectroscopic techniques that have been used in the laboratory are attenuated total reflectance (ATR) analysis (sometimes called frustrated multiple internal reflection [FMIR]) and transmission spectroscopy. Although transmission spectroscopy is, by far, the more commonly used technique for general routine analysis of a wide variety of samples, the ATR technique offers several advantages for RO samples. In many cases, these advantages render ATR more appropriate as the first-choice screening technique for RO membranes.

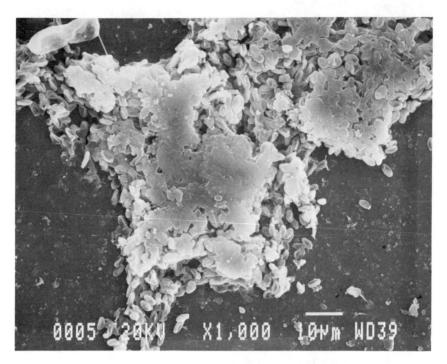

FIGURE 8.27.   SEM of blood cells from membrane in Figure 8.26 (1,000×).

ATR-IR Spectroscopy

Although the technique of ATR-IR spectroscopy is normally explained through the use of quantum mechanics (Harrick 1967), an alternate conceptual approach is presented here. The pertinent information is shown schematically in Figure 8.29.

When a light ray in an optically dense medium strikes an interface between that medium and a less dense medium, whether that ray is transmitted into the rarer medium or reflected into the denser medium depends on (1) the indices of refraction of the two media and (2) the angle of incidence of the incoming ray. The two indices of refraction uniquely define a critical angle, according to Snell's law. Rays striking the interface at less than this angle (i.e., more nearly perpendicular to the interface) will always be at least partially transmitted, although there may still be some reflective loss. Rays striking the interface at greater than the critical angle (i.e., more nearly parallel to the interface) will always be *totally internally reflected*.

Suppose that some means of monitoring the internally reflected ray exists, and examine what happens to a ray that is incident at a value close to, but slightly

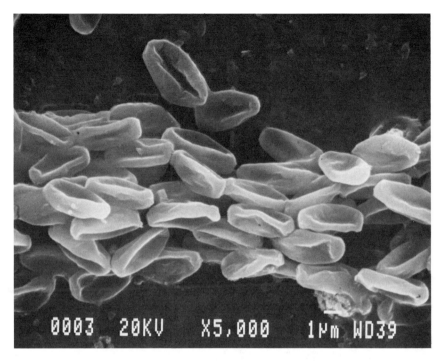

FIGURE 8.28.   SEM of blood cells (5,000×).

greater than the critical angle; it has already been shown that this ray is normally totally reflected. Consider what happens when the rarer medium has an absorption band at or near the energy of the ray. The index of refraction changes drastically around a strong absorption band. The changed index of refraction, in turn, means that there is a different, and usually greater, critical angle for the interface. The net effect is that the ray is now incident at *less* than the critical angle, and there will be at least some transmission from the denser to the rarer medium. The formerly totally internally reflected ray is now *attenuated* by that part that was transmitted. If the internally reflected ray is scanned spectroscopically, the result is a spectrum that is qualitatively similar to the conventional transmission spectrum of the rarer medium. The effective sampling depth of this technique varies as a function of the incident wavelength, but is typically on the order of several microns. Some control of the effective penetration depth can be achieved by appropriate adjustment of either the incidence angle and/or the index of refraction of the denser medium. There have even been papers on "spectroscopic microtoming," in which successively deeper penetration depths can sometimes reveal layers below the surface (Hirschfeld 1977).

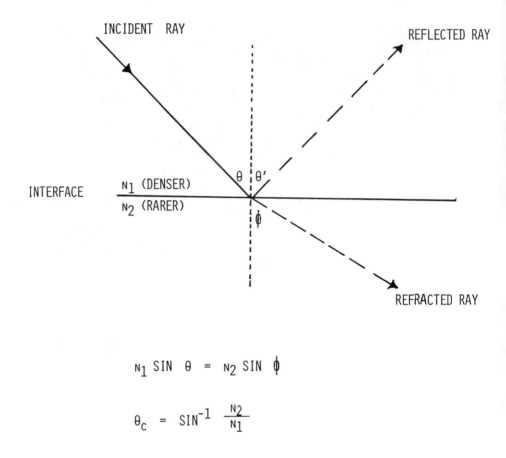

$$N_1 \; SIN \; \theta \; = \; N_2 \; SIN \; \phi$$

$$\theta_c \; = \; SIN^{-1} \; \frac{N_2}{N_1}$$

ATTENUATED TOTAL REFLECTION

FIGURE 8.29.   Schematic illustrating the principles underlying the ATR technique.

One of the advantages of the ATR technique is the minimal sample preparation involved. In general, it is merely necessary to press the sample into intimate contact with the ATR crystal (the higher index medium in Figure 8.29) by means of a C-clamp, or some similar device. Samples can even be run wet, as received, although better results are usually obtained if the sample is dried first. Moreover, unlike SEM-EDX, this technique is essentially nondestructive; hence, the analyzed sample can subsequently be used for other tests. (Although the technique itself is nondestructive, it is usually necessary to cut two small pieces of the sample, normally about 45 × 8 mm, to fit in the ATR unit. There are, however, some ATR units sold for skin analysis and similar applications that can accommodate the entire sample without any cutting; there are also micro-units that require significantly less sample area.) The ATR technique works best on flat, moderately smooth surfaces. ATR also works well for powders, and there are accessories that can be used for the analysis of liquids as well.

Because the ATR technique analyzes only the surface of the material in contact with the crystal, it is ideally suited for the analysis of coatings, as well as surface foulants, on the surface of RO membranes. However, note that in general, ATR is somewhat less sensitive to low-level surface components than is SEM-EDX.

The primary limitation of the ATR technique is that good surface contact between the sample and the ATR crystal is required to obtain a spectrum. Thus, badly etched or abraded surfaces or very coarse powders may be too rough to yield useful spectra with this analysis technique. (The flatness required, however, is significantly less than that necessary for backscattered SEM imaging.) Similar problems have been observed with RO systems that use fiber bundles rather than flat membranes.

Other limitations are inherent in the basic physics of this technique. Very thin coatings (less than 0.1 to 0.3 μm thick) may not be detected. Conversely, the analyst must always remember that this is a surface technique, which may not reflect the composition of materials below a surface coating. Finally, in a complex mixture, it may be difficult to sort out all the materials present on the membrane surface.

A typical example of a fairly simple case of membrane contamination is illustrated in Figure 8.30. It shows an RO membrane that was used on well water; membrane fouling was observed within hours of start-up. The combination of the strong, broad band near 1420 cm^{-1} and the weaker, sharper bands near 880 and 710 cm^{-1} in the ATR-IR spectrum clearly shows that the bulk of the foulant is a carbonate salt, most likely calcium carbonate (i.e., solids from hard water). Optical microscopic analysis of the same membrane also found evidence of bacterial contamination, which could not be detected by IR. Based on these analyses, additional pretreatment and filtering of the well water was recommended. When those steps were taken, subsequent premature membrane failure was prevented.

A second example is shown in Figure 8.31. In this case, the primary foulant appears to be a crystalline clay or other silicate-containing material (as evidenced by the strong band near 1030 cm^{-1}). There is also evidence of a low-level carbonate component. Complementary data from EDX identified the silicate material as a clay and found sufficient calcium to suggest that the carbonate detected was calcium carbonate. However, the bands in the 3000 to 2800 cm^{-1} region and near 1460 cm^{-1} in this Figure indicate that there is also a significant amount of an oil-like material present, which would not be detected by EDX. In

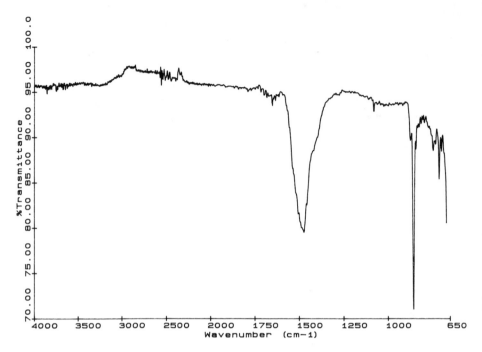

FIGURE 8.30.   ATR-IR spectrum of an RO membrane contaminated with calcium car-bonate.

the third example (Figure 8.32), almost all of the foulant is a fatty acid. If only EDX were used on this sample, the cause of this problem would not be correctly identified and remedial action based on that analysis would be wasted.

The ability to detect chemical changes in the RO membrane is illustrated in Figure 8.33a and b. Figure 8.33a is the control spectrum of a cellulose triacetate membrane. Figure 8.33b is the ATR-IR spectrum of the membrane whose SEM micrographs and EDX spectra were shown in Figures 8.24 through 8.28. This membrane failed after approximately 1 year of an expected 5-year lifetime. In Figure 8.33 there is clearly a difference in the ratio of the relative intensities of the 1225 and 1040 cm^{-1} bands. This difference indicates that an appreciable amount of the cellulose triacetate membrane has hydrolyzed to the diacetate or monoacetate. This membrane cannot be regenerated; moreover, the harsh operating environment that caused the hydrolysis must be modified prior to installation of replacement membranes, or they too will fail prematurely. Based on this analysis, the blood contamination found by SEM appears to be an artifact unrelated to the membrane failure. This is an example of the importance of

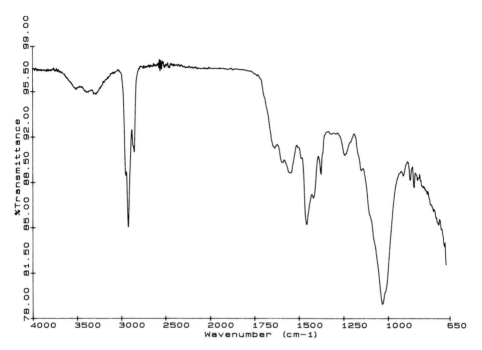

FIGURE 8.31.   ATR-IR spectrum of an RO membrane contaminated with clay, a low level of calcium carbonate, and a hydrocarbon oil-type component.

analyzing the entire sample, preferably by several different analytical techniques.

One of the limitations of the ATR technique can be seen in Figure 8.34a, b, and c. In this sample, the coating of foulant on the membrane was so thin that most of the bands observed in the spectrum are from the membrane itself, rather than from the foulant. In this instance, however, it was still possible to obtain a spectrum of the foulant by using the spectrometer computer data system to subtract Figure 8.34b from Figure 8.34a. The result, shown in Figure 8.34c, indicates that the bulk of the foulant is an inorganic silicate material such as silica or amorphous clay.

The ATR-IR accessory used most often in the laboratory normally samples a total area of almost 700 mm². This large-area sampling can be both an advantage and a disadvantage. The pitfalls of heterogeneous samples have previously been discussed, when it was pointed out that the SEM-EDX results obtained from a small area of such a sample may not be representative of the bulk sample. Whether or not the sample is heterogeneous, the ATR spectrum obtained will be

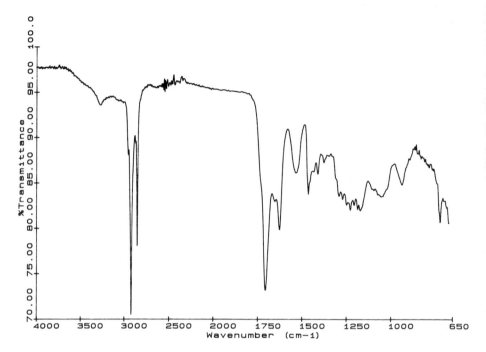

FIGURE 8.32.   ATR-IR spectrum of an RO membrane contaminated with an organic fatty acid.

the *average* of all the material in contact with the analysis crystal. If the membrane surface is heavily fouled, then the area of foulant material in contact with the crystal surface is relatively large and the results obtained are probably representative of the bulk of the sample surface. Conversely, important foulants that cover only small portions of the total membrane surface (or small amounts of a second foulant in the presence of a large amount of a primary foulant) may not be detected, because they are hidden by the other material present on the bulk of the membrane.

Transmission IR Spectroscopy

Transmission IR spectroscopy, often accompanied by an IR microscope (Messerschmidt and Harthcock 1988) or another microsampling device, offers an alternative to ATR-IR. Transmission IR has an advantage in that a spectrum can be obtained of virtually any sample. For example, if contamination of the membrane with an organic solvent is suspected, it is usually possible to identify the solvent from the spectrum of the vapors emitted by residual solvent in the membrane. In addition, transmission IR microanalysis is suitable for hollow

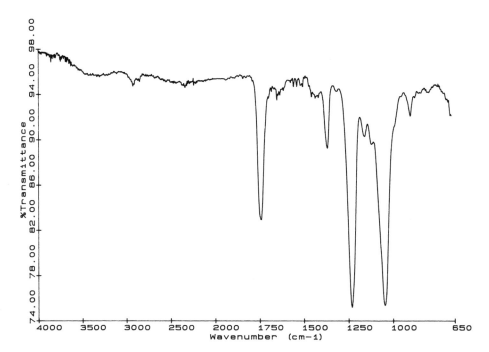

FIGURE 8.33.   (a) ATR-IR spectrum of a control cellulose triacetate RO membrane; (b) ATR-IR spectrum of failed cellulose triacetate-base RO membrane, showing severe hydrolysis of the cellulose triacetate.

fiber RO systems, which cannot be routinely handled by ATR. With an IR microscope it is possible to identify *individual particles,* down to approximately 10 μm in diameter, in contrast to the large-area limitation of ATR.

The transmission IR spectrum of a powder that clogged a hollow fiber RO unit is shown in Figure 8.35. The combination of the broad band near 1150 cm^{-1} and the sharper bands near 3560, 3420, 1680, 1625, 670, and 600 cm^{-1} identifies calcium sulfate (gypsum). This spectrum was taken using a miniature, high-pressure, diamond-anvil, optical cell, where a small amount of the foulant powder was placed directly between the diamond window faces of the optical cell. This technique facilitates identification by avoiding more conventional preparation in a pressed alkali-halide salt pellet, where interaction of the sample with the salt matrix and/or absorption of water by the highly deliquescent alkali halide can be a problem. The high noise level in the region of 2600 to 1900 cm^{-1} is an artifact of the diamond window; however, there are very few bands of analytical interest in this region.

As in SEM-EDX and ATR-IR, problems can arise when attempting to use transmission IR spectroscopy on heterohomogeneous samples. These problems

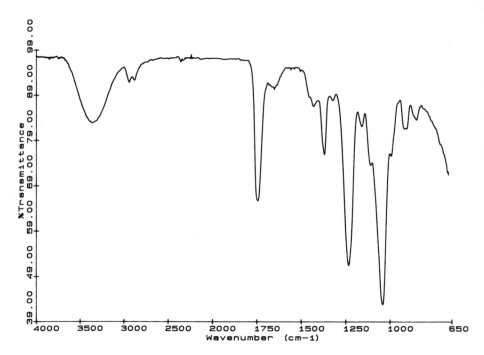

FIGURE 8.33.   *(continued)*

are similar to those that can occur when ATR is used on a heterogeneous surface. The difference in the two is that for ATR the *surface* should be homogeneous, whereas for transmission IR the *entire sample* under analysis should be homogeneous. An illustration of this problem is the diamond cell transmission spectrum of a single RO hollow fiber, shown in Figure 8.36. This fiber was taken from the membrane that was clogged with the gypsum, shown in Figure 8.35. There are bands in the spectrum attributable both to a polyamide fiber substrate and to a reasonably thick polysulfone overcoating. The actual working RO membrane surface is too thin to be easily identified in this spectrum. This spectrum illustrates the point that it is always possible to obtain spectra of heterogeneous samples; however, it can be difficult to obtain meaningful information from such spectra.

Thus, for transmission IR spectra in particular, if a heterogeneous surface coating is suspected, it will usually be necessary to *physically separate* individual components of the coating from the sample prior to performing the analysis. For this reason, the effort involved in sample preparation for transmission IR analysis of RO membrane samples is usually significantly more tedious and time-consuming than for ATR analysis.

There are now some IR microscopes with reflectance capability that can obtain the reflectance spectrum of a material on the membrane surface without re-

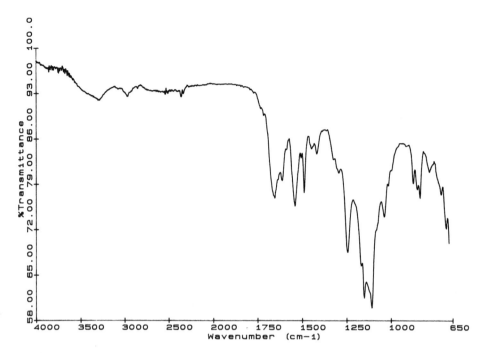

FIGURE 8.34.   (a) ATR-IR spectrum of a lightly fouled RO membrane; (b) ATR-IR spectrum of a control membrane; (c) difference spectrum obtained by digitally subtracting (b) from (a), which enhances the foulant bands.

quiring the physical isolation of that material (Messerschmidt and Harthcock 1988). In this instance, it should be possible to perform parallel IR microscopic analysis on *exactly* the same sample imaged by SEM-EDX, with substantially reduced sample preparation. Other advantages of such a capability are obvious. However, even when reflectance microscopy is available, if a contaminant is embedded within the RO membrane, it will usually be necessary to physically separate the contaminant from the bulk of the membrane matrix before analysis.

Thus, IR spectroscopy can be an extremely valuable adjunct to optical microscopic and SEM-EDX analyses. The equipment necessary to perform IR analysis is significantly easier to operate and maintain than SEM-EDX; moreover, its cost is a small fraction of that of an SEM-EDX system and only slightly greater than that of a reasonable quality optical microscope. Samples can be in virtually any form, although reasonably smooth samples are required for ATR. IR provides information on chemical species present in both foulants and the membranes themselves, and can thus identify organic contaminants and chemically degraded membrane surfaces, as well as many common inorganic contami-

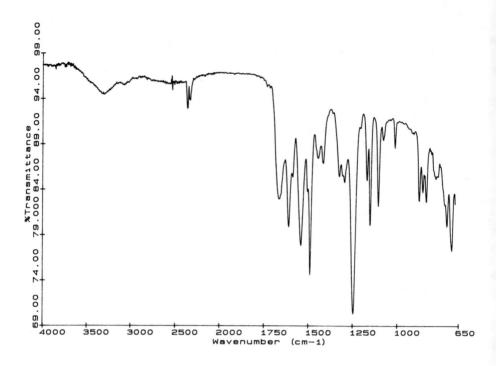

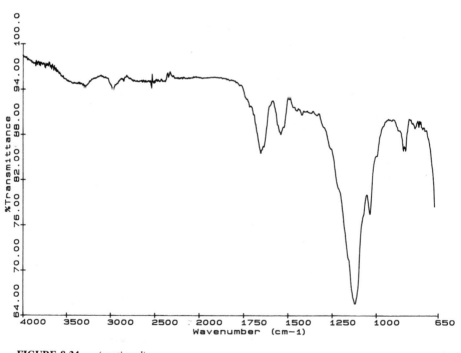

FIGURE 8.34. *(continued)*

270

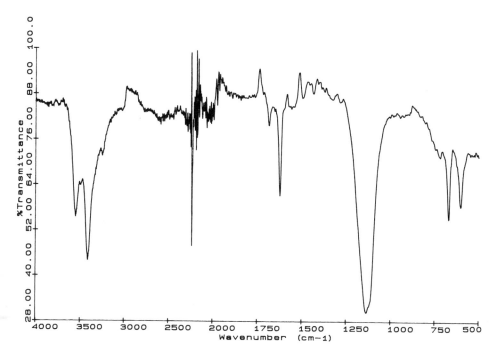

FIGURE 8.35.   **Transmission IR spectrum of gypsum powder clogging a fiber-bundle membrane; spectrum taken using a miniature, high-pressure, diamond-anvil optical cell.**

nants. The principal disadvantage of IR is its insensitivity to elemental sulfur and its general inability to identify metal species present.

Other Analytical Techniques

Many additional methods can be used to analyze RO foulants. This chapter has concentrated on those most commonly employed because of their ease of use and/or the value of the information provided. Two other techniques of interest are Auger and ESCA (electron spectroscopy for chemical analysis) analyses. These techniques are characterized as true surface analyses because they typically examine only the top 10 to 20 Å of the sample.

Auger Spectroscopy

As described in the section dealing with EDX analysis, when inner-shell electrons of the sample are ejected by electrons from the electron beam, the vacancies can be filled by electrons from a higher level. Associated with this process are two possibilities: the emission of a photon (EDX), or the emission of an electron (the basis for Auger spectroscopy) (McCrone and Delly 1973). Because

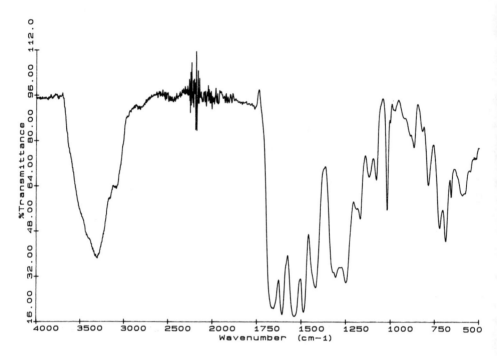

FIGURE 8.36.   Transmission IR spectrum of one fiber of a fiber-bundle membrane.

the escape depth for Auger electrons is only about 10 to 20 Å (see Figure 8.11), only those electrons produced within the first few atomic layers can contribute to the spectrum (Lee 1977).

ESCA

Electrons can also be ejected from inner or outer shells when excited by photons of sufficiently high energy. The kinetic energy of the ejected electron is equal to the photon energy minus the binding energy of the electron. This kinetic energy is characteristic for every atomic species in a particular state of chemical coordination. The binding energies of all elements ($Z = 1$ to 104) have been tabulated. By measuring the kinetic energy of the ejected electron, the elemental composition and chemical coordination of a species can be determined. This is the basis for ESCA. Like Auger spectroscopy, ESCA is sensitive to only about the first 10 to 50 Å of the sample's surface (McCrone 1973).

Both ESCA and Auger techniques have the advantage of being able to analyze a very thin foulant layer without much interference (or much contribution to the spectrum) from the underlying substrate. Unfortunately, the instrumentation is very costly and has not been extensively used for RO analysis.

The acquisition costs of the various techniques discussed in this chapter vary greatly. Optical microscopy is probably the least expensive, followed by IR spectroscopy, SEM-EDX, and Auger or ESCA. The operating costs of some of these instruments may also be appreciable; consumables such as liquid nitrogen, which may not be readily available, may also be required. Finally, the personnel costs involved in these analyses are not negligible. It can take years of experience to interpret correctly some of the analytical data obtained with these techniques. For these reasons, it is likely that one or more of these techniques may not be available "in-house." There are, however, a variety of commercial laboratories available to perform such services on a contractual basis, and a number of directories that identify these laboratories. The directories are published by a variety of sources and are available in many libraries; the American Society for Testing and Materials (ASTM) publishes one such directory, which is fairly comprehensive. As in any contract work, it is important to determine whether the prospective laboratory has sufficient expertise in the client's area of interest (i.e., in this instance, the analysis of RO membranes) by means of a detailed discussion with laboratory personnel prior to the submission of any samples.

SUMMARY

There are several techniques that can be used to analyze complex RO foulants. In choosing a method of analysis, the important roles played by energy-dispersive X-ray analysis and elemental mapping in diagnosing the nature and composition of foulants should be kept in mind. The utility of IR spectroscopy for detecting chemical changes in the RO membranes must also be considered.

ACKNOWLEDGMENT

The authors thank The BFGoodrich Company, Specialty Polymer & Chemicals Division, for support in publishing this work.

References
Amjad, Z., J.D. Isner, and R.C. Williams. 1988. The role of analytical techniques in solving reverse osmosis fouling problems. *Ultrapure Water* 5(5):20–26.
Colthup, N.B., L.H. Daly, and S.E. Wiberly. 1990. *Introduction to infrared and raman spectroscopy*. 3rd ed. New York: Academic Press.
Duncumb, P., and P.K. Shields. 1963. Calculation of fluorescence excited characteristic radiation in the x-ray microanalyzer. *British Journal of Applied Physics* 14:617.
Dyer, J.R. 1965. *Applications of absorption spectroscopy of organic compounds*. Englewood Cliffs, N.J.: Prentice-Hall.
Goldstein, J., and H. Yakowitz, editors. 1975. *Practical scanning electron microscopy*. New York: Plenum Press.

Harrick, N.J. 1967. *Internal reflection spectroscopy*. New York: Wiley Interscience. (Now available through Harrick Scientific Corp., 88 Broadway, Ossining, NY 10562.)

Hayat, M., editor. 1974. *Principles and techniques of scanning electron microscopy*. Vol. 1. New York: Van Nostrand Reinhold.

Hirschfeld, T. 1977. Subsurface layer studies by attenuated total reflection fourier transform spectroscopy. *Applied Spectroscopy* 31(4):289–292.

Lee, L.H., editor. 1977. *Characterization of metal and polymer surfaces*. Vol. 1. San Diego: Academic Press, pp. 133–154.

McCrone, W., and J. Delly. 1973. *The particle atlas*. Vol. 1. 2d ed. Ann Arbor, Mich.: Ann Arbor Science Publishing, pp. 178–182.

Messerschmidt, R.G., and M.A. Harthcock, editors. 1988. *Infrared microspectroscopy: Theory and applications*. New York: Marcel Dekker.

Silverstein, R.M., G.C. Bassler, and T.C. Morrill. 1991. *Spectrometric identification of organic compounds*. 5th ed. New York: Wiley.

Williams, D.H., and I. Fleming. 1980. *Spectrometric methods of organic chemistry*. New York: McGraw-Hill.

9

RO Application in Brackish Water Desalination and in the Treatment of Industrial Effluents

C. A. Buckley

C. J. Brouckaert

C. A. Kerr

Pollution Research Group, Department of Chemical Engineering, University of Natal

BRACKISH WATER DESALINATION

The term *brackish* is usually applied to natural waters, surface or artesian, with salinities between about 1,500 and 5,000 mg/l. In many arid parts of the world, such waters often constitute a significant fraction of the total available water supply. To meet World Health Organization (WHO) specifications for potable water, the salinity must be reduced to below 500 mg/l (WHO 1984)—a removal efficiency requirement of up to 90 percent. This level of performance was achieved by the first generation of commercial reverse osmosis (RO) membranes. Brackish water desalination was consequently one of the first successful applications of RO, with the first large-scale plants appearing in the late 1960s. The technology is now well established. Nearly all brackish water desalination plants use either RO or electrodialysis (ED) (Wade 1991; Wangnick 1991). So much experience has been gained with plants of capacities ranging from 440 gpd to 198,000 gpd that the pretreatment of feed water and the disposal of concentrated brine are more likely to cause difficulties than the design and operation of the RO equipment itself. The nature and scope of pretreatment depends on the characteristics of the feed water of the membrane system. In regions such as Saudi Arabia and Florida, the problem of brine disposal has become a major consideration as the volume of brackish water processed has become an appreciable portion of the total surface water resources.

Spiral wound membranes are most widely used for desalination of brackish water, but hollow fiber modules, using cellulose acetate (CA) or aromatic polyamide fibers, are also used (Henley 1991). The choice of module may be a matter of the design contractor's preferred technology, but the nature of the feed will also influence the choice of module type. Hollow fibers are susceptible to particulate fouling and unsuitable for waters with high levels of suspended solids. Polyamide membranes exhibit greater resistance to biological attack, high temperature, and pH extremes. They may be preferred for more aggressive waters, as less elaborate pretreatment is needed and stronger cleaning solutions can be used.

Module Array Configuration

A salinity of about 10,000 mg/l in the concentrate stream, corresponding to an osmotic pressure of roughly 100 psi, is a practical limitation on the level of water recovery. This generally corresponds to an overall water recovery of 70 to 90 percent, which can readily be achieved with a single pass in two or three stages, with the recovery per stage limited to about 50 percent. For overall recoveries between 60 and 85 percent, two stages are typically used, increasing to three stages when recoveries of over 85 percent are required (Quinn 1985). The number of modules in the end stages is decreased to maintain the flow velocities. An even distribution of flow between parallel hollow fiber modules in the same stage is usually achieved by routing their discharges through flow-balancing "pigtails" coils of fixed hydraulic resistance. It is generally not economical to provide booster pumps to offset the pressure drop between stages.

Ancillary Equipment

Essential elements of an RO plant include a high-pressure pump, the RO modules, and the back-pressure valve(s). Additional equipment for monitoring, control, and safety is also required.

In large plants, multistage centrifugal feed pumps are invariably used. In small installations, triplex piston pumps are often chosen because of their simple construction, relatively low cost, and easy maintenance. As a safety measure, a prefiltration stage (e.g., cartridge filters) is often installed between the booster pump and the high-pressure pump. It is important that all parts in contact with the feed be highly resistant to corrosion by the saline stream, not only for the sake of reliable and efficient operation of the pump, but also because iron corrosion can lead to precipitation of ferric hydroxide in the RO modules and subsequent fouling of the membranes. Stainless steel (316L or better) should be used for stressed components or, where feasible, polymeric materials may be used. The same considerations apply to the materials of valves and pipe work (Backish 1985).

The feed pressure is controlled by valves upstream of the modules, and the flow rate is controlled by a valve downstream. The downstream valve must not be linked to pressure. If the pump delivery pressure should drop owing to some malfunction, the valve may close completely, resulting in a dead-end filtration situation and precipitation in the modules. For centrifugal pumps, the pressure control valve throttles the flow, whereas for a positive displacement pump, pressure should be regulated by a relief valve.

Some fouling of the membranes is inevitable despite the most rigorous pretreatment. Therefore, provision must be made for periodic cleaning. Cleaning equipment consists of a tank for preparing the cleaning solutions, a low-pressure circulation pump, a filter, piping, and valves.

In most cases, the concentrated brine becomes supersaturated with sparingly soluble salts such as calcium sulfate, but precipitation can be prevented by dosing with crystallization inhibitors. Because these only retard precipitation rather than prevent it, brine must be flushed out of the modules when the plant is shut down. A flushing system, using stored permeate water, must be provided for such shutdowns.

Pretreatment

The nature and degree of pretreatment required depends on the nature of the feed water, which varies widely with location. Removal of particulate and colloidal material is a common pretreatment step, as both spiral wound and, particularly, hollow fiber modules are intolerant of particulates, which can cause plugging of the flow channels. Colloidal material can cause fouling of the membrane surfaces.

The degree of filtration required can be gauged by undertaking a Silt Density Index (SDI) test (ASTM D4189). This is a widely used empirical measure of the colloidal fouling tendency of a water supply. The applicability of this dead-end filtration test to a cross-flow situation is theoretically dubious. However, experience has shown that hollow fiber modules usually operate satisfactorily if the SDI is less than 3. Spiral wound modules are slightly more tolerant to particles in the feed (an SDI limit of 5 is considered to be good practice) but this is unlikely to make a significant difference to the design of a pretreatment plant (Quinn 1985).

Groundwaters usually have very low SDI values, and cartridge filtration is a generally sufficient pretreatment. Filter elements of 5 to 20 μm nominal size are typically used, although Goozner and Gotlinsky (1990) have claimed that cartridge filters with nominal ratings are often ineffective and that absolute-rated filters should be used. Surface waters may have SDI values as high as 200 and require more rigorous pretreatment. A rough guide is that waters with SDI values less than 6 can be passed through deep-bed media filters. SDI values between 6 and 50 require coagulation with cationic polyelectrolytes, followed by clarification. Coagulation with alum or ferric salts, followed by sedimentation and deep-bed filtration, is a suitable pretreatment for waters with SDI values above 50 (Matsuura and Sourirajan 1985).

The fouling tendency of colloids can be reduced in some instances by increasing their zeta potential, thus reducing their tendency to adhere to surfaces. Methods used to achieve this include softening the feed water by means of ion exchange (IX) and/or dosing the feed with a polymeric dispersant.

Prevention of Precipitates

The most common precipitates encountered in brackish water desalination are calcium and magnesium carbonate; sulfates of calcium, barium, and strontium; and silica. Well water may have special problems associated with species that are soluble under reducing conditions but form insoluble substances when oxidized by contact with air. Dissolved ferrous salts and hydrogen sulfide fit in this category.

As brackish waters usually contain high levels of calcium, acid dosing is commonly used to prevent calcium carbonate precipitation. Sulfuric acid is frequently used. This increases the potential for calcium sulfate scaling, which may become the limiting factor in determining the overall water recovery. If the calcium or magnesium concentrations are high, cation exchange or lime-precipitation softening should be employed before acid dosing.

Precipitation of sulfates can be prevented by dosing with polyelectrolyte crystallization inhibitors. Sodium hexametaphosphate (SHMP) is also widely used. Its effect is to hinder the rate of crystallization, rather than to prevent it altogether. Again, the brine must be flushed from the modules before shutting the plant down. SHMP is almost totally rejected by RO membranes and has little or no effect on the permeate quality.

Recently, polymers have been developed (Amjad 1989) that inhibit calcium carbonate precipitation (for example, the Flocon® products from Pfizer, the

Aqua Feed® products from BFGoodrich, the Belgard® products from FMC, and the Aquakreen® products from Kao). Use of these polymers may reduce the need for acid dosing. They have advantages over SHMP in that they are more stable against thermal and hydrolytic degradation, making dilution procedures less critical. These advantages are particularly useful in small installations, but the use of such polymers has become established in large plants as well. The extent to which they have supplanted the sulfuric-acid–SHMP combination in general practice is not clear.

Silica precipitation can be a serious problem. The formation of insoluble silica compounds is not easy to predict accurately, as the solubility is strongly dependent on temperature and pH. These compounds, once formed, are difficult to remove—particularly from CA membranes, which are not tolerant to high pH cleaning solutions. Pretreatment options are limited. Silica, soluble and colloidal, can be removed by the lime-precipitation process, but, because of the relative cost and complexity of the process, this is generally avoided if possible (Rautenbach and Albrecht 1989).

The concentration of ferrous ions that can be tolerated by most membranes depends on the concentration of dissolved oxygen. If the dissolved oxygen is below 0.1 mg/l, the brine may contain up to 4 mg/l ferrous ions, but if the brine is oxygenated, the iron concentration should be below 0.05 mg/l. Either contact with air should be avoided or the water should be aerated and then filtered or clarified (Matsuura and Sourirajan 1985).

Well water may contain dissolved hydrogen sulfide. At pH values above 8, a significant portion of the hydrogen sulfide is present as sulfide ions, which may be oxidized to elemental sulfur by either oxygen or chlorine. This can be prevented by acidifying the water. Under acid conditions, the hydrogen sulfide passes through the membrane into the product water, which then requires degasification. The release of hydrogen sulfide gas can cause corrosion and environmental problems.

Biological Control

Bacterial and algal growth in the system must be prevented to avoid fouling and, in the case of CA membranes, degradation of the membrane. The usual treatment for the prevention of microbial growth in CA-membrane–based systems is chlorination to yield 0.2 mg/l free chlorine. Polyamide membranes are, however, degraded by chlorine, and for these sodium metabisulfite may be used to dechlorinate the water after disinfection. Activated carbon adsorption may also be used for dechlorination, although it has the potential disadvantage of fouling the membranes with carbon particles and of providing sites for regrowth of organisms in the dechlorinated water. It is also a relatively expensive process owing to the low specific capacity for chlorine (Rautenbach and Albrecht 1989).

Standard Methods Applicable to Reverse Osmosis Desalination

The widespread use of RO by public utilities for the production of municipal water supplies has led to the establishment of standard methods that relate to various aspects of RO plants. Recognized standard methods are useful for contractual specifications concerning the performance and operation of such plants.

The following ASTM standards apply specifically to RO plants (American Society for Testing and Materials 1990).

D3739	Standard practice for calculation and adjustment of the Langelier Saturation Index for reverse osmosis
D4582	Standard practice for calculation and adjustment of the Stiff and Davis Index for reverse osmosis
D4962	Standard practice for calculation and adjustment of sulfate scaling salts ($CaSO_4$, $SrSO_4$, $BaSO_4$) for reverse osmosis
D4993	Standard practice for calculation and adjustment of silica (SiO_2) scaling for reverse osmosis
D4194	Standard test methods for operating characteristics of reverse osmosis devices
D3923	Standard practice for detecting leaks in reverse osmosis devices
D4516	Standard practice for standardizing reverse osmosis performance data
D4195	Standard guide for water analysis for reverse osmosis application
D4472	Standard guide for record keeping for reverse osmosis systems
D4189	Test method for silt density index (SDI) of water

The ASTM subcommittee D19.08 on Membrane and Ion Exchange Materials has the responsibility for drawing up these standards.

Current Trends in Brackish Water Desalination

The current status of brackish water desalination now exhibits the characteristics of a mature technology: widespread application; reliable techniques for adapting to special circumstances; efficient and comprehensive ancillary support systems to handle such aspects as monitoring, control, maintenance, cleaning, etc.; and keen competition between manufacturers, contractors, vendors, and specialists. As is typical for a mature technology, opportunities for truly fundamental innovation are increasingly rare, and progress consists largely of incremental refinements and adaptations of known principles and systems.

Steady improvements in membranes, coupled with flexibility of operation, resulted in 85 percent of new desalination capacity in 1990 consisting of RO (Pritchard 1991). This flexibility is illustrated by the range of plant sizes under consideration, from a giant 1860 MGD plant being planned for Beijing, China (Pritchard 1991), to mobile, containerized units for emergency use in Kuwait (Mace 1989), and small plants to supply potable water to remote rural villages in India (Prabahkar et al. 1987, 1989). There is a vigorous mobile desalination industry in the United States based on trailer-mounted ion exchange (IX) and/or RO plants (Slejko 1990).

Membrane improvements have followed two trends: membranes with improved general capability, that is, higher fluxes and rejection and greater chemical resistance; and membranes with properties that are tailored to specific needs. The effect of the general improvement in membranes is illustrated by the lower operating pressures of modern plants, between 200 and 250 psi as opposed to 400 to 600 psi for older installations. This reduces energy requirements from 7.75 kWh/1,000 gallons to between 3.1 and 5.8 kWh/1,000 gallons (Riley 1990). An example of tailoring membranes to specific needs is the emergence of membrane softening as a recognized technique (Comstock 1989; Conlon et al. 1990). This process employs nanofiltration or "ultralow-pressure" membranes that reject divalent ions, particularly SO_4^{2-} and CO_3^{2-}, strongly (± 95 percent), but monovalent ions only moderately (50 to 60 percent). Membrane softening has been used with particular success to produce municipal water from brackish groundwater in Florida.

Membrane fouling is widely accepted as the single largest cause of permanent flux decline and premature failure of RO membrane elements in brackish water desalination (Paul and Rahman 1990). Consequently, a considerable amount of effort has been devoted to effective means of controlling or removing fouling layers. Techniques include polymer dosing (Amjad 1989), special cleaning chemicals, specialist membrane regeneration services (Paul and Rahman 1990), and membrane surface modification to minimize the tendency of foulants to become adsorbed (Riley 1990).

Finally, an important trend affecting the brackish water desalination market is concern over the disposal of the concentrated reject brine. This issue has become especially significant in Florida, which has a large number of RO installations. With production of about 9.8 mgd, they discharge 4.9 to 9.8 mgd of concentrate (Buros 1990). Although all constituents of the concentrate are derived from the feed water, the process may concentrate them to environmentally unacceptable levels.

The widespread use of RO thus gives rise to both technical and regulatory concerns. The approach to the problem depends greatly on the specific circumstances at a particular site, and methods of treatment include surface water discharge (Malaxos and Morin 1990), deep well injection (Muniz and Skehan

1990), blending with irrigation water (Edwards and Bowdoin 1990), solar evaporation (Smith 1990), vapor compression evaporation (Awerbuch and Weekes 1990), and ED reversal (Reahl 1990).

TREATMENT OF INDUSTRIAL EFFLUENTS

Industrial effluents should be considered as a mixture of water, chemicals, and energy. As parts of an effluent, the components are viewed as contaminants and there is a cost associated with disposing of such an effluent. On the other hand, effluents can be considered as a resource whose components frequently have a value associated with them. The true cost of an effluent discharge must be considered to be the sum of the discharge costs and the component values. Conversely, the costs associated with recycling water, chemicals, and energy would be offset by the true savings (the sum of the discharge costs and the component values if no reuse were to be undertaken).

In general, a certain amount of effluent will have to be discharged. A primary objective of a responsible manufacturer is to minimize the waste of resources and the excessive discharge of effluents. A process such as RO is but one device in the toolbox of a process engineer.

Introduction and Overview

RO is a single-stage separation and concentration process. Although compared with a process such as evaporation it is not energy intensive, RO is nevertheless expensive. To be a cost-effective process, it is necessary for RO to achieve some economic benefit. Consideration should be given to modifying the processes that generate the effluent so that segregated effluent streams are produced. The objective should be to separate concentrated from dilute streams, and those containing undesirable components from streams that are innocuous.

The required quality of all process waters should be critically evaluated, inasmuch as potable or municipal water is often of higher quality than required. Where a lower quality water can be accepted, a fraction of the effluent or permeate might be recycled. Further, consideration should be given to cascading relatively high quality water to those processes that can accept a lower quality water. It must be noted that there is no single parameter to describe water quality. For each use of water there is an associated critical quality parameter. As an example, water containing hydrolyzed dyes might be used as a first rinse for newly dyed textiles. The presence of color in the rinse water may at first be thought to preclude its reuse in a textile operation, but the chemical nature of the hydrolyzed dye is such that it does not react with fibers (Townsend et al. 1992).

An aspect to be considered in the recycling of chemicals is that a higher-value chemical that can be recycled might be used in place of a lower-value chemical that is used once and then discharged. Ease of recovery would be a factor in the choice of a substitute chemical. For an RO recycle system, selection of substitute chemicals would be based on such aspects as membrane rejection, osmotic pressure, precipitation potential, process operating temperature, oxidation potential, and pH.

If a recovery process is limited by the fouling potential of constituents in the process feed water, it could be beneficial to pretreat the feed water to remove these constituents.

The safe disposal of effluents is the responsibility of the producers of the effluents. As a consequence, effluent treatment must be considered an integral part of the manufacturing process. Frequently, the least expensive method of effluent treatment is to eliminate the source of the effluent. The techniques of waste minimization have been well documented (Environmental Protection Agency [USA] 1985; Department of Trade and Industry 1990; 3M 1985).

Discharge regulations are specific to each region. In general, the regulations are framed to protect the environment. A publication outlining the various philosophies for water quality management has recently been produced (Department of Water Affairs and Forestry 1991), which sets out the following principal categories of pollutants:

1. Chemicals (including acids, metals, salts, inorganic compounds and organic compounds—both biodegradable and relatively nonbiodegradable)
2. Suspended matter (particles and colloids that are temporarily or permanently suspended in water and that can act as transport media for metals or pathogens)
3. Radiogenic materials, if present at significant levels

The overall goal of water quality management is to maintain water quality in such a state that it remains fit for recognized uses. The concept of fitness for use implies an evaluation of water quality in terms of the requirements of a particular user, or category of users, and is measured in relation to criteria or norms representing ideal quality for a particular use. In general, the principle of "polluter pays" applies.

This goal can be achieved by the "uniform effluent standards" approach, which aims to control the input of pollutants to the water environment by requiring that effluents comply with uniform standards. The underlying philosophy is that minimum pollution (from point sources) is the desirable ultimate goal. The "receiving water quality objectives" approach involves specification of the desired quality of the receiving water environment and the control of sources. This approach recognizes that the receiving water has a certain capacity to assimilate pollution without serious detriment to quality requirements of the users. The "pollution prevention" approach is aimed specifically at control of the handling and disposal of hazardous substances. Toxicity persistence and capacity for bioaccumulation present major threats to the environment. In most cases, legislation includes aspects of all three of these approaches.

RO is a relatively expensive operation and should be considered only after aspects such as waste minimization, segregation of streams, and the review of process water quality requirements have been addressed. Specific considerations for the use of RO are addressed in subsequent sections of this chapter.

RO is particularly suitable for use in water-scarce regions because a high quality permeate can be produced. The purity of the permeate can be controlled by water recovery and the intrinsic membrane rejection. In certain circumstances, the permeate from one RO unit can be fed to a second unit. An advantage of RO over IX is that it causes no net increase in the salinity of the environment. It has been well established that a stand-alone RO plant is more energy efficient than a stand-alone evaporation plant. This argument does not apply to specific combined-cycle or integrated plants where waste heat of a suitable thermal quality is available (Wade 1991).

Evaluation and Design of Reverse Osmosis Effluent Treatment Processes

Industrial effluents are the purge or safety valve in manufacturing processes; they "carry the load" resulting from any process upset or malfunction. Invariably, the harshest conditions, in regard to effluent treatment, occur during plant upsets. General factory practice is that all mistakes go down the drain. These unplanned incidents could be sufficient to destroy a set of membranes, that is they may occur only once in the membranes' lifetime. For this reason, it is often advisable to close-couple the membrane treatment process to a specific piece of effluent-producing equipment. An advantage is that the operators of a manufacturing plant are made aware of the vulnerability of the effluent treatment plant to careless operation of the manufacturing process.

Foulants, Pretreatment, and Cleaning

There are three approaches that can be adopted for handling membrane-fouling effluents. The cautious approach involves the installation of a large buffer capacity and a series of safety barriers (sand filtration, cartridge filtration, and reductant-acid-base addition, as well as alarms and cut-out controls). In this case, the cost of pretreatment would exceed the cost of the basic membrane plant. Furthermore, all controls and alarms must be regularly calibrated and checked. The control strategy must be carefully considered as to its ability to handle any eventuality.

The cavalier approach is to accept the fact that the membranes will foul and that they have a finite life. It is then necessary to balance the cost of appropriate safety measures against the cost of new membranes. Fouling is handled by instituting a scientifically devised cleaning procedure. The nature of the foulant can be determined by analyzing small test membranes mounted in cells in the reject line of the plant (Buckley et al. 1987; Kerr and Buckley 1992).

With formed-in-place membranes, which enable a system to be constructed with minimal pretreatment, fouling is allowed to occur (Spencer et al. 1989). At some stage, the membranes are stripped from the supports and a new membrane-forming solution is passed through the system. This procedure can be automated (Townsend et al. 1992).

Information on the nature of foulants in the feed to an RO plant can be gained by a detailed autopsy of fouled membranes. Specialist companies undertake these investigations. This information can then be used to improve the pretreatment or cleaning procedures.

Bench Scale Tests

Bench scale tests can be used successfully to evaluate and highlight factors such as suitable membrane types, fouling tendencies, and rejection characteristics of different membranes in the presence of a particular effluent (Kerr and Buckley 1992). A bench scale 3-cell flat sheet test rig such as that developed by Patterson Candy International (PCI), can be used (Gutmann 1987). The hydrodynamics of the system are similar to those in a PCI tubular membrane system. Although absolute flux results cannot be compared directly with other module configurations, the relative relationships hold. Thus, the rejection and fouling characteristics can provide a good indication of the phenomena that would be encountered in large-scale applications.

Discs of flat sheet membrane are cut, conditioned by soaking in deionized water for 24 hours, and inserted in the test cells. Membranes from different manufacturers can be used to evaluate the most appropriate membranes for a particular effluent. A pure water flux is determined, initial membrane compaction is quantified, and the appropriate flux and rejection characteristics of the membranes are determined using pure solutions. This data can be used as a base level against which future results can be compared to gauge the state of the membranes and the effectiveness of any cleaning procedures.

Initial test experiments are conducted in total recycle mode with the permeate returned to the feed tank. A schematic diagram of the test rig in total recycle mode is shown in Figure 9.1. Cooling coils or heating elements can be added as required to adjust the temperature of the feed to that which would pertain in a large-scale operation. Permeate flux and species rejections can be calculated. The effects of temperature and pH of the feed on species point rejection and permeate flux can be evaluated. The possible formation of a gel polarization

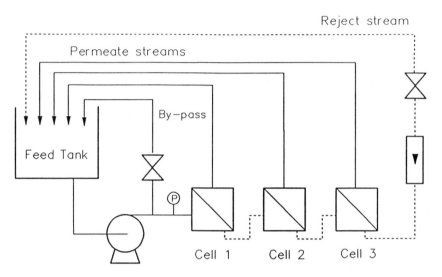

FIGURE 9.1.   Schematic diagram of test rig in total recycle mode.

layer can be checked by varying feed flow rate at a given pressure until the flux is independent of velocity. In total recycle mode, the feed composition remains constant. A continuous decline in permeate flux (after initial decline) could indicate feed-membrane interaction, with physical and/or chemical modification of the membrane by the feed. An increased flux could indicate hydrolysis of the membrane.

A pure water flux should be determined between each total recycle experiment and compared with the initial water flux. A decline in pure water flux compared with the initial measurement could indicate membrane fouling. A comparable pure water flux would indicate no serious membrane fouling or deterioration of the membrane structure.

The osmotic pressure characteristics of an effluent can be found by pre-concentrating the effluent to known levels prior to total recycle experiments. A decline in permeate flux with increasing feed concentration, (but with no decline in pure water flux compared with the original measurement), indicates that the osmotic pressure of the feed solution is becoming the limiting factor to RO.

If no osmometer is available, an indication of the osmotic pressure of the feed can be obtained in the following way: under total recycle conditions and with a constant feed flow rate, the input pressure is varied and the resultant fluxes are monitored. A plot of pressure against permeate flux should give a straight line since:

$$J = \frac{K_w(\Delta p - \Delta \pi)}{A} \tag{9-1}$$

Where:

J = permeate flux
K_w = water permeability constant of the membrane
Δp = pressure differential (feed-permeate)
$\Delta \pi$ = osmotic pressure differential (feed-permeate)
A = membrane area

At $J = 0$, the line will intersect the pressure axis at the osmotic pressure of the feed solution.

Figure 9.2 is a schematic diagram of the test rig in a batch concentration mode. During these experiments, the permeate is continually removed, resulting in an increase in feed concentration. Depending on the size of the feed tank, six to eight batches of fresh effluent are concentrated to 50 percent water recovery. The concentrates are combined and concentrated further.

Serial batch concentration tests are similar, except that instead of removing the effluent at 50 percent water recovery, fresh effluent is added to the feed tank to make up the original volume. Species rejection and permeate flux are monitored throughout the experiment. The pure water flux should be determined frequently, so that the effects of membrane and/or osmotic pressure can be determined. If membrane performance deteriorates below acceptable limits, various cleaning techniques can be implemented and their effectiveness assessed by determining the pure water flux.

Concentration-dependent fouling may occur when, with increasing feed con-

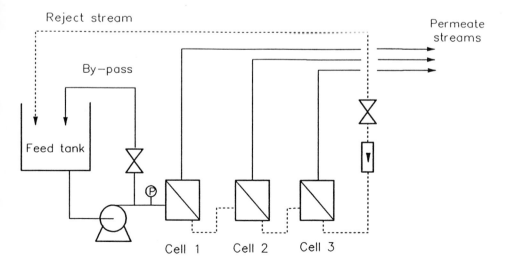

FIGURE 9.2.   Schematic of batch concentration mode.

centration, solubility limits are exceeded and precipitation occurs. Examination of the concentrates at various water recoveries, together with the collection and analyses of any precipitates, can enable the identification of the foulant and the lower limit of water recovery at which fouling occurs.

Generally, concentration-dependent fouling can be minimized or eliminated by pretreatment measures. A more serious form of fouling is progressive membrane fouling. This is somewhat similar to feed-membrane interaction (compaction-hydrolysis) in that it is irreversible. The mechanics of this type of fouling are not well understood, and its existence can be identified with certainty only when all the possible combinations of cleaning and pretreatment sequences have been exhausted.

Batch concentration tests are time consuming and, because concentration-dependent fouling may not become apparent until fairly high water recoveries have been achieved and a significant area of the membrane has become fouled, a rapid screening method for potentially fouling effluents has been developed (Kerr and Buckley 1992). Feed is evaporated under vacuum at a temperature similar to that of the proposed effluent treatment plant. At increasing water recoveries, the concentrate is passed through 0.22 μm filters in order to detect and collect any precipitates formed. These tests can be undertaken simultaneously with or prior to batch concentration experiments, and the threshold of water recovery for the onset of precipitation determined. The evaporation test concentrates can be used as the feed for membrane experiments.

Bench scale concentration and evaporation tests can highlight problems such as flux decline, decline in species rejection, osmotic pressure effects, and fouling These problems can then be prevented at the pilot-plant stage by adequate pretreatment and plant design.

In parallel with the bench scale tests, the tolerance of various membranes to the raw effluent can be determined by suspending test sections of flat sheet membranes in the flowing effluent. Periodic testing of the membranes can indicate any long-term incompatibility between a membrane and an effluent. This test complements the tests involving limited samples of effluent.

Pilot-Plant Tests

Short-term events, such as oil contamination or high chlorine concentrations, do not become apparent during most sampling programs. For this reason, it is necessary to run pilot-scale experiments at the factory by drawing effluent from the process continuously. Through this practice, the membrane plant personnel learn more about the effluent and the process operators become familiar with the characteristics of the membrane systems.

Because industrial effluents are site and process specific, the guarantees offered by equipment vendors, which are often vague and contain many conditions, are frequently meaningless. The operation of a pilot plant then becomes necessary to establish confidence in the place of any guarantee.

The consequences of any changes in the objectives of the investigation, as a result of the interim findings, must be carefully considered. Otherwise, insufficient data relating to the final objective may be obtained, leading to inconclusive results.

Specific objectives of a pilot-plant investigation may include the following:

- Acquisition of design data
- Determination of membrane life
- Production of large volumes of permeate or concentrate to evaluate the long-term effect of its reuse
- Reliability and operability of the process
- Transfer of expertise to the factory management and staff

The following are frequent shortcomings of pilot-plant exercises:

- Insufficient funds to engineer the plant to adequately high standards. In this case, the time a plant is down for maintenance is longer than the time spent processing effluent.
- Insufficient supervision, control, record keeping, and instrumentation, resulting in significant periods of incorrect operation that may cause membrane damage. The cost of adequate instrumentation and control for a pilot plant approaches that of a full-scale plant.
- Operating conditions that are not equivalent to those in the full-scale plant. The feed to the pilot plant should be a continually representative sample of the entire feed. Due consideration should be taken of aspects such as temperature, redox potential, age, dissolved gases, biological activity, and seasonal or market-related changes in effluent characteristics.
- Insufficient funds to modify and rectify any design faults. In this case, the process conditions are known to be incorrect and most of the assessment is spent justifying and extrapolating inappropriate data.

The water recovery of the plant determines the scaling and fouling potential on the membranes at the reject end of the plant. The total volume of fresh effluent handled by the pilot plant will determine the detection limit of low concentration, low solubility chemicals that will gradually foul the system. It is necessary to process sufficiently large volumes of fresh effluent to the required water recovery (that is, ratio of permeate volume to feed volume). Sufficient membrane area must be installed to enable continuous operation at the required water recovery. If this is not feasible, then a feed and bleed configuration (operating at the required water recovery) should be used in preference to a series of batch concentrations. A sequence of batch concentration tests might result in alternate precipitation and dissolution of scalants. A feed and bleed system at the required water recovery provides the most severe operating conditions for the entire membrane surface.

The necessity of a will to succeed cannot be overemphasized, inasmuch as many exercises are carried out with a subconscious objective of management to "prove" that effluent treatment is impractical and uneconomical.

Characteristic Reverse Osmosis Process Configurations in Effluent Treatment

RO applications in industrial effluent treatment fall into four broad process configurations, which are determined by the economic and technical aspects of each situation.

Total Recycle of Reject and Permeate
The ideal application of RO is the total recycle of both the permeate and the concentrate streams (see Figure 9.3). This imposes severe constraints on the system because there are no separate purge streams for impurities. Not many systems fall into this category. The most important and striking example is the treatment of nickel-plating rinse waters.

Reverse Osmosis as a Component of a Chemical Recovery Process
Where RO is unable to effect the required effluent treatment on its own, it may still find application as a concentration stage in combination with other processes (see Figure 9.4). Generally, the fraction that can be recycled is restricted, unless the associated processes do not add extraneous components or the posttreatment is able to remove what has been added in pretreatment. The treatment of chrome-plating effluents is an example of this pattern.

Reverse Osmosis as a Concentration Step Before Disposal
RO can be used as a process to reduce effluent volumes prior to disposal (see Figure 9.5). The effluent is split into a concentrate stream for off-site disposal

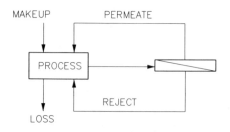

FIGURE 9.3.   Total recycle of reject and permeate.

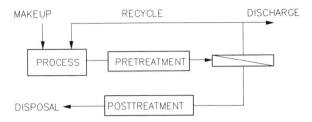

FIGURE 9.4.   RO as a component of a chemical recovery process.

and a permeate stream. Part of the permeate stream can be recycled and the balance discharged.

Reverse Osmosis as a Component of Pretreatment Before Disposal

Finally, RO can be used to purify and concentrate an effluent in conjunction with other treatment steps (see Figure 9.6). An example of this use of RO is in the treatment of dilute toxic effluents. The toxic component in the dilute effluent is concentrated, whereas a fraction of the permeate, which is not totally suitable for reuse, can be blended into the higher quality process water streams. The balance of the permeate is discharged. The concentrate is further treated to destroy or immobilize the toxic components through techniques such as precipitation, oxidation, or incineration.

EXAMPLES OF REVERSE OSMOSIS EFFLUENT TREATMENT PROCESSES

Electroplating Solutions

The treatment of effluents from electroplating processes by RO came under investigation very soon after membrane systems were widely commercialized (Belfort 1984). Several characteristics of these effluents make them ideally suited to treatment by RO. They contain contaminants that are both toxic and

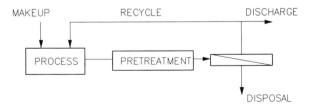

FIGURE 9.5.   RO as a concentration step before disposal.

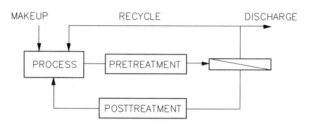

FIGURE 9.6.   RO as a component of pretreatment before disposal.

valuable, such as nickel, chrome, zinc, copper, cadmium, and cyanide. They may be either extremely acid or alkaline, and thus unsuited for direct discharge. The incentive for treatment of these effluents is thus high, from both economic and environmental aspects. The effluents are not usually direct products of the electroplating process itself, but result from the loss of plating bath solution through drag-out by the plated articles. The losses are recovered in a countercurrent series of rinsing baths, but the solution is then too dilute to be returned to the process line. RO offers a means to recover pure water and to return the concentrated chemical stream to the process at a lower energy cost than that associated with evaporation.

The use of RO in electroplating applications has been notably successful only in treating effluents from the so-called Watts nickel-plating baths. To put this in perspective, the electroplating of tin is the largest of the electroplating operations in terms of tonnage of plating metal, followed by nickel, copper, and chrome, which are roughly equal to one another in terms of tonnage of plating metal, and are most often used together in composite coatings, generally referred to as "chrome plate." RO has been used to treat copper (McNulty, Goldsmith, and Gollan 1977a) and chrome plating wastes, but not with the same wide general acceptance as in treating nickel.

A Watts nickel solution is suitable for recovery by RO. Divalent nickel salts are strongly rejected by membranes, and the pH of 3.0 to 4.0 does not cause rapid degradation of CA membranes. The concentrate is recycled at a lower concentration than the plating bath to compensate for evaporative losses.

The use of RO has made it possible to virtually eliminate any Watts bath effluent. The concentrate is returned to the plating bath, and the permeate to the rinsing system. Because this constitutes an almost total recycle, makeup water is deionized to minimize the buildup of impurities in the system. The rejection of boric acid, added as a buffer, is variable (McNulty, Goldsmith, and Gollan 1977b) and depends on its degree of ionization. About 40 percent recovery is reported, therefore monitoring and control of its concentration in the plating bath is required (Canning 1982). The recovery of brighteners is problematic where effluents from "bright" and "semibright" baths are treated together. The concentrate can be returned only to the "bright" bath (Cartwright 1982).

The operational costs associated with RO can usually be justified on the basis of the chemical savings alone, although the payback time depends on the size and loading on the plating operation (McNulty, Goldsmith, and Gollan 1977b). RO takes up less space, is easier to operate and maintain than an evaporator, and requires less energy (Rautenbach and Albrecht 1989).

One of the major difficulties hindering the more widespread use of RO in other electroplating processes is the extremes of pH at which they are operated. A number of systems, for example, those for plating copper, zinc, and cadmium, use cyanide as a complexing agent. The solutions must be alkaline to prevent evolution of cyanide gas. Sodium hydroxide is often used in excess to promote high conductivity and to prevent absorbed atmospheric carbon dioxide from evolving cyanide gas. The pH values of these baths may be as high as 11 to 13, too high even for polyamide-based membranes to operate continuously (Peterson, Larson, and Cadotte 1982).

Chrome plating is carried out under strongly acidic conditions. The solutions require neutralization before treatment with RO. This reduces the economical attractiveness of RO, particularly as additional processing is needed to remove the neutralizing agent from the RO concentrate prior to its return to the plating bath. Moreover, because chromium is removed from the bath, makeup is required, reducing the incentive for chemical recovery. At the Rock Island Arsenal, (Illinois) RO is used successfully to recover chromium from the rinse water. The pH is raised with sodium hydroxide prior to RO, and the sodium is removed from the concentrate by a cation exchange process (Matsuura and Sourirajan 1985).

In addition to the traditional plating systems discussed thus far, there are a number of alternative systems in which RO may find application. For example, cyanide solutions at pH values of approximately 10 are used for copper plating of steel. These solutions can be treated with polyamide membranes. Copper can be plated from pyrophosphate solutions at pH values ranging from 8.6 to 9.2 (Canning 1982). The types of plating effluents that are treatable by RO have been surveyed by Donelly et al. (1976). It is to be expected that the list will grow as more chemically resistant membrane systems are proven reliable.

Although the ideal application for RO in effluent treatment involves chemical recovery, it has also been used in situations where recovery is not economically feasible, such as the treatment of the mixed plating effluent at the Rock Island Arsenal (Matsuura and Sourirajan 1985). This effluent includes the permeate from the chromium recovery process (containing significant levels of chromium), as well as effluents from other plating and related processes. Here the RO is part of a complex of treatment steps and is used as a preconcentrating step to reduce the loading on the final evaporators. Its use is justified by the high rejection of multivalent ions and the low energy consumption, as compared with evaporation.

Power Station Effluents

The water quality requirements of a coal-fired power station vary from the ultrapure water needed for the boiler-turbine circuit to the low quality water used for slurry transport of ash and for ash conditioning. The greatest volume of water is used in the cooling circuit. This has an intermediate quality requirement, which is determined by the need to avoid corrosion and fouling in the condensers and cooling towers. Because water is continuously lost from this circuit by evaporation, cooling tower water is maintained close to its quality limit. This is achieved by a combination of blowdown and treatment. Cooling tower blow-down is the largest volume effluent stream generated by a power station.

Since the 1970s, "zero discharge" of liquid effluents has become an accepted concept in the design of coal-fired power stations. This was made possible by the presence of ash dumps, which must absorb substantial quantities of water (about 15 percent by mass) to stabilize them (Schutte et al. 1987).

Zero effluent discharge requires an integrated management of the station's water balance, whereby the amount of water reporting to the ash dumps plus that lost by evaporation must be equal to the amount of intake water. Two large installations that have recently reported the use of RO to concentrate cooling circuit blowdown are the Bayswater/Liddell complex in New South Wales, Australia (Stuart et al. 1989), and the Lethabo Power Station in the Republic of South Africa (Schutte et al. 1987). The large volumes involved (9.2 mgd at the Bayswater/Liddell complex and 2.4 mgd at Lethabo) make RO economically viable. Permeate from the RO plant is recycled to the cooling circuit, and the concentrate is evaporated to recover more water and to further reduce its volume.

The Bayswater/Liddell plant has a standard design approach, with spiral wound CA elements (Hydranautics). The pretreatment plant consists of a lime-soda softener-clarifier, dual-media filters, cartridge filters, acid-SHMP dosing and chlorination. The Lethabo installation uses a different concept. Here, tubular CA modules (Membratek), fitted with a flow reversal and a sponge-ball cleaning mechanism, are used. The only pretreatment is pH adjustment with acid injection. The plant has had difficulties with mechanical problems and the loss of rejection by the modules. It is currently being recommissioned after a number of mechanical improvements (Aspden and Swanepoel, 1990).

The success of a "zero discharge" philosophy in both these examples arises from the fact that the concept was integrated into the design and management of the whole plant. In the case of the relatively new Lethabo Power Station, which was commissioned in 1986, zero discharge was incorporated into the original design. At the older Bayswater/Liddell stations it was a retrofit concept, which involved the redesign of the entire water supply and effluent management systems.

The spent regeneration solutions from the IX plant producing the ultrapure boiler circuit water are a second source of effluent generated by power stations. RO has the potential to, at least partially, replace IX in the demineralization process. A substantial reduction in the volume of regeneration effluents can be realized. McAfee et al. (1990) reported that RO as a pretreatment for the IX demineralizers at the Permian Basin Steam Electric Station (Monahans, TX) reduced the frequency of regeneration from three times a week to three times a year.

Pulp and Paper Bleach Effluents

Bleach effluent from a pulp and paper mill was successfully treated by RO in a pilot plant designed to handle a continuous flow of 18.5 gpm of pretreated effluent (Groves and Bindoff 1985). A schematic diagram of the pilot plant is shown in Figure 9.7. Pretreatment of the feed to the RO plant consisted of lime

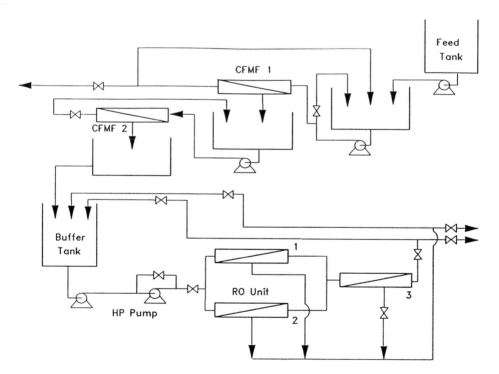

FIGURE 9.7.   Schematic diagram of pilot plant for RO of bleach.

addition to remove magnesium and organic material from the effluent. The precipitates were removed by cross-flow microfiltration. Sodium metabisulfite was used to remove chlorine from the effluent, and dosing was carefully controlled to limit the calcium sulfate scaling potential of the RO feed. Despite these precautions, fouling of the RO membranes occurred. The foulant was identified as calcium oxalate, and an additional pretreatment stage of sodium carbonate addition, followed by further cross-flow microfiltration, was implemented to remove excess calcium from the RO feed.

The RO plant consisted of a 2:1 array with four FilmTec elements per module. The RO membranes rejected between 96 and 98 percent of all species. At 75 percent water recovery, the pilot plant was capable of producing permeate of suitable quality for recycle within the mill.

Mine Drainage

The Seeded Precipitation and Recycled by Reverse Osmosis (SPARRO) process has been developed for desalinating mine water, using CA tubular RO membranes (Membratek) in conjunction with a crystallizer. The RO is operated in a supersaturated mode, and seed crystals are introduced to act as preferential sites for crystal growth, to prevent scaling of the membranes. Because of the seed crystals' relatively short residence time in the RO tubes, most of the crystallization occurs in the external crystallizer vessel, which also provides the seed crystals to be recycled to the RO modules. Seeded-slurry RO pilot plants have been operated to remove calcium sulfate from power station cooling tower blowdown (Hess and Jones 1988) and mine drainage water (Harries 1985). As yet, no information is available concerning commercial examples of this process.

References

American Society for Testing and Materials. 1990. *Water and environmental technology,* section 2. vols. 11.01 and 11.02, Annual book of ASTM standards. Philadelphia.

Amjad, Z. 1989. Deposit control agents for reverse osmosis applications. *Ultrapure Water* 6(5):57–60.

Aspden, J.D., and D.A. Swanepoel. 1990. Minimizing liquid discharges from large power plants. *Water SA* 16(4):287–292.

Awerbuch, L., and M.C. Weekes. 1990. Disposal of concentrates from brackish water desalting plants by means of evaporative technology. *Desalination* 78:71–76.

Backish, R. 1985. "Materials for RO systems." In *Theory and practice of reverse osmosis,* edited by R. Backish. Englewood, N.J.: International Desalination Association, pp. 313–336.

Belfort, G. 1984. "Desalting experience by hyperfiltration (reverse osmosis) in the United States." In *Synthetic membrane processes,* edited by G. Belfort. Orlando: Academic Press, pp. 221–280.

Buckley, C.A., A.L. Bindoff, C.A. Kerr, A. Kerr, A.E. Simpson, and D.W. Cohen.

1987. The use of speciation and x-ray techniques for determining pretreatment steps for desalination. *Desalination* 66:327–337.

Buros, G. 1990. Introduction to the problem of disposal of desalting concentrates in Florida. *Desalination* 78:3–4.

Canning handbook. 1982. *Surface finishing technology*. Birmingham: W. Canning Plc. p. 973.

Cartwright, P.S. 1982. The status of RO and UF in metal finishing recovery and reuse. Paper presented at Water Pollution Control Federation Conference and Exhibition, St. Louis, October 3–8, 1982.

Comstock, D.L. 1989. Desal-5 membrane for water softening. *Desalination* 76:61–72.

Conlon, W.J., C.D. Homburg, B.M. Watson, and C.A. Kiefer. 1990. Membrane softening: The concept and its application to municipal water supply. *Desalination* 78:157–176.

Department of Trade and Industry. 1990. *Cutting your losses*. The Business and Environment Unit, London.

Department of Water Affairs and Forestry. 1991. *Water quality management policies and strategies in the Republic of South Africa*. April. Department of Water Affairs and Forestry, Pretoria, Republic of South Africa.

Donelly, R.G., R.L. Goldsmith, K.J. McNulty, D.C. Grant, and M. Tan. 1976. *Treatment of electroplating waters by reverse osmosis*. EPA-600/2-76-261. Washington, D.C.: U.S. Environmental Protection Agency.

Edwards, P., and P. Bowdoin. 1990. Irrigation with membrane plant concentrate: Fort Myers case study. *Desalination* 78:49–58.

Goozner, R., and B. Gotlinsky. 1990. Field results for the reduction of SDI by pre-RO filters. *Ultrapure Water* 7(2):20–28.

Groves, G.R., and A.L. Bindoff. 1985. "Closed loop recycle and treatment of bleach effluents." In *Proceedings of the Symposium on Forest Products International: achievements and the future*, at Council for Scientific and Industrial Research, Pretoria, South Africa.

Gutman, R.G., 1987. *Membrane filtration: The technology of filter-driven crossflow processes*. Bristol: Adam Hilger.

Harries, R.C. 1985. A field trial of seeded reverse osmosis for the desalination of a scaling-type mine water. *Desalination* 56:227–236.

Henley, M. 1991. Power plants, water re-use and potable water offer growth potential for RO membranes. *Ultrapure Water* 9(2):18–22.

Hess, M.B., and G.R. Jones. 1988. *Field demonstration of waste water concentration by seeded reverse osmosis*. EPRI Research Project 2114-7 Final Report. Palo Alto: Research Report Center.

Kerr, C.A., and C.A. Buckley. 1992. Screening tests for reverse osmosis and ultrafiltration. *Waste SA* 18(1):63–67.

Mace, B. 1989. Design of compact, containerized desalination plants. *Desalination* 76:53–60.

Malaxos, P.J., and O.J. Morin. 1990. Surface water discharge of reverse osmosis concentrates. *Desalination* 78:27–40.

Matsuura, T., and S. Sourirajan. 1985. Fundamentals of reverse osmosis. In *Theory and practice of reverse osmosis,* edited by R. Backish. Englewood, N.J.: International Desalination Association, pp. 225–236.

McAfee, A.L., B. Nowlin, and S. Beardsly. 1990. Pretreatment of an ion exchange demineralizer with reverse osmosis. *Ultrapure Water* 7(7):54–62.

McNulty, K.J., R.L. Goldsmith, and A.Z. Gollan. 1977a. Reverse osmosis field test: Treatment of copper cyanide rinse waters. *US NTIS Report PB-272473.*

McNulty, K.J., R.L. Goldsmith, and A.Z. Gollan. 1977b. Reverse osmosis field test: Treatment of Watts nickel rinse water. *US NTIS Report PB-266919.*

Muniz, A., and S.T. Skehan. 1990. Disposal of concentrate from brackish water desalting plants by use of deep well injection. *Desalination* 78:41–47.

Paul, D.H., and A.M. Rahman. 1990. Membrane fouling: The final frontier? *Ultrapure Water* 7(3):25–36.

Peterson, R.J., R.E. Larson, and J.E. Cadotte. 1982. "Industrial applications of the FT-30 reverse osmosis membrane." In *Proceedings of the World Filtration Congress 3.* Philadelphia, pp. 541–547.

Prabahkar, S., R.N. Patra, B.M. Misra, and M.P.S. Ramani. 1987. Operational experience of reverse osmosis plants for drinking water in Indian villages. *Desalination* 65:361–372.

Prabahkar, S., R.N. Patra, B.M. Misra, and M.P.S. Ramani. 1989. Management and feasibility of reverse osmosis schemes for rural water supply in India. *Desalination* 73:37–46.

Pritchard, C. 1991. Reverse osmosis on top. *The Chemical Engineer* 495:6.

Quinn, R.M. 1985. "Reverse osmosis practice." In *Theory and practice of reverse osmosis,* edited by R. Backish. Englewood, N.J.: International Desalination Association, pp. 257–296.

Rautenbach, R., and R. Albrecht. 1989. *Membrane processes.* Chichester: John Wiley & Sons.

Reahl, E.R. 1990. Reclaiming reverse osmosis blowdown with electrodialysis reversal. *Desalination* 78:77–89.

Riley, R.L. 1990. Reverse osmosis in membrane separation systems: A research needs assessment. Final Report, vol. 2, chap. 5, DOE/ER/30133-H1. Springfield, Va., National Technical Information Service, U.S. Department of Commerce.

Schutte, C.F., T. Spencer, J.D. Aspden, and D. Hanekom. 1987. Desalination and re-use of power plant effluents from pilot plant to full scale application. *Desalination* 67:255–269.

Slejko, F. 1990. Ecolochem Inc.: A pioneer in the mobile DI industry. *Ultrapure Water* 7(4):47–50.

Smith, B.E. 1990. The use of solar ponds in the disposal of desalting concentrate. *Desalination* 78:59–70.

Spencer, H.G., D.A. Jernigan, C.A. Brandon, and J.L. Gaddis. 1989. "Formed-in-place inorganic membranes: Properties and applications." In *Proceedings of the First International Conference on Inorganic Membranes.* Montpellier, France.

Stuart, J., T. Bryant, I. Fergus, and E. Gabrielli. 1989. Review of initial three years operation of waste water management scheme at 4640 MW Bayswater/Liddell Power Station Complex, Australia. *Desalination* 75:379.

3M. 1985. *Pollution prevention pays: An overview by the 3M Company (USA) of low and non-pollution technology.* International Environment and Development Service, a World Environment Centre (WEC) Report. New York.

Townsend, R.B., F.G. Neytzell-de Wilde, C.A. Buckley, D.W.F. Turpie, and C. Steenkamp. 1992. The use of dynamic membranes for the treatment of wool scouring and textile dyeing effluents. *Water SA* 18(2):81–86.

U.S. Environmental Protection Agency. 1985. *Manual for waste minimization opportunity assessments*. EPA 600/2-88/025, Washington, D.C.: Hazardous Waste Engineering Research Laboratory, April.

Wade, N.M. 1991. The effect of the recent energy cost increase on the relative costs from RO and distillation plant. Paper presented at 12th International Symposium on Desalination and Water Re-use, 15–18 April, Malta.

Wangnick, K. 1991. Worldwide desalting plant inventory: The development of the desalination market. Paper presented at 12th International Symposium on Desalination and Water Re-use, 15–18 April, Malta.

World Health Organization. 1984. *Guidelines for drinking water quality*. Vol. 1, *Recommendations*. Geneva.

10

Applications of Reverse Osmosis Technology in the Food Industry

S. S. Köseoḡlu

Engineering Biosciences Research Center
Texas A&M University System

G. J. Guzman

Engineering Biosciences Research Center
Texas A&M University System

INTRODUCTION

Applications of reverse osmosis (RO) in food processing are developing broadly and include processing of meat by-products, fats and oils, milk, beverages, sugar, and fruit and vegetable juices (see Table 10.1). RO technology is utilized either alone or in combination with microfiltration (MF), ultrafiltration (UF), and/or pervaporation (PV) for concentration, purification, or recovery of the valuable components from the feed streams.

The main advantage of RO technology in food processing is the reduction in costs associated with lowering the power requirement or even in the complete elimination of the energy-intensive evaporation process. Figure 10.1 compares primary energy requirements for water removal by selected processes to reach the same concentration level, and includes boiler and electrical generator conversion losses per kilogram of water removed (Kollacks and Rekers 1988). The energy consumed by RO is approximately 110 kj/kg water. On the other hand, the most efficient evaporator found in food processing, the mechanical vapor recompression (MVR), uses 700 kj/kg. Other advantages of RO include (1) the

TABLE 10.1   Some Reverse Osmosis Applications in Food Processing

Industry	Applications
Dairy	Cheese whey concentration
	Milk concentration
	Desalting of salt whey
	Waste treatment
Grain milling	BOD reduction
	Waste water recovery and reuse
	Recovery of by-products from waste water
Beverage	Cold stabilization of beer
	Removal of color from wine
	Removal of alcohol from beer and wine
Fats and oils	Waste water treatment
Sugar	Preconcentration of dilute sugar solutions
	Maple syrup
	Recovery of sugar from rinse water
Fruit and vegetables	Concentration of tomato juice
	Waste water treatment
	Concentrated juices
	Juice flavor and aroma concentration

ability to make separations without heat damage, thus retaining product quality, (2) reduced fresh water requirements owing to reutilization of waste water, (3) potential increased profit margins from the creation of new products, (4) reduced waste treatment volume and costs, and (5) relatively low floor space and capital requirements when establishment of steam-generation facilities can otherwise be avoided. Disadvantages include (1) expenses and time required to document product safety and obtain approval from the Food and Drug Administration for use of new membrane materials in food processing, (2) uncertainty about membrane durability, effective operating life, and replacement costs, (3) concerns about chemical inertness and pH sensitivities, (4) operating pressure range limitations in certain designs, and (5) fouling problems with certain feed stocks.

APPLICATIONS

The Dairy Industry

Continuous improvements of RO membranes and the development of advanced process scale equipment have resulted in extensive application of RO technology within the dairy industry (Morales et al. 1990). RO has enabled concentration of process streams from 10 wt. percent total solids (TS) to 25 percent at lower cost

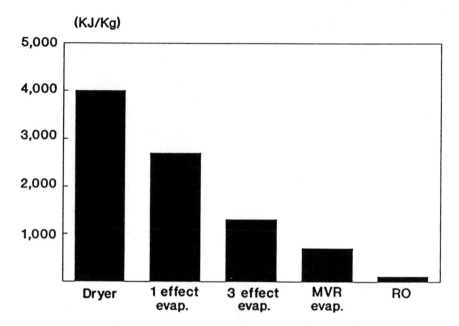

FIGURE 10.1.   Comparison of energy consumption for removing water by selected processes.   (From W.A. Kollacks and C.J.N. Rekers 1988. Reprinted by permission, Starch/(Starke)).

than that incurred by evaporation. Low-temperature processing has minimized losses of volatile flavor components and adverse changes to heat-sensitive components, i.e., protein denaturation. Moreover, RO produces reusable water and reduces discharge volumes to municipal water treatment facilities. Membranes and process conditions used in the concentration of milk and whey are summarized in Table 10.2.

Specific applications of RO in the dairy industry include cheese whey concentration, milk concentration, desalting of salt whey, and treatment of waste streams.

Cheese Whey Concentration

RO has been used in combination with other membrane processes for the concentration of whey and ultrafiltered permeates of whey to produce whey protein concentrates, lactose, and waste streams with reduced biological oxygen demand (BOD) (Spangler and Amundson 1986). A large cheese plant in California uses plate-and-frame RO modules (Dow Chemical, previously Pasilac A/S, Division of De Danske Sukkerfabrikker [DDS], Silkborg, Denmark) to concentrate the lactose present in the UF permeate from production of whey protein concentrate (WPC) (see Figure 10.2) (Morris 1986). UF permeate moves

TABLE 10.2   Major Reverse Osmosis Membranes Used in Concentration of Milk and Whey

Feed	Membrane	Pressure MPa	Temperature °C	Concentration	Flux l/m²h	Reference
Whole	PCI T2/15, CA	3.4	30	2×	4	Glover and Brooker 1974
Skim	PCI T2/15, CA	2.8	25	2×	10	Skudder et al. 1977
Whole	PCI T1/12, CA	4.8	30	2.5×	13	Abbot et al. 1979
Skim	PCI T1/12, CA	4.8	30	3×	14	Abbot et al. 1979
SW	Wafilin BV, CA	4.0	30	2×	17	De Boer and Hiddink 1980
SW	PCI T2/15W, CA	4.5	30	2×	34	Pepper and Orchard 1982
Skim	PCI T2/15W, CA	4.5	30	2×	16	Pepper and Orchard 1982
Whole	PCI T1/12, CA	4	20	2×	—	Agbevavi et al. 1983
Whole	PCI T2/15W, CA	—	30	1.25×	—	Barbano and Bynum 1984
Whole	Ultrapore CA	4.0	30	2×	16	Marshall 1985
Whole	Ultrapore NCA	4.0	50	2×	3	Marshall 1985
Whole	PCI ZF99, NCA	2.7	50	2×	15	Marshall 1985
Whole	Koch NCA	4.2	50	2×	7	Spangler and Amundson 1986
Skim	Koch NCA	4.2	50	2×	14	Spangler and Amundson 1986
SW	Koch NCA	4.2	50	2×	29	Spangler and Amundson 1986
Skim	Ultrapore NCA	3.0	50	2.6×	—	Guirguis et al. 1987

PCI = Paterson Candy International
ZF99 = noncellulosic polyamide membrane
CA = cellulose acetate
NCA = noncellulosic acetate
Whole = whole milk; Skim = skim milk; SW = sweet whey

through two parallel RO lines at 9,200 gallons per line per hour. The system concentrates lactose for alcohol fermentation at a rate of 4,500 gallons per line per hour and returns mineral permeate at 5,300 gallons per line per hour to a second RO system for demineralization. A third RO system (RO1) recovers and returns diafiltration water to the UF.

Preconcentration of whey by RO to approximately 12 wt. percent TS before vacuum evaporation has been reported as standard practice in industry (Kosikowski 1986). In Europe, the standard is in the range of 6 to 24 percent (Pepper 1990). RO-concentrated whey has been found to be a more efficient feed for UF and electrodialysis (ED). For example, RO has been used to concentrate whey to

Whey

Ultrafiltration

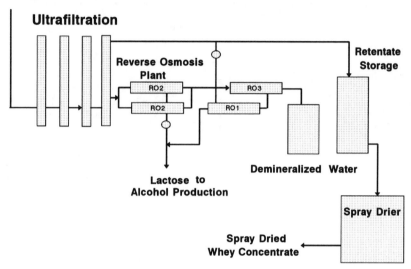

FIGURE 10.2.   Flow diagram for whey protein concentrate processing.   (From C.W. Morris 1986. Reprinted by permission, Food Engineering).

20 to 25 wt. percent TS before demineralization by ED or to 18 to 20 wt. percent TS prior to demineralization by gel filtration (Donnelly et al. 1974). Another application described by Singh (1983) preconcentrates UF permeates to 16 wt. percent TS for use as feedstock in ethanol production. Typical industrial tubular RO systems for whey concentration are operated in recirculation mode at 40°C to 45°C, (104°F to 113°F) at 29.3 to 49.3 atm (430 to 725 psi), and at linear velocities of 201.2 to 274.3 cm/s (6.6 to 9.0 ft/s). Table 10.3 shows the

TABLE 10.3   Composition of Feed, Retentate, and Permeate for Whey Concentrated (4×) by Reverse Osmosis

| | Weight % | | |
Component	Whey	Retentate	Permeate
Total solids	6.5	25.0	0.26
Lactose	4.7	18.6	0.03
Protein	0.7	2.8	—
Salts/ash	0.7	2.5	0.15
Nonprotein nitrogen	0.2	0.7	0.04

From Donnelly et al. 1974. Reprinted by permission, *J. Soc. Dairy Technol.*

effectiveness of this process during which only 0.26 percent of the total solids were permeated through the membrane and their concentration in the retentate fraction increased from 6.5 percent to 25 percent (Donnelly et al. 1974).

The literature contains various examples of this application (Pepper 1981; Pepper and Orchard 1982; Pepper and Pain 1987). Wheys derived from gouda and cottage cheeses, quark, and acid (HCl) casein, have been concentrated commercially to about 28 wt. percent TS using a multistage recycle RO system (Paterson Candy International Ltd.) without affecting the properties of the whey proteins.

Tubular and plate-and-frame type configurations are mainly used in the dairy industry (Pepper 1990). Because of high capital and operating costs of these designs, however, they are being replaced by spiral wound modules for the products requiring lower concentration ranges (<15 percent TS).

The U.S. Department of Energy estimates that energy savings in the pre-concentration of whey by RO to 12, 15, and 20 wt. percent TS before evaporation would be 3, 3.6, and 4 trillion Btu per year, respectively (Mohr et al. 1989). The estimated energy savings for concentration of whey before shipping is 2.4 trillion Btu per year, assuming that one-fourth of the whey processed in the United States is transported from cheese plants to central whey-processing facilities. RO processing of whey can also permit recovery of waste heat and enables use of RO permeate as cleaning water in cheese plants.

Milk Concentration
Concentration of milk before transport to a manufacturing facility decreases fluid volume, and reduces costs of hauling, storage, refrigeration, and processing (Amundson 1970; Donnelly et al. 1974). Most of the research on milk concentration in Europe and the United States has been on preconcentration to increase capacities of existing conventional evaporators or to increase cheese yields. Considerable research has been done in Australia on using RO to reduce milk transportation costs (Snow 1985).

Milk concentration is not a common commercial practice in the United States, owing to (1) limited development of a downstream process for making products from RO milk, (2) high capital costs of RO equipment, and (3) FDA regulations that do not permit mixing concentrated milk solids with water to produce reconstituted milk. However, its use is increasing (Mohr et al. 1989).

RO concentration of milk presents a unique problem: high temperature increases permeation and decreases residence time, but operating at high temperatures is limited by thermal resistance of the available membranes or by the need to avoid whey protein denaturation. Processors are, therefore, forced to operate at intermediate temperature ranges with a consequent bacterial proliferation, or at low temperatures with resulting low permeation and long residence times (De-Boer and Nooy 1980; Hiddink et al. 1980). Processing temperatures are 30°C and 45°C for cellulose acetate (CA) and polyacrylamide thin film composite membranes, respectively.

Heat pretreatment of milk may reduce bacterial counts and improve flux. Heating at 80°C for 20 seconds reduces initial bacterial count and restricts subsequent proliferation of surviving bacteria in whole and skim milk (Abbot et al. 1979). Heat pretreatment at 60°C for 5 to 15 minutes has increased flux about twofold, as compared with nontreated skim milk (DeBoer and Hiddink 1980).

Formations of deposits on membranes reduces flux during RO concentration of milk. Deposits are mainly proteinaceous (Glover and Brooker 1974; Skudder et al. 1977), with casein and calcium phosphate as the primary foulants. Comparisons of RO processing of pure milk protein solutions and skim milk have shown that inorganic ions increase resistance of the deposited protein layer (fouling) and thereby influence the flow rate considerably (Kulozik and Kessler 1988).

During RO processing, high pressure is required to force the permeate through the membrane. The concentrate pressure is reduced to atmospheric pressure by a pressure relief valve. Some physical properties of milk are affected by the sudden reduction of pressure during RO processing. Passage through pressure relief valves acts as a homogenizer and damages fat globules in raw milk (DeBoer and Nooy 1980). The natural enzymes in milk then degrade the milk fat in the damaged globules, causing flavor defects, foaming, and fat losses in cheese and butter manufacture. This rancidity problem can be eliminated either by heat treatment before RO (Abbot et al. 1979; Zall 1984) to inactivate lipase enzymes, by utilization of a technique that involves depressurization of the retentate storage reservoir after the milk is concentrated (DeBoer and Nooy 1980), or by cooling the concentrate below 10°C before pressure release (Versteeg 1985).

The quality of butter made from RO retentate that was not subjected to sudden pressure reductions did not change over 6 months of storage at 18°C (DeBoer and Nooy 1980). In comparison, butter made from RO retentates processed through normal pressure release valves had high rancidity.

RO retentates have been reconstituted with water to produce milks of satisfactory quality (Dixon 1985; Kosikowski 1986; Schmidt and Deeth 1987). RO retentates of whole and skim milk can be stored at 4°C with no adverse effects on quality (DeBoer and Nooy 1980).

The benefits of concentrating milk prior to manufacture of yogurt (Davis et al. 1977; Dixon 1985; Jepsen 1979), cheese (Agbevavi et al. 1983; Barbano and Bynum 1984; Barbano 1985; Jensen et al. 1987; Jepsen 1979; Sward et al. 1989), and ice cream (Bundgaard 1974) have been investigated.

Good quality cheddar cheese can be produced using RO retentates of heat-treated, standardized milk (Barbano and Bynum 1984; Barbano 1985). RO enables the use of milk of consistent composition. Improved fat recoveries and increased retention of whey solids resulted in cheese yield increments of up to 3 percent with a milk volume reduction of 20 percent. RO whole milk concentrate has been made into cheddar cheese, but results in a nonuniform granular product (Agbevavi et al. 1983). This defective texture was a result of the high lactose content in the RO concentrate. Yield increases of 5 percent, compared with the control cheese made from unconcentrated milk, have been obtained in cottage cheese production using RO retentates of skim milk (Barbano 1985).

Ice creams superior in quality to conventionally produced products, and with improved flavor release, better mouth feel, and no cooked flavor have been produced from milk concentrated by RO (Bundgaard 1974). Ice cream made from RO retentates of skim milk was equivalent in quality to standard ice cream (Donnelly et al. 1974). Dixon (1985) has used concentrated whole and defatted-milk to make liquid milk products, butter, skim milk powder, and yogurt. Higher churning losses and increased curd contents were obtained for butters made from cream separated from RO-concentrated milk than for butters prepared from whole milk. A blend of creams prepared by the two methods would result in improved yields. Yogurt prepared from skim milk concentrated by RO has been preferred over yogurt made from fresh skim milk fortified with skim milk powder.

Yogurt is reported to be produced commercially in France from skim milk concentrated at 5°C to 13 wt. percent TS by RO and standardized with 38 wt. percent-fat cream (Jepsen 1979). Good quality yogurt has also been prepared from RO concentrate, fresh milk, and fresh skim milk fortified with whey protein concentrate (Guirguis et al. 1987).

Desalting of Salt Whey
The high salt content (5 percent) of salt whey limits its use in food products. If the salt whey output of a cheese plant is smaller in volume than that of nonsalt whey, the two can be mixed and separate treatment is not required. However, salt whey must usually be discharged to municipal waste-water treatment facilities. Many cheese plants have found that treatment of effluents high in BOD before discharge are necessary to reduce surcharge costs and to avoid other penalties.

Some RO membranes, referred to as nanofiltration membranes, are now being used to remove salt from whey. These usually are thin film composite membranes (FilmTec Corporation, Minneapolis) that permit permeation of monovalent salts but retain divalent salts, sugars, and other components of higher molecular weight. One commercial operation has been reported that processes 10,000 gallons of salt whey per day (gpd) to produce 6,000 gpd of retentate and 4,000 gpd of low salt permeate, which is discharged as low-BOD waste water to municipal waste water treatment facilities (Mohr et al. 1989).

Dairy Waste Water Treatment
Increased municipal treatment costs, combined with increasingly tighter state and federal environmental standards, have had a considerable impact on the

profitability of the dairy industry, especially for the fluid milk and cheese industries. According to Mohr et al. (1989), approximately 95 percent of fluid milk processors discharge waste water into municipal systems. Processors are searching for innovative technologies to cut their costs and, consequently, increase the profitability of their operations.

A PCI tubular system was evaluated by Delbeke (1981). The system recovered 90 and 96 percent of the water with average fluxes of 33.9 and 35.6 liters/m² h, respectively. The permeates did not contain detectable concentrations of fat, protein, lactose, citric acid, or acetic acid and had very low chemical oxygen demand (COD). The process was found to be economically justified.

Another practical application of RO is in the treatment of evaporator condensate to produce high-quality water. One large cheese plant has reduced waste water discharges by 150,000 gpd by treating evaporator condensate with RO (Mohr et al. 1989).

The Michigan Milk Producers Association's milk plant in Ovid, Michigan, uses RO to process evaporator condensate produced during condensed milk manufacture. Every hour, the RO system treats 80,000 lb of evaporator condensate containing 30 ppm organic material to produce 145 gal/min (72,500 lb/hr) of filtered water with only 3 to 5 ppm organic material. Part of the filtered water is used in the plant's clean-in-place systems, and the rest is used to wash down tanks to reclaim solids or used as boiler feed water. The total project cost was $550,000, and the payback period was estimated to be 18 months. Membrane replacement cost was estimated at about $72,000 every 18 to 24 months (Electric Power Research Institute 1991).

Fruit and Vegetable Juice Processing

Orange Juice

Concentration of orange juice by evaporation results in the loss of various volatile alcohols, esters, and aldehydes and causes a significant deterioration in quality, overall aroma, and flavor of the concentrated juice (Brent 1964). In addition, chemical alteration of the aroma and flavor compounds owing to lipid oxidation and Browning reaction causes off-flavors in the orange juice concentrates (Strobel 1983).

Application of RO in orange juice processing was reported by Hedges and Pepper (1986). A two-stage RO plant equipped with PCI ZF-99 membranes increased the capacity of an existing plant by preconcentrating the juice to 14 to 18° Brix before concentration to 40 to 65° Brix in an evaporator. This increased the capacity of the evaporator and reduced the steam usage by 40 to 60 percent. A 20° Brix RO product, rediluted to the feed juice strength of 11 to 12° Brix, retained all the natural juice characteristics of taste and flavor.

The concentration of fresh-squeezed orange juice from 11.75 to 23° Brix with a SEPA-97 (CA) RO membrane was reported by Paulson et al. (1985). They showed that neither sugars nor flavors were detected in the permeate stream. The flux decreased from 13 to <3 l/m^2/hr at ° Brix values from 12 to 23, respectively. Medina and Garcia (1988) also concentrated orange juice up to 50 percent hydraulic recovery with a SEPA polyamide membrane. Permeate flux and solute recovery were reported at various transmembrane pressures.

Watanabe et al. (1978) showed that fouling materials on RO membranes during mandarin orange juice concentration consisted mainly of pectins and other insoluble materials similar to cellulose. Experiments with model solutions of pectin and cellulose showed that a large decline in flux was due to the presence of high molecular weight pectins in the mixture (Watanabe et al. 1979). Intermittent lateral surface flushing restored the flux of the RO membranes to more than 90 percent of its original value. This method is useful for keeping the total flux high; however, it promotes permanent fouling (Watanabe 1979).

A study by Fukutani and Ogawa (1983a) evaluated Satsuma mandarin juice concentration by a plate-and-frame RO unit fitted with different membranes. The efficient concentration of Satsuma mandarin juice was limited to 30° Brix. The high pulp content of the feed juice resulted in lower permeation rates, and clarified juice was found to be the best feed. The rejection rates for acid, potassium, and vitamin C were sightly lower than for sugar. Monoterpene and sesquiterpene hydrocarbon volatiles were retained by all of the membranes examined. However, monoterpene alcohols, aliphatic alcohols, and aliphatic aldehydes permeated partially through a CA membrane but were retained by other (composite and polyacrylonitrile) membranes. A follow-up study (Fukutani and Ogawa, 1983b) evaluated concentration of natsudaidai, apple, and grape juices and showed that sugar composition had a significant influence on osmotic pressure.

Matsuura et al. (1973, 1974) discussed the relevant physicochemical criteria for RO separations and process design, and presented data on osmotic pressures of commercially available lime, lemon, prune, carrot, and tomato juice solutions. Their experimental data show that low-temperature RO treatment of membrane-permeated apple juice water results in significant recovery of aroma compounds.

Limonene and terpenes from citrus-processing waste streams, such as cold-pressed oil, centrifuge effluent, and molasses evaporators, have been concentrated by a combination of UF and RO (Braddock 1982). Limonene rejection rates were 78 to 97 percent and 87 to 99 percent for UF and RO, respectively. Limonene adversely affected the permeation rates of all membranes. This effect was minimal with Teflon-type membranes, but polysulfone membranes were affected more than CA and Teflon-type membranes. In a separate study, dilute aqueous essences from commercial juice evaporators were concentrated by polyamide hollow fine fiber membranes (Braddock 1991). The rejection rates for ethanol and acetaldehyde ranged from 90 to 25 percent. Rejection was greater than 85 percent for aroma molecules.

An upper attainable level of 20° Brix has limited the use of RO in the production of orange juice concentrates. However, newly available membranes (Cross 1988, 1989) have allowed concentration of clarified juices up to 60° Brix without drastic increases in applied pressure. Tests have shown that product quality is comparable to that obtained by freeze concentration at costs similar to those of thermal evaporation. According to the data, RO concentration up to 30° Brix is less costly than three-stage thermal evaporation.

Concentration of orange juice was studied by Merlo et al. (1985) using reconstituted orange juice (10 percent TS). The RO system utilized PCI tubular ZF-99 membranes. The concentration scan from 2 to 20 percent TS at 40.8 atm reduced the flux by half.

Koseoglu et al. (1990a) have described a process (see Figure 10.3) to produce sterile and concentrated citrus juices with improved flavor and reduced acid, using industrial membranes. The juice was fractionated into three streams: (1) pulp, (2) a heat-sensitive solution containing small molecule compounds including flavors, acids, and sugars, and (3) a heat-insensitive solution and dispersion containing the larger color, protein, and other molecules, and microbes. The heat-sensitive fraction was cold concentrated by RO without experiencing a sig-

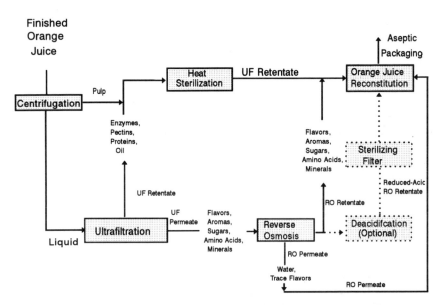

FIGURE 10.3.   Simplified flow diagram for an improved orange juice RO process.   (From S.S. Koseoglu et al. 1990a. Reprinted by permission, *Food Technol.*)

nificant loss of volatile compounds. The heat-insensitive fraction may be cold- or hot-concentrated without the loss of flavor compounds or formation of sugar-protein browning products. Finally, the pulp and concentrated heat-insensitive solution were heat sterilized using scraped surface heat exchange, and aseptically combined and filled with the cold-concentrated, heat-sensitive fraction to produce a concentrated and/or sterile juice. Reduced-acid orange juice and grapefruit juice were prepared by passing a portion of the RO concentrate through a single-stage ion exchange column.

Other Fruits and Fruit Juices
Other applications of RO include concentration of fruit-cocktail dicer rinse water, hot waste stream from a peach-puree screw cooker, container wash water, and evaporator condensate (Merlo et al. 1985); concentration of betalaines (Lee et al. 1982); preconcentration of apple and pear juices (Sheu and Wiley 1983; Chua et al. 1987); and processing "second press" apple juice (Paulson et al. 1985).

The application of UF and RO to the production of high quality concentrate fruit purees was studied (Gherardi et al. 1989a). The pilot plant used in pear and apricot puree trials is described. UF serum, bright in color and with a flavor typical of the original fruit, was concentrated by RO and produced a 25 to 27° Brix clear liquid. For both purees, the color, flavor, and aroma qualities of the final and initial products were similar. In pear puree, UF high molecular weight esters remained bound to the pulpy fraction. In the case of peach puree, eight lactones aroma components were identified in liquid-liquid extraction (Gherardi et al. 1989b).

RO preconcentration of apple juice has been studied by both membrane manufacturers and apple juice processors (Mohr et al. 1989). Single-strength apple juices (10° Brix) were processed by RO to 20 to 25° Brix using RO systems equipped with CA membranes with recoveries of solutes and apple volatiles ranging from very low to 97 and 87 percent, respectively. The flow rate range was 15 to 26.9 l/m^2 hour. A comparative study (Moresi 1988) for combined RO and falling-film evaporation (FFE) indicated that processing costs for concentration of apple juices were greatly reduced by introducing an RO preconcentration step. The overall operating costs of the concentration unit, however, were still more sensitive to live-steam–specific cost than to either membrane-specific cost or progressive reduction of mean permeate flux after a series of cleaning operations.

Cloudy pineapple juice was concentrated from 130 to 250 g/kg soluble solids at 6,000 kPa, 40°C, and a flow rate of 3 m/s, with good quality retention (Bowden and Isaacs 1989). Pilot-scale tubular and plate-and-frame RO units were equipped with ZF99 and HR98/HR95 membranes, respectively. Permeate flux averaged 20 l/m^{2}h, and losses of soluble components into the permeate were very slight. The flavor of the reconstituted juice was comparable to single-strength juice.

Vegetable Juices

Characteristics of vegetable juice processing differ from those of fruit juice processing. For example, tomato processors are far less concerned with the heat damage that occurs during final evaporation. In fact, a slightly cooked flavor in tomato juice is preferred by consumers (Merson et al. 1980). Energy consumption is the second highest cost after raw materials, and has stimulated the search for alternative energy-efficient processing methods. Dale et al. (1982) compared energy consumption of MVR, RO, and centrifugation processing with that of standard triple-effect scraped-surface evaporators. The lowest cost system was found to be one that used a combination of centrifugation, RO, and MVR.

The literature on vegetable juice processing is not as extensive as that for fruit juice processing. Several groups in Japan, Italy, and the United States, however, have reported findings on the concentration of tomato juice and paste by RO (Ishii et al. 1981; Dale et al. 1982; Merlo et al. 1986a, b; Hedges and Pepper 1986; Watanabe et al. 1982; Gherrardi et al. 1986).

Merlo et al. (1986b) processed tomato juice (4 to 6 percent natural tomato soluble solids [NTSS]) using a two-stage tubular RO pilot-plant unit (see Figure 10.4) equipped with 15.6 m^2 of aromatic polyamide thin film composite membrane (PCI ZF 99) to study the effects of temperature, pressure, and flow rate on NTSS, sugars, and organic acids. Fluxes at Stages 1 and 2 were in the range of 48–34 l/m^2hr and 30–20 l/m^2 hr respectively, and concentrations of 7.4 to 9.0 percent NTSS at 8.6 l/min and 6.3 to 6.4 percent NTSS at 18.6 l/min were achieved. All of the sugar and most of the organic acids stayed in the retentate. In another study, Merlo et al. (1986a) examined NTSS rejection during long-term high temperature testing. One membrane had a flux rate of 39.7 l/m^2hr at 72°C and 55.2 bar after 52 hours of operation. Another membrane had a flux of 41 l/m^2 hr at 78°C and 41.4 bar after 717 hours. NTSS rejection rates ranged from 94 to 98 percent. Maintaining acceptable flux and rejection at high temperatures for long periods of operation, in order to preconcentrate tomato juices prior to evaporation economically, was feasible.

Gherardi et al. (1986) reported results of 2-year trials of preconcentrating tomato juice from 5 to 8.5° Brix prior to manufacture of tomato paste. RO concentration of tomato juice was done at 60°C to 70°C and 35 to 45 bar using a PCI ZF 99 tubular membrane. Analyses of tomato juice permeate indicated very low losses of soluble solids. Membrane selectivity for components of tomato juice was high. RO produced concentrates with a lower loss of color, greater retention of characteristic flavor and aroma, and higher viscosity than other conventional methods of processing. The microbial load of the product was reduced, and no contamination was observed. The treated tomato juice could either be packed directly or concentrated to 28 to 30° Brix.

Ishii et al. (1981) evaluated applicability of membrane processing to produce fresh-flavor tomato juice concentrate using semicommercial equipment and CA

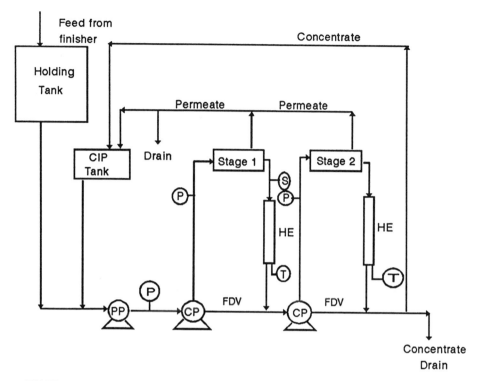

FIGURE 10.4.   Flow diagram of two stage tubular RO pilot plant unit used for tomato juice.   (From C.A. Merlo et al. 1986b. Reprinted by permission, *J. Food Sci.*).

membranes (Daicel, Tokyo, Japan) at 25°C. A sterilized juice (4.5 percent Refractive Index [RI]), cooled to 30°C, was fed to a membrane system containing 20 units, each equipped with 72 tubes. Approximately 97 percent of the original sugars, 92 percent of acids, 96 percent of amino acids, and 62 percent of the vitamin C were retained. This unit produced 20 percent RI condensed fresh tomato juice, superior in quality to the conventional product, with lighter and more brilliant red color and better taste.

Watanabe et al. (1982) reported on concentration of tomato juice with a dual dynamic RO membrane (Zr[IV]-PAA) of hydrous zirconium oxide and polyacrylic acid on a porous ceramic tube. After concentration of the tomato juice for 8 hours, the membrane was regenerated with good reproducibility by washing successively for 30 minutes in water, 0.1 M NaOH, and water. Operating conditions of 40°C to 45°C, 70 kg/cm², and flow rate 5 l/min were recommended for the membrane.

Koseoglu et al. (1991) described a process for the production of concentrated tomato juice (see Figure 10.5) and flavor concentrates from other vegetables such as celery, carrots, and cucumbers. Screened tomatoes were processed by a hollow fiber UF membrane system and permeate (clarified juice), and possessed the delicate flavor of fresh tomatoes because the product had received no heat treatment. Clarified juice was concentrated using a PCI RO system at a flow rate of 295.7 l/m²/day. The tubular system was equipped with 2.6 m² of PCI's ZF-99 noncellulosic membrane.

The Beverage Industry

The beverage industry includes the production of alcoholic beverages and carbonated soft drinks. Most of the membrane technology applications are in the production of beer and wine. Table 10.4 wt. summarizes some of the membranes and process conditions used to produce beverages of low and high alcohol content.

Low-Alcohol Beers

Most regular beers sold in the United States have an alcohol content between 3.1 and 4.0 wt. percent. The objective of low-alcohol beer research is to produce

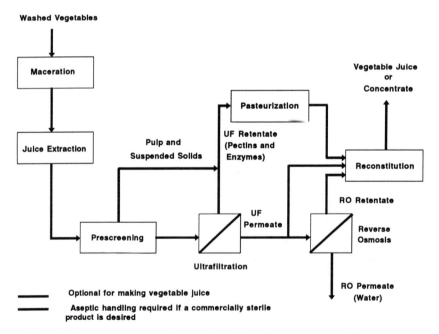

FIGURE 10.5.   Simplified flow diagram of an alternative process to produce concentrated tomato juice.   (From S.S. Koseoglu et al. 1991. Reprinted by permission, *Food Technol.*).

TABLE 10.4   Some Reverse Osmosis Membranes Used to Modify Ethanol Content of Alcoholic Beverages

Feed	Membrane	Pressure MPa	Temperature °C	%EtOH Change	Reference
Beer	PCI ZF99 NCA	4.14	2–19	–30%	Goldstein et al. 1986
Beer	CAMP NCA	2.41	9–11	–50%	Light 1986
Wine	DDS ALCO 95 NCA	4.50	26	–43.3%	Bui et al. 1986
Wine	IRCHA CPMA NCA	8.00	26	+15%	Bui et al. 1986
Wine	DDS CPMA NCA	6.00	26	+14.9%	Bui et al. 1988

PCI = Paterson Candy International
ZF99 = noncellulosic polyamide membrane
CAMP = crosslinked amine-modified polyepichlorohydrin
DDS = De Danske Sukker Fabrikker
ALCO 95 = composite membrane with polysulfone base
NCA = noncellulosic acetate
CPMA = crosslinked polyhydroxy methyl acrylamide

two full-bodied products with typical beer flavor: a low-alcohol beer with an alcohol content of 2.0 wt. percent, and a nonalcoholic beer with alcohol content below 0.5 wt. percent.

There are many ways to produce low-alcohol beer. The most common method is to alter the brewing process so that less alcohol is produced. Other methods that remove the ethanol from beer include dialysis, distillation, and RO. Beer produced by dialysis tends to have a very light body because a high percentage of the beer's small molecular components pass through the membrane. Distillation changes the taste and is very expensive (Mohr et al. 1989).

Commercially available spiral wound FilmTec RO membranes can effectively reduce the ethanol content of beer without loss of desirable aroma (Darnes and Jordain 1987).

Another RO process for lowering the alcohol content of beer employs a semipermeable membrane composed of a thin film of nonporous polymer (amine-modified polyepichlorohydrin cross-linked with toluene diisocyanate) on a porous polysulfone-backed fabric. The system is operated at 5°C to 20°C and 3.4 to 17 atm (Light 1986). A minor portion (1 to 20 percent) of the retentate is recovered as product, and the major portion of the retentate is recycled back to the RO system to admix with fresh beverage and added water. The percentage of the system feed that is recovered as permeate ranges from 1 to 10 percent per pass through the system. The alcohol content of the beer is reduced from 4 to 2 percent.

Beers containing 1.25 and 1.85 percent ethanol have been prepared from a stock beer containing 3.7 percent ethanol (Goldstein et al. 1986). The membrane had a molecular weight cut-off of less than about 100 and permitted about 25 to

30 percent alcohol in the beverage to pass through. The temperature of operation was between 2°C and 19°C, and the recommended pressure was less than 61.2 atm (Goldstein et al. 1986). Table 10.5 shows compositions of the stock beer and the filtered beer. Almost all of the volatiles were rejected by the thin film composite PCI ZF-99 membrane, whereas only 16 percent of the volatiles were rejected by the CA membrane. The ethanol rejection rate of the composite membrane was 73 percent, whereas there was no change in ethanol content of the feed and permeate with use of a CA membrane.

Concentration of Beer
Fricker (1984) devised a process to concentrate lager-type beer from 3.1 to 11.8 percent dissolved solids using PCI ZF-99 and T1/12W membranes. A portion of the permeate from RO was distilled to produce distillate containing about 90 percent ethanol. The final beer concentrate formulated from distillate and RO retentate contained 11.8 percent dissolved solids and 15.5 percent alcohol. There was no significant difference in taste and aroma between the reconstituted beer produced by this process and the original beer.

Wine Processing
Applications in the wine-processing industry, especially in Europe, are influenced by the European Economic Committee regulations that define the wine-making process (Pepper 1990.) The current technological advances in this area are reviewed briefly in the following paragraphs.

Bui et al. (1986) described a process for the production of wines of low and high alcohol content (see Figure 10.6). An alcohol-reducing RO unit equipped

TABLE 10.5   Composition of Feed and Permeate of Beer Treated by Reverse Osmosis

Composition	Cellulose Acetate		Thin Film Composite	
	Feed	Permeate	Feed	Permeate
Acetaldehyde	1.1	0.9	1.2	0.3
n-Propanol	16.2	14.5	11.7	0.0
Ethyl acetate	41.0	40.4	26.0	1.0
Isobutanol	25.3	19.4	14.4	0.0
Amy alcohol	23.0	17.8	20.3	0.0
Isoamyl alcohol	82.9	66.2	42.8	0.4
Isoamyl acetate	4.8	4.3	2.7	0.0
Diacetyl	0.02	0.01	0.02	0.0
2,3-Pentanedione	0.02	0.01	0.02	0.0
Ethanol	5.3	5.4	3.7	1.0

From Goldstein et al. 1986.

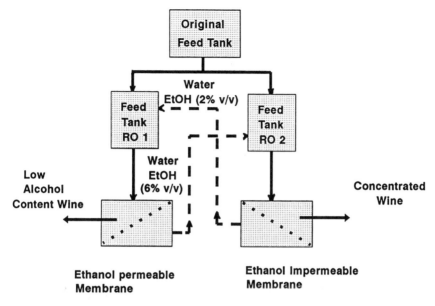

FIGURE 10.6.   Flow diagram of low alcohol content wine made by RO processing. (From K. Bui et al. 1986. Reprinted by permission, *Am. J. Anol. Vitic.*).

with a low-ethanol rejection DDS membrane (ALCO 95, composite polysulfone base, or CAP 995, CA) produced low-alcohol wine (7.1 percent ethanol) from the original 10 percent alcohol feed. A second system, equipped with an ethanol-impermeable membrane made of cross-linked polypropylene or polyester, produced wine with 11.5 percent ethanol.

Grape musts (juices) deficient in soluble solids, particularly sugars, can be ameliorated by the addition of concentrated grape juice or sucrose (chapitalization). RO has alternatively been used to concentrate the must to an optimum sugar concentration before the cold fermentation process (Palmer-Benson 1986). Grape juice concentrated by RO before fermentation yielded a wine superior in quality to wines made from musts with added sugars, or concentrated by evaporation or freezing. Semipilot RO trials enabled the adjustment of sugar concentration in the raw fruit juice up to 18–39 percent range without adding any sugars. The system improved the quality of the wine produced from white *Vitis Vinifera* varieties such as Riesling and Chardonnay. Another important benefit of RO is that the aging time can be reduced from an average of 3 years to about 3 months (Sakaguchi 1976). A pilot-plant RO process using a Separem 400 T polyamide membrane for preparation of low-suger musts suitable for making low-alcohol wines is described by Ambra (1988). The membrane excluded approximately 51 percent of the sugars, and the naturally enriched concentrate was suitable for normal wine making. Organoleptic tests showed that the most acceptable wines were those without added aromas.

In a recent study, Duitschaever et al. (1991) compared the compositions of grape (Riesling variety) must concentrated by RO to three different levels of soluble solids and the resulting wines against must-wine ameliorated by addition of sugar. RO treatment of the must increased the soluble solids, including the acids. pH was not affected. Volatile acidity increased linearly over the concentration series for RO, but remained constant for the sugar-addition series. RO was found to be an excellent alternative method for amelioration of grape must within limits of concentration (26 to 28° Brix).

The possibility of concentrating wines by RO has also been investigated. French red wines have been concentrated with an RO system using membranes with an ethanol exclusion rate of 85 percent. Total ethanol loss was 2 percent when the ethanol content in the wines was increased 1 percent (Bui et al. 1988). The permeate of one concentrated wine contained only traces of wine constituents, except for methanol, which could cross the membrane freely. RO is being investigated as an alternative method to induce tartrate precipitation before packaging. The wine is concentrated by removing ethanol and water until the desired amount of tartrate crystallizes. The wine is then filtered and remixed with the RO permeate to produce a wine with reduced tartrate concentration (Amundson 1986).

A method to produce low alcohol (<0.5 percent) and nonalcoholic (0.01 percent) wines using Millipore (Millipore Corporation, Bedford, Massachusetts) polyamide membranes has been described by Gnekow (1989).

Soft Drink Industry

RO is used in the soft drink industry mainly to purify the feed water (Kieninger and Reicheneder 1977; McFiggans 1984; Cross, J. 1988). Soft white flocs in clear lemonade and other soft drinks can be prevented by RO-filtering the water supply (McFiggans 1984). Analysis showed that the flocs were nontoxic carbohydrate-protein aggregates with traces of iron and aluminum. The flocs were caused by polypeptides or polysaccharides secreted by an organism of the Microcystis family from the water reservoir.

Cereal Processing

Treatment of Thin Stillage and Recovery of By-Products

Wu and Sexson (1985) and Wu (1988) recovered corn proteins from stillage solubles of dry milled fractions of corn, grits, flour, degerminated meal, and hominy feeds (see Figure 10.7). The solubles fraction was membrane processed using UF and RO (MWCO 200) membranes. After production of alcohol, the stillage was separated into soluble and insoluble fractions by filtration and

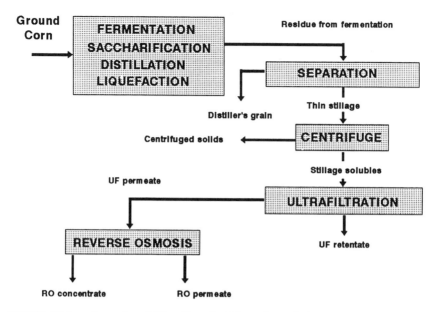

FIGURE 10.7.   Processing of thin stillage by RO.   (From Y.V. Wu 1988. Reprinted by permission, *Cereal Chem.*).

centrifugation. The soluble fraction contained 0.036 to 0.08 percent nitrogen and 1.4 to 7.2 percent total solids. The recoveries of permeate, nitrogen, and solids are summarized in Table 10.6. The RO permeate contained 70 to 86 percent of the original twin stillage volume, 0.1 to 30 percent of the total solids, and 0.3 to 18 percent of the total nitrogen. The combination of UF and RO processes has potential advantages for fractionating proteins with distinct functional properties. Calculations based on the work of Gregor and Jeffries (1979) showed that UF and RO processing of stillage solubles is 67 percent less expensive than evaporation processes. This process produced food-grade by-products and reduced the volume of waste streams.

Corn Whey Treatment
The wet extraction of protein from whole corn or ground degermed endosperm by the UF process is described by Lawhon (1986). Germ separated by dry milling can also be solvent defatted, and the meal can be converted into protein concentrate and isolate by the same process.

In this process, the permeate from the UF process—or the whey if the acid precipitation process is used—is then processed by RO to concentrate the sugars, salts, and other constituents. This fraction can be mixed with nonprotein residue from the extraction and used in production of alcohol, or it can be mixed with

TABLE 10.6   Volumes and Composition of Thin Stillage from Different Sources Processed by RO

Stillage Solubles Membrane (pressure)	Permeate Volume %	Nitrogen %	Solids %	Ash %
Corn				
RO (5,440 k Pa)	80	0.3	0.2	0.2
RO (1,360 k Pa)[1]	71	18.0	30.0	37.0
Corn grits				
RO (5,440 k Pa)	86	0.6	0.3	0.1
RO (1,360 k Pa)[2]	85	12.0	9.2	5.8
Corn flour				
RO (5,440 k Pa)	77	1.3	0.1	0.0
RO (1,360 k Pa)[2]	85	12.0	9.2	5.8
Degerminator meal				
RO (5,440 k Pa)	83	0.3	0.4	0.3
RO (1,360 k Pa)[2]	85	12.0	9.2	5.8
Hominy feed				
RO (5,440 k Pa)	80	0.5	0.3	0.2
RO (1,360 k Pa)[2]	70	4.6	4.5	3.4

[1]Wu et al. 1983
[2]Wu and Sexson 1985

other food ingredients. The purified water from RO (permeate) can be reused in subsequent extractions. This reduces water usage and treatment costs considerably.

Corn Wet-Milling Operations

The simplicity, energy efficiency, and versatility of adaptation to existing plants made RO technology an attractive candidate for removing the major load on evaporators in wet corn milling operations.

A simplified schematic of a corn wet-milling process with and without membrane systems is given in Figure 10.8. Fresh water enters the system in the starch-washing step and leaves as a dilute suspension containing all proteins and some starch. This stream, called "light middlings," then flows countercurrent to the corn fractions toward the front of the process. Finally, the used waste leaves the system as light steep water. The process evaluated by Kollacks and Rekers (1988) used RO technology to reduce fresh water consumption, increased the number of water washing steps, and decreased the load on the steep water evaporator. A single-stage recirculation Stork-Wafilin RO unit equipped with

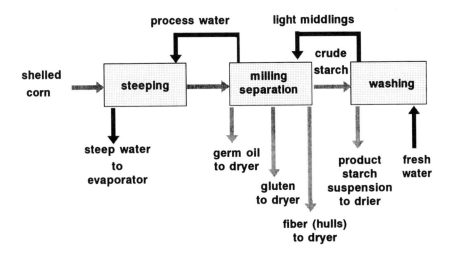

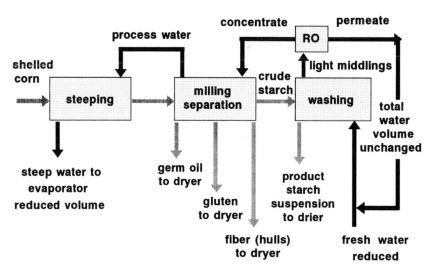

FIGURE 10.8.   Use of RO in wet corn milling processing.   (From W.A. Kollacks and
C.J.N. Rekers 1988. Reprinted by permission, *Starch/(Starke)*).

399 m² of CA tubular membrane type WFR 0950 was installed in a 300 ton/day grind wet corn milling plant. The following advantages were claimed: (1) reduced fresh water consumption, (2) reduced waste water load from reduced steep water condensates and reduced cooling-water demand at the evaporator, (3) reduced electrical power requirements at the evaporator, (4) improved starch quality, (5) increased washing ratio, at the same net fresh water intake, (6) capacity expansion without evaporator and boiler expansion, (7) less additional power needed as compared with conventional hydrocyclone wash station expansion, and (8) less additional power needed as compared with MVR evaporator expansion. An additional advantage of RO is that it can be installed in an existing plant without an operation shutdown. Membrane life expectancy is about 1 year.

RO-Evaporation of Steep Water
The feasibility of using membranes to preconcentrate steep water before evaporation was evaluated by Ray (1987). Cost of the separation process was reduced by preconcentration of the steep water. The energy requirements for the RO-evaporator process were considerably lower than for the conventional evaporator. The calculation was based on processing 3.6×10^5 gal/day steep water and recovering 57 percent of the water with the use of RO system. The electrical and thermal energy requirements for the RO-evaporator process were 500 kw and 40 Btu, respectively. On the other hand, a conventional evaporator requires 1,800 kw of electrical energy and 100 Btu thermal energy to process the same volume of steep water.

Polishing of RO Permeate and Evaporator Overhead
This is applicable to RO permeates and evaporator overheads that have high COD concentrations. In the feasibility study by Ray (1987), the RO permeate and the evaporator overhead had COD contents of 5,000 and 1,500 ppm COD, respectively. In tests done with feeds containing 2,000 to 2,500 ppm COD using spiral wound membranes, 98 percent of the water was recovered, which contained 100 to 300 ppm COD that could be recycled and used as fresh water. Comparison of carbon adsorption and sewage treatment costs showed that commercial feasibility at the present time is dependent on sewage treatment costs.

Concentration of Dilute Sweet Waters
In dilute purification of high-fructose corn syrup by chromatographic separations, large amounts of dilute sugar solutions, called "sweet water," are obtained in the elution of the fructose from the column. Normally, the dilute sugar solution containing 5 percent sugar is concentrated up to 30 percent sugar by evaporation. Osmotic pressures of solutions containing 20 percent solids limit the use of conventional RO membranes in this area. However, newly available membranes (Cross 1989) can allow concentration of dilute sugar solutions up to 60° Brix without drastic increases in applied pressure.

Vegetable Proteins

Production of low-, intermediate-, and full-fat protein products from full-fat soybean or peanut flours, using UF and RO, has been described (Rhee and Koseoglu 1991). In this process, defatted flours are first extracted with warm water at alkaline conditions. After solid-liquid phase separation, the aqueous extract containing dissolved proteins is processed by UF to separate the smaller protein molecules that are tightly bound to the flavor and color components. The permeate is treated by RO to recover these smaller molecular weight proteins and soluble carbohydrates as by-products, and the water is returned to the front end of the process. In this study, RO effluents were lower in total solids than area tap water, indicating suitability for reuse without further treatment. The spray-dried RO product essentially consisted of sugar and ash, and contained only 11.5 percent protein. Inclusion of RO in protein-recovery processes improves the overall economics of the process by reducing waste treatment costs, reutilizing the extraction water, and recovering by-products that can be used as food ingredients.

Oils and Fats Processing

Waste Water Treatment
Waste waters from vegetable oil extraction plants and refineries are sent to municipal treatment systems or are treated on-site and discharged directly to rivers or streams. On-site treatment typically includes pH adjustment to an acidic state, gravity separation, pH readjustment to near neutrality, chemical complexing or coagulation, biological treatment, and, in some cases, filtration through granular media. Until recently many plants have found that despite high costs, discharge to municipal systems is more economical than on-site treatment. However, increasingly tight state and federal environmental standards are being imposed on industrial discharges to municipal treatment facilities. As regulations become more stringent and municipal treatment costs increase, processors are looking for alternative technologies for on-site waste water treatment.

The most common application of membranes follows a conventional waste treatment plant process, including solids screen, settling tank, and fat skimmer. It consists of a UF and RO membrane system that separates oils from the waste streams. Several membrane companies now offer systems for handling oily waste streams.

Chemical characteristics of a typical waste water from an edible oils refinery are as follows: 500 to 2,000 ppm BOD; 1,500 to 10,000 ppm COD; 400 to 1,600 ppm TSS; 1,100 to 2,900 ppm TDS; 200 to 700 ppm oil and grease; pH 4 to 6; 1,000 to 2,500 ppm total phosphorous as PO_4; 1 to 6 ppm iron; 1 to 6 ppm silicon as SiO_2; and 0.01 to 0.25 ppm manganese.

The waste treatment problem is more pronounced in olive oil processing, as compared with other common oils such as soybean, cottonseed, and peanut. The amount of vegetation water accumulated for each ton of oil produced can range from 2 to 5 cubic meters, depending on the type of continuous or discontinuous process used. The vegetation water is discharged after sedimentation and recovery of superficial oil (Massignan et al. 1989) and contains high amounts of organic and inorganic impurities. COD and BOD contents of waste water are 30 to 200 and 20 to 60 kg/m^3, respectively. Treatment methods based on RO technology are suitable for concentrating the organic and inorganic substances and enabling recovery of some valuable components. Technical difficulties associated with this application include the presence of gelling substances, such as pectins, that foul the membranes. Pretreatment with calcium chloride (1,435 ppm) or natural aging for 4 months was found to be an effective method to produce good quality feed for the RO system. The rejection rate of chemically treated waste water was 98 percent, as compared with the 80 percent rate of naturally aged waste water. A semipilot continuous DDS (De Danske Sukker-fabrrikker) module with DDS-HR$_2$95 membrane gave a flow rate of 380 to 390 l/m^2/day. A new type of inorganic tangential microfiltration membrane combined with RO membranes has been successfully applied to the treatment of "vegetable water" and vegetable oil refining waste (Serafin 1988). The overall process achieves 100 percent removal of suspended solids and colloids. In combination with RO, it archieves 95 percent reduction of COD, 99.9 percent reduction of fat, 99.5 percent reduction of dry residues, and complete elimination of turbidity and color. COD is reduced to 200 to 300 ppm.

Commercial applications of Koch polymeric membranes in this area include coolant streams treatment in the metals industry, wash water from mayonnaise and oleo packaging, and industrial laundries. A typical commercial system can process 25,000 to 100,000 gal/day with a membrane life of 3 to 5 years, depending on waste stream impurities, pretreatment systems, and RO process conditions.

Miscella Distillation
The recovery of solvents used in the extraction step of edible oil processing is required for economic, environmental, and safety considerations. The miscella (mixture of extracted oil and solvent) exits the extractor at 70 to 75 wt. percent solvent content, and currently the solvent is recovered by distillation.

Koseoglu et al. (1990 b, c) have evaluated the ability of various RO membranes to separate vegetable oils from commercial extraction solvents, such as mixtures of hexane, ethanol, or isopropanol and 25 wt. percent cottonseed oil. Industrial scale membranes procured from Osmonics and Paterson Candy International (PCI) were evaluated (see Table 10.7).

Osmonics's Sepa O (PA) membrane did not deteriorate in hexane. The flux achieved in the larger unit, 7.3 l/m^2/hr, was lower than the laboratory unit, 53.4

TABLE 10.7   Characteristics of Membranes Evaluated for Removal of Solvent from Edible Oil Miscellas

Company	Membrane	Estimated MWCO	Polymer
Romicon, Inc.	PM 1	1000	Polysulfone
Osmonics, Inc.			
	OSMO Sepa 0 (PA)	500–1000	Polyamide
	OSMO Sepa S-20 (VF)	20,000	Fluorinated polymer
	OSMO 192T-89 (PA)	300–400	Polyamide
	OSMO 192T-0 (PA)	500	Polyamide
	OSMO 192T-97 (CA)	150–200	Cellulose acetate
Paterson Candy International Ltd.	PCI ZF-99	<400	Composite
	PCI RO (CA)	<400	Cellulose acetate

From Koseoglu and Engelgau 1990b. Reprinted by permission, *J. Am. Oil Chem. Soc.*

$l/m^2/hr$. One reason for this was that the Sepa membranes were housed in PVC tubes that withstand pressures of only 17 atm in pilot plant tests, whereas the laboratory test cell was subjected to 54.4 atm. A second cause of lower flux from the larger unit may have been the higher concentration of oil in the ethanol-oil miscella fed into the system. When processing a hexane-oil mixture, the larger Sepa O membrane unit also passed less oil in the permeate than the laboratory-size membrane.

The PCI ZF-99 membrane separated ethanol well, while concentrating oil in the feedstock miscella from 22.38 to 63.86 percent. Average oil content in the permeate was only 0.082 percent. The flux varied from 3.2 to 48.4 $l/m^2/hr$ and back down to 3.8 $l/m^2/hr$ as temperature increased from 82°F to 144°F while oil concentration in the feed rose to 63.86 percent. Rising temperatures decrease viscosity and raise the oil solution capacity of ethanol. The ratio of soluble oil to undissolved oil in the feed appeared to affect the flux. In a second test processing ethanol-oil miscella, the pressure was varied from 27.2 to 54.4 atm while the temperature remained at 93°F or below. Under these conditions, flux rose from 3.75 to 16.3 $l/m^2/hr$; again, the feed contained a large percentage of nonsolubilized oil at 93°F.

The PCI ZF-99 membrane permeated hexane from a hexane-oil miscella at a very low rate, and the low percentage of oil in the permeate indicated that a membrane of the same composition but with larger pores might be satisfactory. However, this membrane began to show signs of failure when allowed to sit in contact with alcohol for several days. It was not determined whether failure was due to deterioration of the membrane or of the support materials (i.e., adhesives, seals, etc.).

The PCI CA RO membrane showed good oil retention in a series of tests with ethanol-oil mixtures. In the initial test, system pressure was held constant at 54.4 atm with the temperature rising only slightly, and oil content of the retentate increased from 25.08 to 64.31 percent. A mean flux of only 11.7 l/m²/hr was obtained; however, the percentage of oil in the composite permeate was desirably low at 0.15 percent. In a second test of the same system with an ethanol-oil mixture, the temperature was held constant and the pressure was varied from 13.6 atm to 54.4 atm. Increased pressure caused an increase in flux from 11.1 l/m²/hr to 31.4 l/m²/hr at the original oil concentration of 25.08 percent.

Another potentially useful membrane is Allied Signal's polysulfone membrane that was developed for separation of hexane, pentane, or heptane from hydrocarbons obtained in the deasphalting process. In one study (Li et al. 1984, 1985), a spiral wound element, aided by the formation of a dynamic gel layer, recovered pentane from the hydrocarbon mixture with fluxes of 17 to 25.5 l/m²/hr. Allied Signal has suggested that this concept could be extended to food processing. Further on-site testing is needed to verify that it is applicable to edible oil processing. However, a conceptual design of a combined membrane-distillation process proposed by Allied Signal for solvent recovery is shown in Figure 10.9, which could be applicable to miscella recovery in edible oil processing. The solvent-rich stream that exits the membrane unit is recycled to the extractor, and the oil-rich stream is processed by distillation to recover the remaining solvent. Energy savings of 2 trillion Btu per year is estimated for the

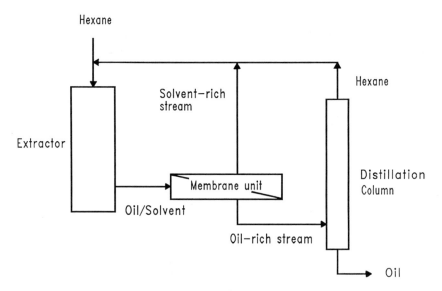

FIGURE 10.9.   Membrane solvent recovery process for edible oil miscellas.   (From S.S. Koseoglu and D.E. Engelgau 1990. Reprinted by permission, *J. Am. Oil Chem. Soc.*).

combined-membrane–distillation-recovery operation in the domestic oil seeds extraction industry.

The Sugar Industry

Tragardh and Gekas (1988) summarized the applications of membrane technology in the beet and cane sugar industry. Experiments with a DDS plate-and-frame module showed that at 40 bar and 75°C to 80°C, thin juice can be concentrated from 15 to 20 percent TS at an average capacity of 40 $l/m^2/hr$. RO of thin beet juice has been suggested as an alternative to enlargement of an existing evaporator.

Experiments were performed by the Hawaiian Sugar Planter's Association to demonstrate the use of RO to concentrate cane juice. Tubular PCI RO membranes capable of operating at 80°C (Hsu 1986) were used, and tests were run on mixed juice and clarified juice. No significant differences in permeate flux were noted. Salt rejection, sugar loss, and total dissolved solids (TDS) rejection were all unaffected by the presence of suspended solids. Birch sap was concentrated by RO in a pilot-scale recirculating system using ZF-99 tubular membranes (Kallio et al. 1985). Retentions of glucose, fructose, sucrose, and malic acid were practically 100 percent with the use of the ZF-99 membrane, whereas with the CA membrane the retentions fell to 92 to 99 percent at 30°C and 4 MPa. Concentrating mixed juice rather than clarified juice contributes favorably to the overall process economics by decreasing the juice volume and, consequently, reducing demand for juice heaters, clarifiers, and filters, as well as decreasing evaporation requirements.

Other Applications

A soluble or dissolved tea product with acceptable flavor quality is prepared by extraction of leaf tea with hot water, pH adjustment, addition of soluble caseinate, and cold concentration by RO or freeze-concentration (Smallwood 1985).

Concentration of black tea extracts by CA membranes in an RO process was studied by Schreier and Mick (1984). Darjeeling orange pekoe extract was concentrated with an RO membrane from 46 l of extract to 5 l of concentrate. Results showed that basic compounds were completely retained by the membrane. The authors claim that optimization of membrane properties to suit aroma volatile retention should be feasible.

References
Abbot, J., F.A. Glover, D.D. Muir, and P.J. Skudder. 1979. Application of reverse osmosis to the manufacture of dried whole milk and skim milk. *Journal of Dairy Research* 46:663–672.

Agbevavi, T., D. Roleau, and R. Mayer. 1983. Production and quality of cheddar cheese manufactured from whole milk concentrated by reverse osmosis. *Journal of Food Science* 48:642–643.

Ambra, S. 1988. Reverse osmosis applied to grape musts. *Vini d'Italia* 30(2):39–46.

Amundson, C.H. 1970. "Some recent technical advances and probable trends in processing, manufacture and packaging of condensed milks." In *Proceedings of 15th International Dairy Congress* 11:642–643.

Amundson, C.H. 1986. Membrane technology and its application in the dairy and food industry. *Food Science* 33:98–102.

Barbano, D.M. 1985. "Reverse osmosis prior to cheese making." In *Proceedings of IDF Seminar,* held at Atlanta, 8–9 October. Brussels, Belgium: International Dairy Federation, pp. 31–53.

Barbano, D.M., and D.G. Bynum. 1984. Whole milk reverse osmosis retentates for cheddar cheese manufacture: Cheese composition and yield. *Journal of Dairy Science* 67:2839–2849.

Bowden, R.P., and A.R. Isaacs. Concentration of pineapple juice by reverse osmosis. *Food Australia* 41(7):850–851.

Braddock, R.J. 1982. Ultrafiltration and reverse osmosis recovery of limonene from citrus processing waste streams. *Journal of Food Science* 47:946–948.

Braddock, R.J., G.D. Sadler, and C.S. Chen. 1991. Reverse osmosis concentration of aqueous-phase citrus juice essence. *Journal of Food Science* 56:1027–1029.

Brent, J.A. 1964. Orange concentrate and method of making. U.S. Patent No. 3,140,187.

Bui, K., R. Dick, G. Moulin, and P. Galzy. 1986. A reverse osmosis for the production of low ethanol content wine. *American Journal of Enology and Viticulture* 37(4):297–300.

Bui, K., R. Dick, G. Moulin, and P. Galzy. 1988. Partial concentration of red wine by reverse osmosis. *Journal of Food Science* 53(2):647–648.

Bundgaard, A.G. 1974. Hyperfiltration of skim milk for ice cream manufacture. *Dairy Industries* 39:119–122.

Chua, H.T., M.A. Rao, T.E. Acree, and D.G. Cunningham. 1987. Reverse osmosis concentration of apple juice: Flux and flavor retention by cellulose acetate and polyamide membranes. *Journal of Food Process Engineering* 9(3):231–245.

Cross, J. 1988. Reverse osmosis as purification method. *Soft Drinks Management International* (Jan.):pp. 22–24.

Cross, S. 1988. Achieving 60 Brix with membrane technology, no. 590. Paper presented at the IFT Annual Meeting, 19–22, June at New Orleans.

Cross, S. 1989. Membrane concentration of orange juice. *Proceedings of the Florida State Horticultural Society* 102:146–152.

Dale, M.C., M.R. Okos, and P. Nelson. 1982. Concentration of tomato products: Analysis of energy saving process alternatives. *Journal of Food Science* 47:1853–1858.

Darnes, P., and P. Jordain. 1987. How Elgoods reverse osmosis to brew low-alcohol beer. *Brewing and Distilling International* 17(11):47–48.

Davis, F.L., P.A. Shankar, and H.M. Underwood. 1977. Recent developments in yoghurt starters: The use of milk concentrated by reverse osmosis for the manufacture of yogurt. *Journal of Society of Dairy Technology* 30:23–28.

De Boer, R., and J. Hiddink. 1980. Membrane processes in the Dairy Industry. *Desalination* 35:169–192.

De Boer, R., and P.F.C. Nooy. 1980. Concentration of raw whole milk by reverse osmosis and its influence of fat globules. *Desalination* 35:201–211.

Delbeke, R. 1981. Recovery of milk solids by hyperfiltration. *Milchwissenschaft* 36(11): 669–672.

Dixon, B.D. 1985. Dairy products prepared from reverse osmosis concentrate: Market milk products, butter, skim milk powder and yogurt. *Australian Journal of Dairy Technology* 40(3):91–95.

Donnelly, J.K., A.C. O'Sullivan, and R.A.M. Delaney. 1974. Reverse osmosis: Concentration applications. *Journal of Society of Dairy Technology* 27:128–140.

Duitschaever, C.L., J. Alba, C. Buteau, and B. Allen. 1991. Riesling wines made from must concentrated by reverse osmosis. I. Experimental conditions and composition of musts and wines. *American Journal of Enology and Viticulture* 42(1):19–25.

Electric Power Research Institute. 1991. *Membrane separation in food processing.* Technical Application 3(1). Palo Alto, Calif.

Fricker, R. 1984. Concentration of alcoholic beverages. European Patent Application 84300797.2.

Fukutani, K., and H. Ogawa. 1983a. Comparison of membrane suitability and effect of operating pressure for juice concentration by reverse osmosis. *Nippon Shokuhin Kogyo Gakkai-Shi* 30(11):636–641.

Fukutani, K., and H. Ogawa. 1983b. Juice concentration by reverse osmosis and membrane properties relating to permeability of juice components. *Nippon Shokuhin Kogyo Gakkai-Shi* 30(12):709–715.

Gherardi, S., L. Bolzoni, M. Careri, U. Rognoni, and A. Trifiro. 1989a. Possibilities of using membrane processes in the preparation of fruit purees. *Industria Conserve* 64(2)93–101.

Gherardi, S., M. Careri, L. Bolzoni, R. Aldini, and A. Trifiro. 1989b. Influences of ultrafiltration and reverse osmosis on the aroma composition of pear and peach purees. *Industria Conserve* 64(3):199–206.

Gherrardi, S., R. Bazzarini, A. Trifiro, A. Lo Voi, and D. Palamas. 1986. Preconcentration of tomato juice by reverse osmosis. *Industria Conserve* 61(2):115–119.

Glover, F.A., and B.E. Brooker. 1974. The structure of the deposit formed on the membrane during the concentration of milk by reverse osmosis. *Journal of Dairy Research* 41:89–93.

Gnmekow, B.R. 1989. Simultaneous double reverse osmosis process for production of low and nonalcoholic beverages (particularly wines). U.S. Patent 4,888,189.

Goldstein, H., C.L. Cronan, and E. Chicoye. 1986. Preparation of low alcohol beverages by reverse osmosis. U.S. Patent 4,612,196.

Gregor, H.P., and T.W. Jeffries. 1979. Ethanolic fuels from renewable resources in the solar age. *Annals of New York Academy of Sciences* 326:273–276.

Guirguis, N., K. Versteeg, and M.W. Hickey. 1987. The manufacture of yoghurt using reverse osmosis concentrated skim milk. *Australian Journal of Dairy Technology* 42(1–2):7–10.

Hedges, R.M., and D. Pepper. 1986. Reverse osmosis and ultrafiltration: Advances in the fruit and vegetable juice industries. Paper presented at the 19th International Symposium of the International Federation of Fruit Juice Producers. 12–15, May at the Hague, Holland.

Hiddink, R., R. De Boer, and P.F.C. Nooy. 1980. Reverse osmosis of dairy liquids. *Journal of Dairy Science* 63(2):204–214.

Hsu, T. 1986. Prospects of applying reverse osmosis in sugar cane processing. Paper presented at the Third World Congress of Chemical Engineering, at Tokyo.

Ishii, K., S. Konomi, K. Kojima, and M. Kai. 1981. "Development of a tomato juice concentration system by reverse osmosis." In *Synthetic membranes: Hyperfiltration and ultrafiltration uses,* edited by A.F. Turbak, ACS Symposium Series 154. Washington, D.C.: American Chemical Society, pp. 1–15.

Jensen, L.A., M.E. Johnson, and N.F. Olson. 1987. Composition and properties of cheeses from milk concentrated by ultrafiltration and reverse osmosis: A review of literature. *Cultured Dairy Production Journal* 22(2):6–10.

Jepsen, S. 1977. Membrane filtration in the manufacture of cultured milk products. *American Dairy Reviews* 39:29–33.

Jepsen, S. 1979. Membrane filtration in the manufacture of cultured milk products: Yogurt, cottage cheese. *Cultured Dairy Production Journal* 2:5–8.

Kallio, H., T. Karppinen, and B. Holmbom. 1985. Concentration of birch sap by reverse osmosis. *Journal of Food Science* 50(5):1330–1332.

Kieninger, H., and E. Reicheneder. 1977. Wasseraufbereitung durch umgekehrte Osmose. *Brauwelt* 117(42):1629–1633.

Kollacks, W.A., and C.J.N. Rekers. 1988. Five years of experience with the application of reverse osmosis on light middlings in a corn wet milling plant. *Starch (Starke)* 40(3):88–93.

Koseoglu, S.S., J.T. Lawhon, and E.W. Lusas. 1990a. Sterile orange juice (or concentrate) with improved flavor by commercial membrane technology. *Food Technology* 44(12):90, 92–97.

Koseoglu, S.S., and D.E. Engelgau. 1990b. Membrane applications and research in edible oil Industry: Assessment. *Journal of American Oil and Chemical Society* 67(4):239–245.

Koseoglu, S.S., J.T. Lawhon, and E.W. Lusas. 1990c. Membrane processing of crude vegetable oils. II. Pilot scale solvent removal from oil miscellas. *Journal of American Oil and Chemical Society* 67(5):281–287.

Koseoglu, S.S., J.T. Lawhon, and E.W. Lusas. 1991. Vegetable juice or flavor concentrates produced with commercial membrane technology. *Food Technology* 45(1):124–130.

Kosikowski, F.V. 1986. Membrane separations in food processing In *Membrane separations in Biotechnology,* edited by W.C. McGregor. New York: Marcel Dekker, pp. 201–254.

Kulozik, U., and H.G. Kessler. 1988. Permeation rate during reverse osmosis of milk influenced by osmotic pressure and deposit formation. *Journal of Food Science* 53(5):1377–1383,1437.

Lawhon, J.T. 1986. Process for recovery of protein from agricultural commodities prior to alcohol production. U.S. Patent 4,624,805.

Lee, Y.N., R.C. Wiley, M.J. Sheu, and D.V. Schlimme. 1982. Purification and concentration of Betalaines by ultrafiltration and reverse osmosis. *Journal of Food Science* 47:465–475.

Li, N.N., E.W. Funk, Y.A. Chang, S.S. Kulkarmi, A.X. Swamikannu, and L.S. White. 1984. Membrane separation process in the petrochemical industry, Phase I. Department of Energy Report DOE/ID/12422-T1 (DE85017030), December 15.

Li, N.N., E.W. Funk, Y.A. Chang, S.S. Kulkarmi, A.X. Swamikannu, and L.S. White. 1985. Membrane separation process in the petro-chemical industry, Phase II. Department of Energy Report DOE/ID/12422-1 (DE87009326), December 30.

Light, W.G. 1986. Production of low alcoholic content beverages. U.S. Patent 4,617,127.

Marshall, S.C. 1985. Reverse osmosis plant operation. *Australian Journal of Dairy Technology* 40:88–90.

Massignan, P.D. Leo, and D. Traversi. 1989. Use of membrane technology for processing wastewater from olive-oil plant with recovery of useful components. In *Future Industrial Prospects of Membrane Processes,* edited by L. Cecille and J.-C. Toussaint. Essex, England: Elsevier Applied Science, pp. 179–189.

Matsuura, T., A.G. Baxter, and S. Sourirajan. 1973. Concentration of fruit juices by reverse osmosis using porous cellulose acetate membranes. *Acta Alimentaria* 2(2):109–113.

Matsuura, T., A.G. Baxter, and S. Sourirajan. 1974. Studies on reverse osmosis for concentration of fruit juices. *Journal of Food Science* 39:704–711.

Medina, B.G., and A. Garcia, III. 1988. Concentration of orange juice by reverse osmosis. *Journal of Food Process Engineering* 10(3):217–230.

Merlo, C.A., L.D. Pederson, and W.W. Rose. 1985. Hyperfiltration/reverse osmosis. Department of Energy Report DOE/CE/40691-1 (DE86006737), July 15.

Merlo, C.A., W.W. Rose, L.D. Pederson, and E.M. White. 1986a. Hyperfiltration of tomato juice during long term high temperature testing. *Journal of Food Science* 51(2):395–398.

Merlo, C.A., W.W. Rose, L.D. Pederson, E.M. White, and J.A. Nicholson. 1986b. Hyperfiltration of tomato juice: Pilot plant scale high temperature testing. *Journal of Food Science* 51(2):403–407.

Merson, R.L., G. Paredes, and D.B. Hosaka. 1980. Concentrating fruit juices by reverse osmosis. In *Ultrafiltration Membranes and Applications.* New York: Plenum Publishing, p. 405.

McFiggans, A. 1984. Snowstorm in a lemonade bottle. *Soft Drinks Trade Journal* 38(1): 14–15.

Mohr, C.M., D.E. Engelgau, S.A. Leeper, and B.L. Charboneau. 1989. Membrane applications and research in food processing. Park Ridge, N.J.: Noyes Data Corporation, 256–275.

Morales, A., C.H. Amundson, and C.G. Hill, Jr. 1990. Comparative study of different reverse osmosis membranes for processing dairy fluids. I. Permeate flux and total solids rejection studies. *Journal of Food Processing Preservation* 14:39–58.

Moresi, M. 1988. Apple juice concentration by reverse osmosis and falling-film evaporation. In *Preconcentration and drying of food materials,* edited by S. Bruin. Conference at Eindhoven, Netherlands. Amsterdam: Elsevier Science Publishers, pp. 61–76.

Morris, C.W. 1986. Plant of the year: Golden cheese company of California. *Food Engineering* 58(3):79–90.

Palmer-Benson, T. 1986. Improving Ontario wine through reverse osmosis. *Food in Canada* 46(8):20–21.

Paulson, D.J., R.L. Wilson, and D.D. Spatz. 1985. Reverse osmosis and ultrafiltration applied to the processing of fruit juices. In *Reverse osmosis and ultrafiltration*. ACS Symposium Series, edited by S. Sourirajan and T. Matsuura. 281:325–344.

Pepper, D. 1981. Reverse osmosis using multi-stage recycle designs. *Australian Journal of Dairy Technology*. 36(3):120–122.

Pepper, D., and A.C.J. Orchard. 1982. Improvements in the concentration of whey and milk by reverse osmosis. *Journal of Society of Dairy Technology* 35(2):49–53.

Pepper, D. 1990. RO for improved products in the food and chemical industries and water treatment. *Desalination* 77:55–71.

Pepper, D. and L.H. Pain. 1987. Concentration of whey by reverse osmosis. *Bulletin of International Dairy Federation* 212:25–26.

Ray, R.J. 1987. The use of membranes in hybrid industrial separation systems. Electric Research Power Institute Report EPRI EM-5231. Bend, Ore.: Bend Research Inc.

Rhee, K.C., and S.S. Koseoglu. 1991. Oilseed protein separations by membrane processing. Paper presented at the *Membrane Separations in Food Processing: Applications Short Course*. 18–21 Aug. at the Food Protein Research and Development Center, Texas A&M University, College Station, Texas.

Sakaguchi, Y. 1976. Method for the manufacture of fruit wines from fruit juices of low sugar content. U.S. Patent 3,979,521.

Schmidt, D., and H.C. Deeth. 1987. Reconstitution of reverse osmosis milk concentrate and sensory evaluation of the reconstituted milk. *New Zealand Journal of Dairy Science Technology* 22:38–47.

Schreier, P., and W. Mick. 1984. Aroma of black tea: Production of a tea concentrate by reverse osmosis and its analytical characterization. *Zeitschrift fuer Lebensmittel-Untersuchung und-Forschung* 179(2):113–118.

Serafin, G., F. Camurati, and E. Fedeli. 1988. Advances in vegetable oil effluents treatment. *Rivista Italiana delle Sostanze Grasse* 65(1):7–11.

Sheu, M.J., and R.C. Wiley. 1983. Preconcentration of apple juice by reverse osmosis. *Journal of Food Science* 48:422–429.

Singh, V. 1983. Fermentation processes for dilute food and dairy wastes. *Process Biochemistry* 18(2):13,16,17, 25.

Skudder, P.J., F.A. Glover, and M.L. Green. 1977. An examination of the factors affecting the reverse osmosis of milk with special reference to deposit formation. *Journal of Dairy Research* 44:293–307.

Smallwood, K.C. 1985. Soluble or dissolved tea product. European Patent (EP) No. 0 133 772 A1.

Snow, N.S. 1985. Reverse osmosis: National seminar. *Australian Journal of Dairy Technology* 40(3):83.

Spangler, P.L., and C.H. Amundson. 1986. Concentration of milk and whey using composite, spiral wound, reverse osmosis membranes. *Journal of Dairy Science* 69(6):1498–1509.

Strobel, R.G. 1983. Stable aroma flavor and aroma and flavor products from aroma and flavor substances. U.S. Patent No. 4,374,865.

Sward, G.J., T.E.H. Downes, N.L. van der Merwe, and R. Holton. 1989. Reverse osmosis concentration of whole milk for cheddar cheese manufacture. *Suid Afrikaanse Tydskif vir Suiwelkunde* 21(3):69–71.

Tragardh, G., and V. Gekas. 1988. Membrane technology in the sugar industry. *Desalination* 69:9–17.

Versteeg, C. 1985. Process and quality control aspects of reverse osmosis concentration of milk. *Australian Journal of Dairy Technology* 40(3):102–107.

Wanatabe, A., T. Othani, S. Kimura, and S. Kimura. 1982. Performance of dynamically formed Zr(IV)-PAA membrane during concentration of tomato juice. *Nippon Nogei Kagaku Kaishi* 56(5):339–344.

Wanatabe, A., S. Kimura, and S. Kimura. 1978. The fouling on the membrane during reverse osmosis concentration of mandarin orange juice. International Congress of Food Science and Technology, p. 109 (abstract), Tokyo.

Wanatabe, A., Y. Ohta, S. Kimura, K. Umeda, and S. Kimura. 1979. Fouling materials on the reverse osmosis membranes during concentration of mandarin orange juice. *Nippon Shokuhin Kogyo Gakkai-Shi* 26(6):260–265.

Wu, Y.V. 1988. Recovery of stillage soluble solids from corn and dry-milled corn fractions by high pressure reverse osmosis and ultrafiltration. *Cereal Chemistry* 65:345–348.

Wu, Y.V., and K.R. Sexson. 1985. Reverse osmosis and ultrafiltration of stillage solubles from dry-milled corn fractions. *Journal American Oil Chemists* 62:92–96.

Wu, Y.V., K.R. Sexon, and J.S. Wall. 1983. Reverse osmosis of soluble fraction of corn stillage. *Cereal Chemistry* 60:248–251.

Zall, R.R. 1984. Trends in whey fractionation and utilization: A global perspective. *Journal of Dairy Science* 67:2621–2629.

11

High Purity Water Production Using Reverse Osmosis Technology

Gregory A. Pittner

Arrowhead Industrial Water, Inc.
A BFGoodrich Company

INTRODUCTION

Water is in a class by itself. It consists of a simple combination of two elements, hydrogen and oxygen, which are among the most common in the universe. Having a molecular mass of 18 units, water is unusual in its tendency to be liquid between 0°C and 100°C at atmospheric pressure. Most compounds of similar weight require much lower temperatures or higher pressures to liquify. The high polarity of the water molecule contributes to its ability to dissolve a range of materials, from salts to aliphatic hydrocarbons.

These unusual characteristics of water contribute to its popularity as an industrial fluid for washing and rinsing manufactured goods, for diluting personal-care products and other household chemicals, and for driving turbines in power plants. Advanced electronic products developed over the past 20 years have required high purity water as one of their most important raw materials. Future advances in the electronics industry will, to some extent, be linked to advances in water purification and the economics thereof.

The purpose of this chapter is to examine the significance of reverse osmosis (RO) as a tool for water purification, to discuss how this tool is properly designed and incorporated into high purity systems, and, finally, to conclude with an examination of the present and future role of RO in high purity water production.

It is customary to speak in terms of parts per million (ppm), or the almost equal units, milligrams per liter (m/l), as the unit of measure of contaminants in high purity water. Increasingly, parts per billion—or the metric equivalent, micrograms per liter—is coming into common use. Furthermore, parts per trillion is expected to be used with greater frequency in the future. These units are usually expressed in numbers with too many zeros to be understood by most persons. In fact, one could consider relatively impure water containing 500 ppm to be quite pure in comparison with most commercial products and materials. The standard for pure gold is 0.999, which translates to an impurity content of 0.1 percent. Water containing 500 ppm of dissolved solids has only 0.05 percent impurities, many times less than water that is considered to be pure. High purity water contains less than 0.0001 percent impurities, making its purity among the highest of any materials ever encountered. Only the vacuum of outer space approximates this low level of contaminants.

HIGH PURITY WATER APPLICATIONS AND QUALITY REQUIREMENTS

The major industrial applications for high purity water are discussed in the following paragraphs in order of decreasing economic importance.

Boiler Feed Water

High purity water is produced for boiler feed to be converted to steam for production of electricity and process heating, with lesser amounts used for space heating. Power utility companies are the major users of high purity water for electric power generation. In these applications, water in liquid form is fed to a steam generator (boiler), which converts the water to a gas through the use of oil, coal, natural gas, or nuclear energy sources. The steam generated at a pressure of 600 to 3,000 lb is then fed to a turbine, which removes a portion of heat energy in the steam, converting it to mechanical energy in the form of a rotating shaft. That mechanical energy then is transferred to an electric generator and, in turn, converted to electrical energy. Exhaust steam from the turbine, still primarily gaseous, is condensed for repumping into the steam generator. The efficiency of the power generation cycle improves as the temperature and pressure of the steam increase, and therein lies the basic reason for the water to be of high purity in this application. Increasing temperature and pressure are more demanding, both chemically and physically, on the boiler and the turbine. Small amounts of contaminants such as sodium, chloride, or silica in the steam can cause corrosion or fouling of turbine blades, both of which can weaken the turbine and can result in its destruction (Burris 1987). At the same time that the turbine becomes more sensitive to contaminant levels, the high pressure and temperature of the steam increase the volatility of salts in the feed water, especially silicon dioxide, increasing carryover of those contaminants into the steam phase. Therefore, under conditions of higher temperature and pressure, the distillation process is not able to separate the feed-water contaminants from the steam as effectively, thus requiring the contaminants to be removed upstream of the steam generator.

As a direct consequence of these relationships, the quality of the water that must be maintained in the steam generator is determined by the operating pressure. Makeup water quality is, in turn, determined by an economic analysis of makeup and blowdown costs. The generator concentrates contaminants in the feed water because the steam, even at high pressures, tends to have a lower level of contaminants than the water from which it was produced. A portion of the water in the steam generator must therefore be blown down to maintain the generator water quality at the desired level. To ensure that the blowdown stream is only a small portion of the total feed-water stream, the feed-water maximum contaminant level must be considerably less than the contaminant level tolerated in the steam generator. For example, in boilers with pressures of 600 or more pounds, it is not uncommon to have a blowdown stream of 1/100 of the feed-water rate, requiring feed-water qualities that are 100 times better than the water maintained in the steam generator.

Table 11.1 lists guidelines for the maximum level of boiler water impurities necessary to assure continued, reliable operation of modern steam generators. This should be used as a rough guide only, and as an indication of the extent to which boiler water quality must change with pressure. The particular features of each system, including boiler design and presence of a turbine, will also have a great impact on the quality limits that are appropriately maintained.

Many steam generators, especially those found in chemical process plants, such as oil refineries, pulp and paper production plants, or food processing systems, are used to produce steam exclusively for heating. The required steam quality is far different in these applications than in power generation. Precipitation of silica and corrosion of turbine blades are not an issue; rather, low pH

TABLE 11.1   Steam Generator Water Quality

Drum Pressure	Silica (ppm SiO_2)	Total Alkalinity (ppm $CaCO_3$)	Specific Conductance (micromhos/cm)
0– 300	150	350	3,500
301– 450	90	300	3,000
451– 650	40	250	2,500
651– 750	30	200	2,000
751– 900	20	150	1,500
901–1,000	8	100	1,000
1,001–1,500	2	0	150
1,501–2,000	1	0	100

From ASME 1979. Reprinted by permission.

corrosion of the condensate return lines is of most importance. This is dealt with by removal of bicarbonate and carbonate alkalinity from the feed water, thus avoiding disassociation of these compounds in the steam generator and the resultant carryover of carbon dioxide into the condensate lines.

Electronics

The electronics industry encompasses the production of a large variety of products ranging from cathode ray tubes, to discrete devices, such as diodes and transistors, to the most sophisticated integrated circuits having nominal feature sizes of less than 0.5 μm. Integrated circuit production, especially state-of-the-art fabrication, holds the distinction of requiring water of the highest purity of any industry or application. In addition, whereas steam generator requirements are demanding in terms of ionic levels, there is no concern for particulates or bacteria. However, the electronics industry requires attention to all water contaminants, including ions, organics, particulates, and silica (Hashimoto et al. 1990). Generally speaking, ultrapure water is used in the electronics industry as a cleansing rinse for surfaces. During production of integrated circuits, bare silicon undergoes a number of steps, perhaps as many as 30 or 40, in which layers of conductive or insulating materials are added to the silicon surface. A portion of the surface is then etched away by aggressive chemicals, such as sulfuric acid and hydrofluoric acid, prior to addition of the next layer. Ultrapure water is used between all of the chemical etching steps to ensure a complete rinse and removal of the chemical from the wafer surface. Other electronics manufacturing processes involve surface washing; however, the features on an integrated circuit are so small that even minute amounts of contamination in the water can add a sufficient amount of impurities to render the circuit inactive. Ions such as sodium and chloride can absorb into some layers of the circuit and thereby change the electrical characteristics of the device and the properties of the end product. Organic materials in the water tend to be surface active, and thus migrate toward and become associated with surfaces. Even low levels of dissolved organics can attach to the integrated circuit and disrupt placement of the following layer (Chu 1989). Particles such as bacteria can easily exceed the size of smaller features, disrupting additional layers or creating electrical shorts between adjacent circuits (Green 1983).

Table 11.2 shows a widely accepted specification for ultrapure water for manufacturing of 1 megabit DRAM products. This should be used only as a guide, inasmuch as the appropriate water quality will vary with the exact nature of the product, as well as the nominal feature size of the individual devices. A semiconductor product that utilizes design rules comparable to those required by a 4 megabit DRAM would require improved water quality.

TABLE 11.2   Ultrapure Water Specifications for IM DRAM Manufacturer

	Attainable	Acceptable
Resistivity 25°C	18.2	18.0
TOC (ppb)	Less than 10	Less than 30
Particles/liter		
By SEM		
0.2–0.3 μm		Less than 2,000
0.3–0.5 μm	Less than 200	Less than 200
Greater than 0.5 μm	Less than 1	Less than 1
Particles/liter		
By laser		
Greater than 0.5 μm		Less than 100
Bacteria/100 ml		
By culture	0	Less than 6
By SEM	Less than 1	Less than 10
By EPI	Less than 5	Less than 50
Silica, dissolved (ppb)	Less than 0.4	4
Boron (ppb)	Less than 0.05	2.0
Ions (ppb)		
Na^+	Less than 0.05	0.1
K^+	Less than 0.1	0.1
Cl^-	Less than 0.05	0.1
Br^-	Less than 0.1	0.1
NO_3^-	Less than 0.1	0.1
SO_4^-	Less than 0.05	0.2
Total ions	Less than 0.5	Less than 0.7
Residue (ppm)	Less than 0.1	0.1
Metals (ppb)		
Li	Less than 0.03	0.05
Mg	Less than 0.02	0.05
Ca	Less than 2	Less than 2.0
Sr	Less than 0.01	0.05
Ba	Less than 0.01	0.05
B	Less than 0.05	2.0
Al	Less than 0.05	0.05
Cr	Less than 0.02	0.05
Mn	Less than 0.02	0.05
Fe	Less than 0.02	0.1
Ni	Less than 0.02	0.05
Cu	Less than 0.02	0.05
Zn	Less than 0.02	0.05
Pb	Less than 0.05	0.05

From Balazs 1988. Reprinted by permission.

Metal Finishing

Many of the devices encountered in daily life are furnished with bright and colorful finishes designed to make them more attractive or more durable. Items ranging from doorknobs, to lighting fixtures, to electrical relay contacts are regularly plated with gold, copper, cadmium, chrome, or any of a variety of other metals. High purity water, ranging in quality from 1 megohm to 10 megohms, is customarily used to rinse the starting surface clean of any interfering dirt or chemicals prior to the plating operation. This ensures a better, more uniform bond and more uniform application of metal. Following the plating operation, high purity water is used to rinse off any excess plating solution before drying. Improper rinsing, or rinsing with poor quality water, can reduce the luster of the finish or spot the surface.

Medical and Pharmaceutical Applications

Water is the solvent of choice for most medical preparations, including lotions and creams, and is, of course, the principal constituent of intravenous fluids for replacing natural body fluids in the event of serious illness or injury. In these applications, the presence of any contaminants can lead to unwanted side effects of the medication, interfere with chemical characteristics of the medication, or be directly harmful to the patient. Pharmaceutical products requiring high purity water include both prescription and over-the-counter items such as eye drops or contact lens cleaning solutions. Smaller amounts of high purity water are also used by the medical industry in various laboratory testing procedures.

Of particular significance is the large quantity of water needed for hemodialysis. In hemodialysis, the patient's blood flows through a machine that brings it into close proximity with water, separated only by a semipermeable membrane. Contaminants in the blood diffuse through the membrane into the water, and are thus purged from the patient's body. The water used in such a device is normally from the municipal supply, carefully treated to remove most ionic materials that could interfere with the dialysis procedure. Of special concern is the presence of chloramines, which can readily diffuse through the dialysis membrane and cause illness in the patient.

There are several standard specifications for water quality that apply to medical and pharmaceutical uses, depending on the particular application. For instance, there are designations such as "sterile water for injection," "purified water," and various grades of sterile water. The specifications do not simply exert control over objective water quality parameters. They are also concerned with accidents that may occur during the water preparation. Water for injection is the most difficult and demanding to produce because of extreme quality control requirements and must be purified by either distillation or RO. Only these two processes are considered capable of reliably producing nonpyrogenic water. Purified water, which is commonly used for a variety of applications including preparation of topical and oral medications, may be produced by ion exchange (IX) in addition to distillation and RO.

Packaging

Many common household products are diluted with high purity water. High purity is needed because ions in most municipal supplies may precipitate with other ingredients in the product, thus rendering either performance or appearance unsatisfactory. Some examples of products diluted in this fashion include fabric softeners, window cleaners, and shampoos. Collectively, this industry is referred to as "packaging," because the purchasers of the high purity water are often facilities that concentrate on packaging, rather than product development or marketing. The specifications for most users are specific to the product; however, quality comparable to a single-distilled water containing less than 5 ppm of dissolved solids is generally satisfactory.

WATER PURIFICATION WITH REVERSE OSMOSIS

For many years, the goal of water purification was to reduce ionic materials that were capable of being measured with the use of conductivity instrumentation. Both distillation and IX were well suited for this task. During the 1950s and 1960s, IX began to dominate the industry because it could achieve much greater levels of separation, especially with the use of the mixed bed concept, than could be attained with distillation (Envirex 1968). Mixed bed quality is, in fact, comparable to water distilled 28 times in quartz (Rohm and Haas 1974).

When RO first appeared as a commercial water treatment process, its principal purpose was an alternative to IX for source waters containing high concentrations of dissolved solids. The operating cost of IX equipment is almost directly proportional to the concentration of dissolved solids in the feed-water stream. RO, however, is relatively insensitive to the dissolved solids concentration up to a level of 1,000 or 2,000 ppm. There was an immediate recognition that, as the dissolved solids concentration in the feed water increased, the economics of purification would eventually dictate conversion from a system using only IX to a system with RO polished by IX. A rule of thumb soon emerged which suggested that at 350 or 400 ppm of dissolved solids, RO was economically attractive (Whipple et al. 1987). The levels of dissolved solids in water supplies in southern California are generally above 400 ppm, thus RO first gained greatest support for industrial applications in the West. Water supplies of other areas of the country typically have less than 200 ppm, thus reducing the economic need for RO.

During the 1980s, the perception of RO as a water treatment process underwent substantial change (Benedek and Johnston 1988). This has been driven by the recognition that water quality is a function of a variety of contaminants, of which ionic materials are but one type. Other constituents in water that can seriously affect suitability for many applications include organics, silica, and particulates. Bacteria are also of concern, especially for electronic and pharmaceutical applications (Balazs and Walker 1983).

Once the criteria for high purity water encompassed other contaminants, RO truly came into its own. Whereas IX has the ability to remove a portion of the particulates and organic material present in the feed water, RO is intrinsically a broad-spectrum process that is effective in removal of virtually all particulate matter, and one of the most effective means of organic compound removal. RO is widely used today to obtain a logarithmic reduction of all incoming water contaminants. In many cases it has become a necessary pretreatment for downstream processing designed to polish the water with respect to individual quality parameters.

Ionic Removal

RO continues to be widely used solely for removal of dissolved salts, such as sodium chloride and calcium bicarbonate. This reduces the ionic load on polishing IX equipment and, in some cases, depending on the water purity required, eliminates the need for IX completely. Less sophisticated applications for high purity water, including some segments of the metal finishing industry and the majority of the packaging industry, could utilize RO in this fashion.

Organics Removal

The presence of organics is critically important in both electronics and steam generation, although for vastly different reasons. In electronics, organic material in the water tends to migrate toward the interface between the water and the product, reducing the quality of the surface. Organics in water systems also increase the food supply for bacteria, commonly present in the piping distribution systems. Increased availability of food increases biological growth rates, generating more biological particulates in the high purity water. Organics can thus play an indirect role in particulate quality and quantity.

Organic control in steam generation feed water is needed to control sodium, hardness, and various other ionic impurities. Organics, although nonionic and noninteractive with IX systems, generally contain some sodium, chlorine, sulfur, and nitrogen, which can, under the conditions present in a steam generator, break down to liberate the respective ions and seriously degrade water quality. Organic removal is thus necessary to indirectly control ionic levels (Miller and Fredricks 1988).

Silica Removal

Silica has properties that distinguish it from other inorganic contaminants in water. Silica is nonionic under normal conditions and is found at appreciable concentrations in most water supplies. It can be present as a discrete molecule, or polymerize into colloids with a range of molecular weights. Under certain conditions of concentration and pH, silica will precipitate to form discrete particles. In this particulate form, IX is largely ineffective in its removal; however, RO is totally effective for particulate silica removal while also removing dissolved silica.

RO may have some other, less well understood, advantages over IX for the removal of silica, especially for some applications (Ammerer and Dahman 1990). IX functions by concentrating the feed-water contaminants within the resin during service. Silica is the least tightly held of all the common water contaminants. In the event that the IX equipment is improperly operated, silica tends to be the first ion that is returned to the process water stream or "leaked." If leakage occurs, the silica concentration in the product will normally exceed the concentration in the feed, and will approach the total anionic concentration. RO is a continuous process and thus does not store the removed contaminants during service. In the event of a total failure of the RO system, the silica level in the product will never exceed that of the feed water. This feature, combined with the fact that a total failure of the process is far less likely with RO than with IX, increases its intrinsic reliability for this process (Pittner et al. 1986).

Particles and Bacteria

In cases where particulate matter must be removed, no known process is more effective than RO. Because RO operates through a mechanism akin to filtration, but at the ionic level, it is necessarily effective for removal of virtually 100 percent of all discrete particulate matter, including bacteria, silt, and colloidal solids.

PROBLEMS IN HIGH PURITY
WATER SYSTEMS

Because RO systems are sophisticated, highly engineered, mechanical processes with moving parts, they are subject to the same frailties as all equipment systems. Some features of high purity systems, however, include specific problem areas that must constantly be watched during the design and operation stages to avoid serious negative effects on water quality.

Recontamination

Sophisticated analysis of particulate levels in the product emanating from an RO unit can indicate an appreciable level of particulates. These contaminants do not, under normal circumstances, result from passage of the feed-water contaminants through the membrane to the product, but from contamination of the product water on the downstream side of the membrane. Such contamination is due to the release of construction fragments from the RO module, or to bacteria sloughage.

This issue must be addressed in both the design and startup phases. To ensure the best particulate performance, modules should be manufactured from materials that are maintained in the cleanest possible form throughout processing. It is extremely difficult to rinse particles from a membrane that has been improperly manufactured. Similarly, the RO piping and pressure vessel unit should be carefully rinsed prior to membrane loading. The addition of a sanitizing agent, such as hydrogen peroxide, during the initial rinse can help remove and dislodge some of this particulate material. Of course, the concentration of these oxidizing agents, as with any chemical agents, should be kept well below that which would degrade the construction materials of the RO unit.

Once an RO unit is operating, it should, if possible, be maintained in continuous operation. Prolonged shutdowns aggravate the particle problem by allowing bacteria to flourish in the permeate channels. Upon restart of an RO unit there will be sharp increases in bacterial contamination of the product, which will take many hours to rinse out. Where the best possible water quality is desired, it may be necessary to recycle the RO product back to the inlet of the unit to ensure that the system operates 100 percent of the time.

Membrane Bypass

The many benefits of RO are obtained when the water flows through the membrane, and not around it. Where removal of ionic material is concerned, membrane bypass may degrade the quality of the finished water and increase the cost of any polishing steps; yet this is not a critical issue. Where RO is used to reduce particulates, especially in semiconductor manufacturing, however, the issue is of key importance. In cases in which only a small percentage of the water bypasses the membrane, the particulate concentration of the treated water will be increased manyfold.

Membrane bypass primarily results from two conditions. The first is where integrity of the seam seal between sheets of membrane has been lost and water is allowed to flow from the brine side, between the membrane sheets, into the permeate channel without first passing through the membrane. This condition is, fortunately, unusual and results most often from a flow of permeate water from the product back into the permeate spaces within the membrane. If this backflow occurs with sufficient pressure, the seams between the membrane will be split. If the backflow is significant and many modules are affected, the degradation and performance will be immediately apparent. However, some loss in seam integrity may occur and remain invisible to the end user unless the RO unit is carefully monitored.

Some situations that can cause this to occur include the following. When an RO unit discharges into a storage tank that is either tall or placed at higher elevation, the head of water in the storage tank can cause backflow at sufficient pressure to destroy the membrane seams. It is customary to place a check valve on the permeate discharge of the RO; however, check valves often fail to operate properly and, in any event, may allow leakage of sufficient volume to destroy the modules.

The second, and probably more common cause of backflow, is direct feed of the RO permeate into IX vessels. It is common for IX units to develop an air space or bubble in the upper head. If, when the RO unit is turned off, the discharge of the demineralizers is also closed, the air gap will exert a back pressure equal to the pressure during normal service of the RO unit. The air will then expand, displacing water from the vessel backward into the RO permeate line, destroying the modules. Some have erroneously assumed that it is necessary to maintain the permeate pressure of an RO unit below a maximum level, sometimes given as a 50 psig. This is incorrect. The pressure of the permeate from an RO unit must only be less than the pressure of the brine side to maintain seam integrity. Permeate pressures can, in fact, be several hundred pounds. Seam disruption occurs during shutdown because the feed pressure ordinarily drops to zero; thus, any back pressure upsets the necessary balance. It does not matter what the discharge pressure from the RO is, as long as the permeate pressure drops at the same time as the feed pressure.

The most commonly encountered problem causing membrane bypass is O-ring seal leakage. Both ends of each membrane element have connectors that are sealed with O-rings. This is an economical method of sealing and very effective under the proper circumstances. For O-rings to seal properly, however, the parts making up the seal must be stationary. This is not actually the case in an RO pressure vessel. When the pump starts and pressure in the pressure vessel increases, the vessel stretches. At the same time, forces are exerted on the membrane elements, arising from drag of the water flow within the brine channel. This causes movement of the membrane elements in the direction of brine flow. The shifting of the membrane elements and the adjacent vessel end cap are thus in opposite directions, creating a considerable amount of displacement at the O-ring seal. This movement causes excessive wear, at the very least, requiring O-ring replacement every 3 to 6 months. In some unusual cases, the movement can be sufficient to roll the O-ring out of its groove, completely eliminating the O-ring seal in one start-up cycle.

In cases where the RO unit is used primarily for dissolved solids removal, regular maintenance of O-rings is generally sufficient to overcome any difficulties. Where the removal of particulate matter is critical, as in some electronic systems, double O-rings may be desirable.

Detection of membrane bypass must receive adequate attention. The vast majority of system users rely on gross measurements of conductivity in the feed and product to estimate ionic levels and rejection. In the event that leakage occurs, the rate of leakage has to exceed the inherent accuracy in these measures to be noticed. Fortunately, there are more accurate means to detect leakage, which should be applied where leakage is critically damaging. First, it is necessary to look at the concentration of one particular ion rather than a collective measure of ions, such as conductivity or total dissolved solids (TDS). By measuring one ion, inasmuch as there is no holdup of ionic material inside the RO unit, most of the sources of interference are automatically eliminated. Second, simple calculations show that smaller amounts of bypass can be detected if the ion of greatest rejection is measured.

Consider two cases. The first involves using sodium as the measured ion, which for this example is assumed to be rejected to a level of 97 percent. The second case uses calcium as the indicator ion, for which a rejection of 99 percent will be assumed. Further assuming that both can be measured to within 1 percent accuracy, it is possible to calculate the minimum amount of water leakage that each ion can be used to detect.

Given the following variables:

S_i = feed-water sodium concentration
S_p = product-water sodium concentration
C_i = feed-water calcium concentration
C_p = product-water calcium concentration
F = feed-water flow rate
B = brine flow rate
L = leakage flow rate

Assuming that the rejection is given on the basis of a feed-brine average, the zero bypass sodium concentration in the product can be calculated by the following expression:

$$.015\ S_i(1\ +\ F/B)\ =\ S_p \tag{1}$$

The weight of sodium appearing per unit of time in the product is given as

$$S_p(F\ -\ B) \tag{2}$$

Combining with equation 1, the product sodium at zero bypass is

$$.015\ S_i(1\ +\ F/B)(F\ -\ B) \tag{3}$$

If a portion of the feed water bypasses the membrane and flows directly into the product stream, the sodium level of the product will be increased. Assuming the leakage results in a 1 percent increase in sodium concentration, the leakage flow rate can be calculated as follows:

$$S_iL\ =\ .01(.015)S_i(1+F/B)(F-B) \tag{4}$$

$$L\ =\ .00015(1\ +\ F/B)(F\ -\ B) \tag{5}$$

The same mathematical procedure applied to calcium yields the following result:

$$L\ =\ .00005(1\ +\ F/B)\ (F\ -\ B) \tag{6}$$

Examining the ratio of equations 5 and 6, the leakage that would cause a 1 percent increase in calcium concentration is more than three times less than the necessary leakage for an equally significant change in sodium concentration.

One difficulty inherent in this procedure is that calcium is normally present at a much lower concentration, thus measurement of a 1 percent change in calcium concentration is harder to perform than for comparable change in sodium concentration. This problem can be eliminated by spiking the feed water with calcium to increase its concentration and, thus, the ability to measure small differences in the product. Depending on the resources available to the user, other ions such as sulfite or magnesium may be employed with similar effectiveness.

Differential Passage of Silica and Carbon Dioxide

RO, even in a double-pass configuration, does not in most cases provide adequate ionic removal to satisfy most high purity water requirements. Under normal circumstances, the water is polished with IX to achieve much higher levels of ionic quality. Unfortunately, carbon dioxide passes readily through RO membranes and silica is poorly rejected. Both these considerations lead to higher anion levels in RO product water and, thus, to greater loads on the anion resin than on the cation resin. Because anion resin tends to be more expensive to purchase and regenerate, and it degrades faster and has less capacity than cation resin, differential passage aggravates one of the weaknesses of IX.

The normal response to this imbalance is to use more anion resin than cation resin in the downstream IX system. Because these polishers are most often mixed beds, there are limitations to the maximum anion ratio that can be used, resulting from the need to provide intimate contact between the water and both types of resin. In the extreme, it would not be desirable to use more than two-thirds anion resin and one-third cation. With maximum anion resin, because of the low anion capacity, the system may still be anion limited. Further aggravating this prob-lem, the first anion that passes through the resin is silica, which cannot be detected through simple conductivity measurements. It is easy to understand why silica has become one of the more common problems in ultrapure water treat-ment. Depending on the exact operating conditions of the mixed bed, this situation may lead to colloidal silica formation in the low pH exhaustion wave front passing down the mixed bed column.

DOUBLE-PASS REVERSE OSMOSIS

Any discussion of RO for the production of high purity water would not be complete without examining the advantages of double-pass RO. Whereas the typical RO unit passes water through one RO membrane, there are certain advantages to treatment with two RO membranes in series (Comb and Schneek-loth 1990). RO is not typically considered a stand-alone demineralization pro-cess. A simple, two-bed IX system consisting of strong acid and strong base units is generally capable of removal of at least 95 percent of the incoming dissolved solids and production of water containing less than $\frac{1}{2}$ ppm of silica. RO membranes typically reject 95 to 99 percent of the feed-water dissolved solids and are typically less effective for the removal of silica. RO has, therefore, been considered as a roughing demineralization step, generally to be followed by IX polishing to obtain water quality appropriate for high purity applications.

The advent of thin film composite membranes has allowed for the simple construction of a double-pass RO unit that can provide treated water superior to that of a two-bed demineralizer, without subsequent polishing (Pittner et al. 1986). Thin film composite membranes use a thin rejecting membrane that decreases the amount of pressure necessary to drive the water through the membrane without sacrificing rejection of dissolved solids. Improved silica rejection is also characteristic of thin film composite membranes. Because the required pressure to operate a thin film system is 200 to 250 psi, a double-pass

RO can be operated below 600 lb differential and provide greatly enhanced ionic separation. If the rejection of the single RO membrane is 95 percent, the salt passage is 5 percent. Under these circumstances, the salt passage of a double-pass RO is $(.05)^2$, or .0025. The rejection calculated across both passes would thus be 99.75 percent. Thin film composite membranes are also capable of enhanced silica rejection; thus the effluent silica level of a double-pass RO unit with thin film membranes is equal to or better than the best expected from a two-bed IX system.

There are many possible mechanical designs for a two-pass system, including arrangements in which only one pump is used to drive the water through both passes. In this case, the first-pass permeate pressure is 200 to 300 psi, sufficient to drive the water through the second RO pass. Alternatively, two individual RO units, both equipped with pumps, may be used in series with exactly the same process results.

The overall recovery of a two-pass RO unit can be identical to that of a single-pass RO unit, because recovery is a function of the degree to which the feed water can be concentrated, not of the configuration of the RO unit. To achieve this recovery, however, the brine from the second-pass RO is recycled to the feed of the first-pass RO. Figure 11.1 shows a diagram of a typical double-pass RO unit. This system includes caustic injection between the passes to enhance removal of carbon dioxide in the second stage. Since an RO membrane cannot reject smaller, uncharged molecules, carbon dioxide in the feed water passes freely through the first-pass membranes. Without injection of sodium hydroxide prior to the second pass, the same concentration of carbon dioxide would pass to the second-pass product, in which it would constitute the majority of the dissolved solids present in the product. The imbalance between the carbon dioxide and bicarbonate in the second-pass product also depresses the pH, making

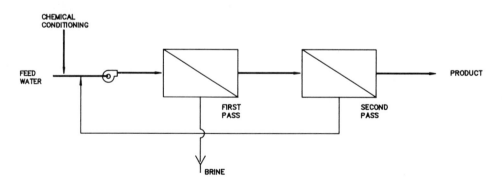

FIGURE 11.1.   Double-pass reverse osmosis system.

the water relatively corrosive. Although sodium hydroxide increases the dissolved solids concentration of the water entering the second pass, the sodium hydroxide is taken up by the carbon dioxide to form bicarbonate and carbonate, causing the rejection of the second pass to be increased to the extent that all added sodium hydroxide is rejected in addition to the carbon dioxide present from the first pass. This situation is unique in that it may be the only circumstance in which the product water quality from an RO system is improved by increasing the dissolved solids level in the feed water. The only situation in which caustic addition, or the addition of a chemical capable of increasing the pH, would not be appropriate, is one in which the water contains no free carbon dioxide. A water with only sodium chloride, or a negligible concentration of carbon dioxide, would not benefit from interstage injection of a pH-increasing chemical.

In some applications the use of two-pass RO may be desirable as compared with IX or with a combined RO and IX system. In the production of water for medical uses, for preparation of packaged products, or for human consumption, it is desirable to avoid adding foreign chemicals directly to the water. There is constant danger that the chemical supply could be contaminated, adding harmful materials to the finished water and thus increasing product liability risk.

Double-pass RO is also simpler to operate because it is a continuous process and not subject to the fluctuations that are characteristic of the batch process of IX. The amount of operating labor required per gallon of produced water is generally less with continuous processing.

PRETREATMENT AND POSTTREATMENT

It is beyond the scope of this chapter to discuss all of the considerations related to pretreatment of water for RO. However, some pretreatment issues take on more importance when using RO as part of a high purity system.

Particle Stabilization

The physical configuration of an RO membrane, either spiral wound or hollow fiber, unavoidably tends to cause some areas of low water velocity against the membrane surface. In order to pack a large amount of surface into a small volume, the spaces through which the water flows must also be very small, providing the opportunity to trap particulates. Both requisites cause particulate accumulations in the brine channels and against the surface of the membrane. Although, in a properly operating system, this accumulation does not degrade the particulate quality of the product water, it does influence pressure drop and concentration polarization.

As previously mentioned, forces are exerted on the membrane elements owing to drag of the water on the membrane surface. Solids accumulation on the brine side of the membrane decreases the cross-sectional area available for flow and increases these drag forces. This increases the amount of movement of the membrane elements with respect to the pressure vessel and increases wear on the O-ring connectors. In the event that sufficient pressure drop develops, the force exerted on the membrane elements will cause mechanical damage. Figure 11.2 presents a view of the end of an RO element damaged by excessive pressure drop. The outer layers of membrane have moved in the direction of the brine flow, whereas the inner layers of the membrane have remained stationary. This condition, known as telescoping, creates a conical profile at both ends of the module. Depending on the extent of differential membrane movement, one or more of the layers can tear, providing a direct path for the passage of brine into the permeate. Because production of high purity can be accomplished only if the RO unit is in good mechanical working condition, excessive particulate fouling can hamper water quality control.

FIGURE 11.2.   Physically damaged (telescoped) membrane module.

Ionic Stabilization

The comments on ionic stabilization in other sections of this book certainly apply to RO when used in the production of high purity water. If precipitation should occur, the result will be the same as described for particulate contamination, diminishing the quality of the product water.

Biological Control

Biological control is a subject of considerable dispute. The controversy arises because there is disagreement on the priority given to control of biological populations. All agree on the desirability of reducing bacterial populations; however, improperly applied biological control programs can adversely affect downstream processing and, in fact, negatively affect the quality of ultrapure water.

A certain amount of biological contamination of the feed water is virtually unavoidable. Similarly, the seeding of the membrane surfaces with microbes of various sorts occurs rapidly. If the circumstances existing within the membrane element are conducive to multiplication of the organisms, such will rapidly occur, establishing a biological film within days or even hours.

Many users prefer to have a sanitizing agent present at all times during operation of an RO unit. For CA modules, the preferred agent is chlorine, which is readily available, easy to measure in the feed water, and highly effective. Sufficient chlorine also passes through the RO membrane to provide bactericidal or bacterial static conditions on the downstream side of the membrane. In applications where the RO unit is used to provide low-particle water, permeate side sanitization helps greatly to reduce the product water particle level. In the absence of sanitizing agents, permeate contamination with biological particles grown on the downstream side of the membrane can be the greatest source of product water particles.

Whereas the presence of a sanitizing agent may be desirable for improved performance of an RO unit, it can have a negative effect on downstream processes. The great majority of ultrapure water systems use IX to polish the RO product. IX resins are degraded by chlorine; it shortens the resin life by cleaving the polymer and oxidizing active groups. Depending on the length of time and concentration of chlorine, sufficient degradation of the resin may occur to release low molecular weight organic contaminants into the water, thus reducing water purity.

Another consequence of chlorination is the formation of organic and inorganic chloramines, produced from ammonia or amine compounds. Reaction of these compounds with chlorine to form chloramine converts polar species to a state of low polarity in which they pass more readily through the RO membrane. Being nonionic, chloramines can also escape removal by IX and pass into the product water. The presence of chloramines can be beneficial, but may also result in negative effects.

Over the last 10 years there has been increasing interest in disinfection of municipal potable water supplies with chloramines instead of chlorine. This treatment reduces the production of trihalomethanes (THMs), most of which are known carcinogens. Although chloramine is present to some extent in all chlorinated water supplies, the increased levels of chloramine resulting from this sanitization approach can negatively affect pure water systems. The presence of chloramines in water used in hemodialysis is especially undesirable, because chloramine compounds can diffuse through the hemodialysis membrane into the patient's blood. Yet removal of chloramines in all applications should not be automatic, because their presence may be desirable for sanitization and not a disadvantage in the downstream processing.

If chloramines in the RO product water are undesirable, they can be removed by activated carbon or sulfite addition. Both processes create ammonia, or an organic amine, as one of the reaction products. Water so treated cannot then be rechlorinated prior to RO, inasmuch as the chlorination process will again create the chloramines that were just destroyed.

Polyamide membranes have different sanitization requirements than CA membranes. The amide bond that is the backbone of this membrane is very sensitive to any halogen in the +1 oxidation state. Free chlorine, bromine, or iodine—all common sanitizing agents—will rapidly destroy this bond, thus weakening the membrane and reducing its rejection characteristics. Exposure to chlorine of even short duration is inadvisable. Although a well-constructed polyamide membrane can tolerate some chlorine without an immediate deterioration in performance, it is a certainty that any chlorine exposure causes damage.

The incompatibility of polyamide membranes with most common sanitizing agents requires that most polyamide systems be operated without residual sanitizing agents. This consideration dramatically increases the cleaning frequency necessary for polyamide membranes, as compared with CA membranes. Many cases are documented in which water supplies were treated very successfully with CA modules, but fouled badly when converted to polyamide membranes.

Although it may seem contrary to conventional wisdom, pretreatment with softening is highly effective in reducing the population of microorganisms, thereby reducing the fouling rate of polyamide membranes. In cases where softening is not necessary to eliminate hardness ions, softening may be necessary to reduce particulate levels.

Polishing Treatment

The following sections discuss ionic, and other polishing processes.

Polishing with Ion Exchange

Included in virtually every ultrapure water system is a polishing device designed to reduce the residual concentration of ionic contaminants. Often there will be two treatment steps, the first of which is customarily designated as *makeup ion exchange,* and a second system most often designated as *polishing ion exchange*.

The makeup IX unit customarily consists of a mixed bed system, but separate bed cation-anion units are sometimes used. In cases of all but the most concentrated feed-water supplies, the RO permeate is sufficiently low in dissolved solids that mixed bed treatment is practical. If the dissolved solids level exceeds approximately 50 ppm, the difficulties and inefficiencies involved in regeneration of a mixed bed unit outweigh the benefits, and separate bed cation and anion exchangers should be used for the makeup system.

The second IX system is always a mixed bed type, which is necessary to obtain the lowest possible level of ionic impurities (Gottlieb 1989).

IX units have the capacity to remove substantial amounts of organic and particulate matter in addition to ions. The removal of organics is a complex process. Some organics, notably carboxylic acids and amines, are exchanged in the same fashion as inorganic ions. However, nonionic organic materials can bind to the resin polymer backbone through Van der Waals forces and, in this manner, can be removed from the water stream. Organics that are ionically exchanged also interact with the polymer structure and thus tend to bond more tightly with the resin under normal circumstances; a reduction in organic levels, as measured by TOC (total organic carbons) instruments, is observed after treatment with IX. Difficulties arise on regeneration, however, because the acid or caustic typically used for cation and anion resins is of little benefit in removing absorbed organics. After many cycles, the IX resin may then lose effectiveness for removal of additional organics.

The same chemical characteristics of IX resin that give it the capacity to remove ions also cause it to develop an electrical surface charge. This charge attracts fine particles, making IX resin effective for removing particles considerably smaller than the spaces between the resin beads. Levels of micron-size particles are reduced by 20 to 90 percent on treatment with IX, depending on the influent level.

The extent to which IX resins can remove ionic materials is controlled by a number of factors, including the resin chemistry, the concentration of the regenerant, the concentration of the impurities in the regenerate, and the concentrations of hydrogen and hydroxide ions in the water (Gottlieb 1989). Levels of cations and anions below 0.1 part per billion are theoretically achievable. RO also has use in ionic polishing. Generally, RO will continue to remove ions when present even at minute concentrations. There are a number of full-scale plants in the United States in which RO is used to polish the effluent from polishing mixed bed IX units, thereby improving the ionic purity of the water.

Other Polishing Processes
Depending on the nature of the water supply, the TOC level in the permeate from an RO system can vary from 50 to 500 parts per billion. For production of the highest purity waters, various techniques are used to remove these residual organic contaminants.

IX polishing and ultrafiltration (UF) will both remove some organic materials; however, organics that escape removal in RO do not tend to interact well with IX resins. The most effective techniques are based on the principle that organics can be oxidized, causing conversion to inorganic carbon dioxide, or to a carboxylic acid, which is ionized and interacts with IX. Many oxidization processes are in common use. These include irradiation with 185 nanometer ultraviolet light, with and without the addition of agents such as hydrogen peroxide, to stimulate production of oxidants. Direct feed of ozone into water, followed by irradiation with 254 nanometer light, has been found to be very effective in reducing organic carbon levels to less than 5 parts per billion.

Regardless of the method used for oxidation of the organic material, the TOC changes only slightly before and after the oxidation step. The primary impact of oxidation is conversion of organics to carboxylic acids, which are then removed with IX. The most dramatic improvement resulting from the oxidation process thus occurs after subsequent treatment with IX.

FIELD EXPERIENCE

Figure 11.3 is the flow diagram for an ultrapure water system recently constructed to service a facility for fabrication of fine-geometry integrated circuit products. It deviates from the typical ultrapure water system in only minor respects and, in fact, highlights the direction in which the industry is moving to combat known problems.

Process Design Considerations

In Figure 11.3, the pretreatment system prior to the RO unit includes a carbon filter, a softener, and a chemical feed of a scale inhibitor. This pretreatment has been tailored to thin film composite, polyamide membranes in the RO unit, which are intolerant to chlorine but can operate at a wide pH range. The carbon

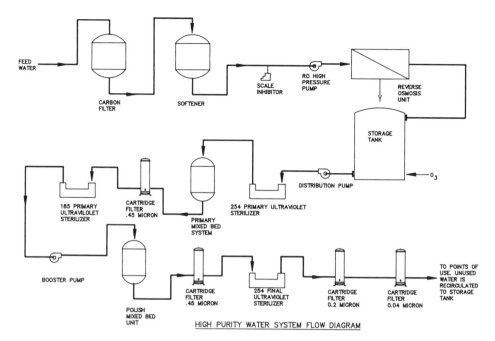

FIGURE 11.3.   High purity water system flow diagram.

filter is the preferred method of dechlorination of the feed water because of its
reliability. The vast majority of systems that use sulfite for chlorine removal
experience several episodes per year in which chemical feed is interrupted.
Although the damage caused by each episode is, in most cases, too small to be
measured, a combination of incidents results in accumulated harm to the membranes. Elimination of chemical handling is also an environmental and safety
benefit.

The presence of a carbon filter is sometimes criticized as a source of bacterial
particulates, which, it is thought, can increase the particulate level of the finished
water. If the RO unit is operated properly, that is, without bypass of the raw
water around the membranes, changes in the particulate level of the incoming
water will have no effect. The presence of the softener tends to enhance the
particulate quality of the water. Although not generally known for this feature,
the slight charge on the surface of the softener resin beads improves the softener's activity as a filter, as compared with granular media beds having the
same particle size. The softener pretreatment thus tends to reduce particulate
fouling of the RO unit, reversing the particle generation features of the carbon
filter.

Because it is a source of bacterial contamination, water storage in an ultrapure
system has received widespread attention. This system's storage tank is posi-

tioned directly downstream of the RO unit, so that downstream treatment with IX and cartridge filtration is performed after storage. This maximizes the opportunity for removal of particulates formed in the storage tank. With storage, fluctuations and demand for the finished water can be addressed economically. The RO unit can be designed for the average demand, and the storage designed to fill the gap between average and maximum. If storage were eliminated, which is technically and operationally feasible, the RO unit would have to be designed to treat at the maximum consumption rate, recycling during periods of low consumption so that operation is continuous. This practice would measurably increase the capital and operating costs of the system.

The storage tank is, in this case, ozonated, which reduces the potential for particulate formation and provides oxidant contact time for the conversion of organics to carboxylic acids. Downstream of storage is a 254 nanometer ultraviolet (UV) light to eliminate the ozone residual, protecting downstream IX and cartridge filters from ozone degradation. The system also comprises two IX steps, each of which are followed by submicron cartridge filtration and UV light. The .45 μm cartridge filters are required as resin traps, but the cartridges are chosen to remove micron-size particulates that might be shed from the mixed bed units. The UV lights downstream of these cartridge filters are designed to kill bacteria, reducing biological seeding of the downstream process. The UV light unit downstream of the primary mixed bed system has been chosen at 185 nanometers to provide some oxidation of organics, in addition to bacterial kill. The additional cost of the 185 nanometer unit, as compared with a 254 system, was minimal, and it provides some organic reduction benefit. Because there is downstream IX polishing, the resistivity drop normally associated with 185 nanometer light irradiation does not create difficulty in meeting resistivity specifications for the final treated water. Last in this system are two stages of submicron filtration. In both cases, the filters are positively charged to minimize particle shedding during minor fluctuations in flow.

Layout

Figure 11.4 shows the layout of the system depicted in Figure 11.3. There is no perfect layout for an ultrapure water system, as layout is perhaps one of the more difficult problems to be resolved in system design. Operating personnel should participate in layout decisions because they can contribute valuable information concerning details of access and serviceability, issues with which most design engineers are not as acutely aware.

In regard to its operability, this system uses portable exchange units for mixed bed polishing, softener pretreatment, and carbon filter pretreatment. Portable units must be replaced on a frequent basis, thus easy access is important. Easy access to the RO unit for cleaning, replacement of interconnectors, O-rings, and

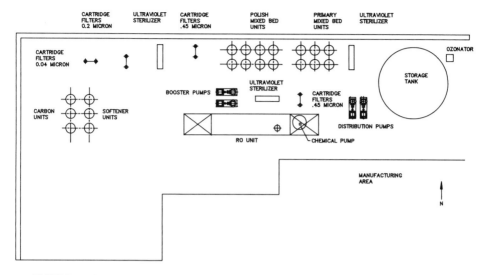

FIGURE 11.4.   High purity water system plot plan.

membrane elements is also desirable. In the layout depicted in Figure 11.4 the space constraints required that some equipment be placed against the wall on the north side of the system. This unavoidably reduced access to some of the mixed bed units, but allowed placement of the RO unit in a clear location.

If a high purity system must generate water low in bacteria and particles, high purity piping materials will be used downstream of the RO unit and in the polishing and distribution areas. For reasons of both cost and reduced water contamination, the length of piping running from the RO unit to the storage tank and through the polishing and distribution systems should be minimized. This requires that equipment be more closely packed. Thus, even if the real estate is readily available, access for maintenance and operation can be reduced. Fortunately, in all but the most unusual cases, an adequate layout can be designed so that access does not create major difficulty.

Start-up

The advantages of a proper system start-up cannot be overstated. Properly organized and conducted, a good start-up can reduce the time required to produce maximum quality water and minimize the overall cost. Improperly conducted start-ups can damage new equipment and make it difficult to identify performance problems.

A good start-up begins with a start-up plan. Because start-up occurs only in systems that have never been operated, both the design engineering and operating personnel should contribute to the plan. Design engineers know best how the units are intended to operate, whereas the operating personnel must understand the operating features of each unit.

Perhaps the most important rule in formulating a start-up plan is to start each individual process unit by itself and establish the proper operation of that unit before proceeding to the next. Moreover, unit operations should be started in the order in which each is used to treat water. For example, one would never attempt to start a polishing IX system before the primary RO unit is started and working properly.

For the system described thus far, the start-up was conducted in the following steps.

1. Operation of the RO pretreatment system was begun prior to loading the RO membranes. Pretreated water was used to flush completely the RO pressure vessels and piping.
2. The RO membranes were then loaded, and the RO unit was operated with the product water going first to drain. This was continued until one or more of the following conditions were met.
   a. The RO unit had run continuously, with stable ionic performance, for 48 hours.
   b. The TOC concentration in the RO product water was not more than 10 percent of the TOC in the feed.
   c. The TOC in the RO product was less than 100 ppb.
3. The storage tank was filled and drained, completely, twice.
4. The storage tank was filled to one-half of its volume with RO product water, and the ozone generator was started. The water in the storage tank was ozonated to a concentration of 200 to 500 ppb.
5. Ozonated water from the storage tank was recirculated throughout the entire polishing and distribution system for a period of 24 hours.
6. The system was emptied and refilled with fresh RO product water, the storage tank ozone level was reestablished, and the system was recirculated for 1 hour.
7. The UV light downstream of the storage tank was started, recirculation was continued for not less than 2 hours, and the ozonator was adjusted to achieve a stable ozone residual in the storage tank. Zero ozone concentration was confirmed at all points downstream of the UV light.
8. A set of primary mixed bed IX units was placed into service, with filters of 5 μm downstream to protect the system against resin fines.
9. An ozone-compatible tubing was established from the discharge of the distribution pumps to a point downstream of the resin trap filters. This allowed continuous ozonation of the system while recirculation through the IX polishing units continued. In this manner, the newly installed piping in

the polishing and distribution systems could be continuously ozonated while using IX resin to remove the heavy particulate load associated with initial operation. Flow rates were adjusted so that the concentration of the ozone in the water returning from the distribution system was 40 to 80 ppb. This recirculation mode, which required in excess of 1 week, was continued until the TOC concentration in the loop was less than 10 ppb and the 0.2 μm particle level downstream of the DI units was less than 30 per milliliter.

10. The entire process in paragraph 9 was repeated with a new set of IX units and 0.45 μm roughing filters.
11. Ozonation of the distribution loop was discontinued, and polishing IX units and permanent .45 μm roughing filters were installed.
12. After 1 day of operation, the first set of final filters, rated at 0.2 μm, were installed.
13. Two days after placement of the 0.2 μm filters, the polishing final filters, rated at 0.04 μm, were installed.

At each point in the start-up, which required more than 30 days, each process unit was started individually and then tested to confirm that it was operating at expected performance levels before proceeding with the next. In this manner, any problems that developed could readily be identified and dealt with.

Table 11.4 shows a summary of the system operation for the first 84 days. (This is not comprehensive of all the information recorded during the first 3 months of operation.)

During the period of first operation, most attention should be given to two points. Process units should be operating as designed and stable. In this system, as is typical of most ultrapure designs, the RO unit must provide water of adequate quality to enable the downstream units to polish that water to the final quality required. The RO unit is the pivotal process in the system. If it is not stable, fluctuations will be transmitted to downstream processes and will likely appear in the final product water. Moreover, if the RO product water is not of sufficient purity, downstream process units will not be adequate for polishing with enough reserve capacity to meet minor swings in demand.

Table 11.3 shows a sharp change in the RO system pressure occurring after the 21st day. This change was purposeful, to reduce the product flow from the RO unit to better match the rate at which water was being consumed in the semiconductor plant, and thus operate the RO unit for a greater percentage of the time. For the rest of the period, the difference in system and concentrate pressures was nearly constant, indicating that the ionic and particulate quality of the water fed to the unit was adequate and not causing excessive fouling. The product water and feed water/TDS (total dissolved solids) ratio also varied in a very narrow range during the entire operating period, indicating good system stability. Comparing the resistivity measurements from both the primary and

TABLE 11.3 High Purity Water System Initial Operating Results

	2/07	2/14	2/21	2/17	3/06	3/15	3/20	3/30	4/06	4/11	4/30	4/27
Days after start	7	14	21	28	35	42	49	56	63	70	77	84
Temperature (°F)	68	70	70	60	68	69	77	72	72	75	75	75
RO pressure (psig)												
System	310	310	330	250	250	260	250	255	250	250	250	250
Concentrate	240	250	250	210	210	210	200	205	210	200	200	200
Feed-water TDS (ppm)	110	110	140	150	150	130	130	140	155	160	135	160
Product water TDS (ppm)	2	2	3	2	4	2	2	2.5	3.5	3.5	3.0	3.5
Product/feed ratio	.018	.018	.021	.013	.027	.015	.015	.018	.023	.022	.022	.022
TOC by on-line monitor (ppb)	6.8	14.1	5.8	4.6	9.4	4.8	4.6	4.1	3.5	4.0	6.3	9.5
Silica (ppb)	0.72	1.6	3.3	5.8	4.7	4.3	4.9	3.8	2.7		6.6	3.4
Primary resistivity (Mega Ohms-Cm)	17.0	17.2	17.2	17.1	17.1	17.2	17.1	16.9	17.1	17.0	17.0	17.1
Polish resistivity (Mega Ohms-Cm)	18.0	17.6	17.8	17.7	18.0	18.0	18.0	18.0	18.0	17.9	18.0	18.0
Distribution loop resistivity (Mega Ohms-Cm)	17.5	17.2	17.7	17.2	17.7	17.9	17.9	17.7	17.8	17.7	17.9	17.8

polishing units, fluctuations in the RO product quality resulting from changes in the influent TDS were small enough to be eliminated by the IX system. Also of interest is the difference in resistivity between the outlet of the polishing unit and the return from the distribution loop. Initially, this difference was 0.5 megohm, but during the 90-day period was reduced to .2 megohm, as the piping in the polishing and distribution systems was rinsed clean of ionic contamination. The magnitude of the resistivity drop in a distribution system tends to be a good indication of the quality of the design.

MARKET SIZE AND GROWTH PROJECTIONS

The market for RO systems used for production of high purity water will increase for two reasons. The demand for high purity water will increase, and RO will provide a larger percentage of current demand. Four of the five applications discussed earlier in the chapter, boiler feed, metal finishing, medical and pharmaceutical, and packaging, will probably expand at approximately the same rate as the economy in general. Assuming a 2.5 percent compounded annual growth rate in these applications would be reasonable. The demand for ultrapure water in the electronics industry will expand faster, perhaps 10 to 12 percent annually, although not as fast as in the early days of the industry.

The most significant growth of RO in high purity water production will be derived from increased use of RO in place of IX. In those applications where dissolved solids removal is the primary objective and economics is the sole factor governing the comparison of alternatives, the percentage of applications handled by RO will continue to expand. At one time it was considered economical to use RO only at TDS levels in excess of 350 ppm. Advances in membrane technology have reduced the cost of the membrane elements, and improvements in the applications skill of the equipment manufacturers have reduced the incidents of misapplication and the associated costs. The economic crossover point for employment of RO is therefore dropping. At this time the crossover point probably falls at a level of 100 to 150 ppm. This makes RO a viable demineralization technology in far more than 50 percent of the water supplies in the continental United States.

Other factors that will influence use of RO arise from concerns for health, safety, and the environment. IX uses strong acids and chemicals to accomplish separation, requiring on-site chemical handling and storage and the associated health, safety, and environmental risks. RO has probably far less negative impact on the environment than IX. Therefore, if all of the costs associated with the two technologies were taken into account, including the negative cost of any residuals produced in the processing, RO would likely be more attractive. Within the next 10 years, this should become an increasingly important factor.

In 1989 it was estimated that 2 million pounds of antiscalant were consumed in the conditioning of RO feed water within the continental United States. Based on a dosage level of 4 ppm, installed RO capacity is approximately 160 million gallons per day. Because approximately 805 of RO applications are for production of high purity water, approximately 130 million gallons per day of RO capacity is in place for this purpose. This is expected to expand by 15 to 20 percent in each of the next 5 years.

CONCLUSION

RO has established itself as a strategic process for the production of high purity water, especially in those applications where particulate, bacteria, and TOC levels must be closely controlled. The quality of the RO membranes available is improving, and the cost per gpm of capacity is decreasing, thus making this process increasingly attractive to all end users. Cost improvements are also derived from a greater understanding of RO application, design, and operation. RO use will expand as a result of expansion of the economy in general, but primarily because of replacement of IX.

References

Ammerer, N., and G.B. Dahman. 1990. Curing silica problems with double-pass reverse osmosis. *Ultrapure Water* 7(4):41–46.

ASME. 1979. American Association of Mechanical Engineers research committee on water in thermal power systems. Pamphlet—Library of Congress 79-54898. "Consensus on operation practice for the control of feedwater and boiler water quality in modern industrial boilers".

Dalazs, M.K., and S. Walker 1983 Solutions to some troublesome pure water. Paper presented at 2d Annual Pure Water Conferences, 13–14 January, San Jose, CA, pp. 110–122.

Balazs, M.K. 1988. *Balazs semiconductor pure water specifications and guidelines.* Balazs Analytical Laboratories. 1380 Borregas Avenue, Sunnyvale, Calif. 408-745-0600.

Benedek, A., and L. Johnston. 1988. Considerations in choosing RO versus ion exchange for demineralization. *Ultrapure Water* 5(7):51–62.

Burris, M.K. 1987. Boiler water treatment principles. *Ultrapure Water* 4(2):60–63.

Chu, T.S. 1989. Investigating THM's, as a cause of RO/DI system problems: The summertime blues revisited. *Microcontamination* 7(7):35–38.

Comb, L., and P. Schneekloth. 1990. High purity water using two-pass reverse osmosis. *Ultrapure Water* 7(3):49–53.

Envirex. 1968. Bulletin No. 335–501. Water Conditioning Division, Rex Chainbelt, Inc.

Gottlieb, M.C. 1989. Equilibrium and kinetic factors in high purity mixed bed demineralizers: Cations effects. *Ultrapure Water* 6(1):32–34.

Green, B.L. 1983. "Microbiological monitoring of electronics grade water." In *Proceedings of the Second Annual Semiconductor Pure Water Conference*, 13–14 January, San Jose, CA, pp. 42–48.

Hashimoto, N., H. Satou, T. Shinoda, and K. Takino. 1990. "Manufacturing equipment for ultrapure water for 16m devices." In *Proceedings of the Ninth Annual Semiconductor Pure Water Conference,* 17–18 January, San Jose, CA, pp. 2–25.

Miller, W.S., and T.L. Fredericks. 1988. Reverse osmosis treatment of demineralized makeup supply to improve steam generator chemistry. *Ultrapure Water* 5(1):44–49.

Pittner, G.A., R. LeVander, and J. Bossler. 1985. Unique double-pass reverse osmosis system eliminates ion exchange for many demineralization applications. Paper presented at 46th Annual International Water Conference, 4–7 November, Pittsburgh, Pa.

Rohm and Haas. 1974. *If you use water.* Publication No. IE-1-54-C. Philadelphia.

Whipple, S.S., E.A. Ebach, and S.S. Beardsley. 1987. "The economics of reverse osmosis and ion exchange." In *Proceedings of First Annual Ultrapure Water Conference and Exposition.* Philadelphia, Pa.

12

An Overview of U.S. Military Applications of Reverse Osmosis

S. A. Choudhury

U.S. Army Belvoir Research, Development and Engineering Center
Fuel and Water Supply Division

INTRODUCTION

The driving principle behind the military land-based water program is that no matter where deployed, the land-based military forces of the United States must be able to locate, purify, store, and distribute water. This chapter discusses the use of reverse osmosis (RO) technology by the military to purify water for the deployed land-based forces. RO technology used at military installations is not considered here, as it is similar to that of municipal water systems.

The armed forces have two standard Reverse Osmosis Water Purification Unit (ROWPU) systems that are currently in use. One is nominally rated at 600 gallons per hour (gph), used in processing fresh water, and the other at 150,000 gallons per day (gpd) for seawater. Under production is a system with a nominal rating of 3,000 gph for use with fresh water. Table 12.1 summarizes these systems. A smaller ROWPU is, at present, under development.

BACKGROUND

The basic need for a military water purification system is the ability to purify water from any available source. The raw water may be fresh, brackish, or seawater; it may also be highly polluted, turbid, or contaminated by nuclear, biological, or chemical (NBC) warfare agents. In the post-World War II era, the Army developed different types of equipment to treat various types of raw water supplies. As the structure for a mobile Army was evolving in the 1960s, it was

---

The statements in this paper are the author's own, and do not reflect official Department of Defense positions. Inclusion of trade names are for identification purposes only.

TABLE 12.1   Summary of ROWPUs

	600-gph ROWPU	3,000-gph ROWPU	150,000-gpd ROWPU
Design accepted	1979	1987	1981
Production rate (gpd):			
Fresh water	12,000	60,000	—
Seawater	8,000	40,000	125,000
Crew size	3	3	7
Elements used:			
Size	6-inch	8-inch	6-inch
Quantity	8	12	80
Weight (pounds)	16,975[1]	—	—
	11,380[2]	39,800[3]	—
Size (feet)	19 × 8 × 8[1]	30 × 13 × 8[4]	—
	10 × 8 × 8[2]	—	
Planned quantity	1,600	400	28
Total installed capacity (seawater)	12.8 mgd	16.0 mgd	3.5 mgd

[1]Trailer-mounted version
[2]Skid-mounted version
[3]ROWPU container only
[4]Includes trailer

apparent that a single multipurpose water treatment system would be needed to replace the various types of equipment currently in use. In the early 1970s, after evaluation of several alternatives, RO was selected as the most appropriate technology for military use. RO could not only treat all waters, but was also found to be capable of successfully treating water contaminated by NBC warfare agents. Work on RO-based water treatment equipment was initiated in the mid-1970s, and the first ROWPU design was implemented in 1979.

DESIGN CONSIDERATIONS

In addition to standard design requirements such as safety, reliability, and ease of maintenance, use by military forces imposes additional requirements on a water treatment system. These relate to battlefield survival and logistical support, such as availability of spare parts and repairs.

ROWPUs, like most military systems, have to meet stringent weight and size limitations to allow for movement by all modes of transportation. ROWPUs have to be designed to function effectively in the modern battlefield, which may include the use of NBC warfare agents. This implies that ROWPUs have to be mobile or easily transportable, and must have the ability to be put into operation and packed up quickly. A 2-hour limit is usually allowed to put a ROWPU into full operation from a completely packed configuration. As ROWPUs can be deployed anywhere in the world, in varied climates, they must be designed to function in air temperatures from $-25°F$ to $+110°F$.

The military RO systems are designed to be in continuous operation—usually in two 12-hour shifts. During a 24-hour day, the ROWPU is expected to be operated for at least 20 hours, the remaining time being used for maintenance activities such as cleaning the media filter or setting up or tearing down the equipment.

Another important factor is that the military personnel operating the equipment are not highly trained chemists or engineers, nor do they have a fully functioning laboratory to advise or support them. The equipment and operating instructions must be simple, yet cover all possible contingencies.

MILITARY WATER TREATMENT

A unique aspect of the military application of RO technology is that a standard design must have the ability to treat all water sources without any modifications to the system. ROWPUs are not site specific—they must be capable of handling all waters. The raw water can be from a surface or subsurface source; it can be fresh, brackish, or seawater. Based on total dissolved solids (TDS) concentration, waters are defined as follows:

Fresh water   TDS: up to 1,500 mg/l
Brackish water   TDS: 1,501 to 15,000 mg/l
Seawater   TDS: greater than 15,000 mg/l

The raw water may also be polluted or contaminated with NBC warfare agents. The only limitations placed on the water is that the raw water turbidity must be less than 150 NTU (Naphelometric Turbidity Unit), and water temperature must be between $+32°F$ and $+110°F$. As an example of the diversity of raw waters that military systems can encounter, the characteristics of some of the sites at which ROWPUs have been operated are shown in Table 12.2.

The RO flow train is designed as a single-pass system using seawater RO elements. The recovery rates for seawater and fresh water are approximately 30 and 50 percent, respectively. The operating pressure is set to obtain the desired recovery. The typical upper limits are 500 psi for a fresh or brackish water source, and 950 psi for seawater.

Drinking water supplies must meet the stringent field water-supply standards established by the military services, and must be approved by military medical authorities. One of the requirements is that drinking water supplies contain a free

TABLE 12.2   Raw Water Characteristics at Some ROWPU Sites

	Pacific Ocean	James River, Virginia	River, Overseas	Panama City, Florida
pH (pH units)	7.8	7.8	7.6	8.4
Temperature (°F)	64.0	58.0	—	86.0
TDS	31,950.0	13,000.0	200.0	18,000.0
TSS	—	33.0	—	—
Turbidity (NTU)	—	—	19.0	0.4
TOC	—	4.1	3.9	4.9
Tot Alk (as $CaCO_3$)	—	66.0	13.0	67.0
Hardness (as $CaCO_3$)	2,920.0	2,600.0	109.0	2,830.0
Calcium	250.0	205.0	27.3	180.0
Magnesium	560.0	508.0	10.0	580.0
Sodium	11,000.0	4,110.0	22.1	5,100.0
Chloride	17,745.0	7,500.0	10.0	9,600.0
Sulfate	2,577.0	910.0	17.0	1,400.0
Silica	3.0	—	—	—

Units are mg/l unless otherwise specified.
TDS   = total dissolved solids
TSS   = total suspended solids
TOC   = total organic content
Tot Alk = total alkalinity

chlorine residual. Consequently, the RO permeate has to be chlorinated before it is considered potable.

Associated with military water treatment equipment are test kits that give operators the capability to conduct limited water quality analyses. These test kits are used by operators to monitor the treatment process and product water quality.

There are some generic problems associated with ROWPUs. Owing to the size and weight limitations, the pretreatment system may not be adequate at certain sites. This may result in filter breakthrough and fouling of the RO elements. As ROWPUs are used only when needed, downtime between operations is quite common. This makes the disinfection and short-term storage of elements crucial. Improper disinfection, preservation, or storage may lead to the degradation of membrane performance.

600-GPH ROWPU

The 600-gph ROWPU was originally developed at the U.S. Army Research, Development and Engineering Center at Fort Belvoir, Virginia, (BELVOIR) as a light-weight, mobile water purification unit for use by highly mobile parachute units. Since then it has come into use as a standard water treatment unit for the

Army, the Marine Corps, and the Air Force. The ROWPU was designed for use in a forward combat area, behind front-line forces. It has been used extensively in training exercises both in the United States and overseas. The 600-gph ROWPU has also been used to purify water in actual operations and for disaster relief.

There are three standard variants on the basic design. In the Army version, shown in Figure 12.1, the skid containing the ROWPU is mounted on a trailer along with a 30 kW diesel generator. The generator provides all the power needed by the ROWPU. All ancillary equipment and a minimum of a 100-hr supply of operating chemicals, expendables, and spare parts are packed on the trailer. This allows for self-contained, independent operations. The Marine Corps and Air Force versions do not include the trailer or a generator and have minor variations in the electrical and piping arrangements. All versions of the 600-gph ROWPU are capable of being moved by highway, railroad, and marine modes of transportation, as well as by military airlift aircraft. If needed, the 600-gph ROWPU can be air-dropped by parachute.

The open skid design limits the temperature range in which the 600-gph ROWPU can operate. When operating at air temperatures below 32°F, the ROWPU and associated equipment have to be inside a heated enclosure, such as a tent, to prevent water from freezing and to avoid damage to the RO elements.

A flow diagram for the 600-gph ROWPU is shown in Figure 12.2. The pretreatment process consists of coagulation by addition of a cationic polymer,

FIGURE 12.1.   600-gph ROWPU at a field site.

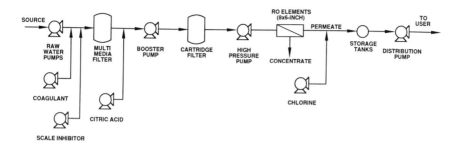

FIGURE 12.2.   Flow diagram—600-gph ROWPU (normal operation).

followed by filtration through a multimedia filter and cartridge filters. Chemicals are also added to the RO feed to reduce scale formation. The eight 6-inch elements in series provide the final purification. Calcium hypochlorite is added to the RO permeate, prior to its leaving the ROWPU, to render the water potable.

Purifying water contaminated by NBC warfare agents provides a further challenge. Studies have indicated that the bulk of the contamination is removed by the pretreatment and RO elements, reducing the quantity of NBC warfare agents to nonhazardous levels. As an additional precaution, however, a posttreatment step is added to the process when operating in an NBC environment. The posttreatment consists of activated charcoal and ion exchange (IX) filters to remove any residual contamination.

During initial setup the concentrate stream is collected for later use in backwashing the media filter and for cleaning the RO elements. Backwash of the multimedia filter is performed after a predetermined change in the differential pressure or every 20 hours, whichever comes first, or at shutdown. The cartridge filters are changed based on an increase in pressure drop.

The operating procedures for the 600-gph ROWPU allow for two types of RO element cleaning—cleaning with citric acid and cleaning with a surfactant. The type of chemical cleaning is left to the discretion of the operator in charge. The RO elements have to be cleaned when there is a 20 percent increase over the initial differential pressure readings. Chemical cleaning is also required when a drop in permeate flow or quality is noted or when the maximum operating pressure has been exceeded. Operational parameters and other information on the 600-gph ROWPU are summarized in Table 12.3, and data from two separate operations are given in Table 12.4.

3,000-GPH ROWPU

The newest member of the family of water purification equipment is the 3,000-gph ROWPU. The unit was accepted as standard in July 1987 and is undergoing final testing prior to being used. At this time, the system will be used only by the

TABLE 12.3   Characteristics of the 600-gph ROWPU

Flow rates (at 77°F)	Feed	32 gpm
	Permeate	
	Fresh/brackish	10 gpm
	Seawater	8 gpm
Normal operating pressures	RO Feed	
	Fresh/brackish	≤500 psi
	Seawater	≤960 psi
Operator actions	Cartridge filter change	ΔP > 20 psi
	Multimedia backwash	ΔP > 10 psi
	Element chemical clean	ΔP > 100 psi
Cartridge filter	8 each, 40-in long, 5 μm (nominal)	
Multimedia filter	Flow rate	6.5 gpm/sq ft
RO vessels and elements	4 pressure vessels × 2 elements (6-inch)	
Chemicals	Coagulant	
	Antiscalant	
	Calcium hypochlorite	
Operating temperature	−25 to 110°F air temperature	
range	+33 to 110°F water temperature	

TABLE 12.4   600-gph ROWPU—Operational Data

	Fresh Water		Brackish Water	
	Raw	Product	Raw	Product
pH (pH units)	7.9	6.0	8.0	5.8
Temperature (°F)	86.0	—	57.0	—
TDS	3.0	2.0	13,000.0	120.0
TSS	—	—	34.0	1.0
Turbidity (NTU)	1.8	<0.2	—	—
TOC	8.2	<0.5	3.8	<1.0
Tot Alk (as CaCO₃)	53.0	2.0	68.0	2.4
Hardness (as CaCO₃)	58.8	<1.0	2,510.0	7.2
Calcium	19.9	<1.0	221.0	<1.0
Magnesium	2.2	<0.2	476.0	1.7
Sodium	5.7	<1.0	4,070.0	36.0
Chloride	6.4	<1.0	13,000.0	61.0
Sulfate	5.7	<1.0	980.0	<2.0
Hours of operation	320		17	
Cartridge filter changes	2		0	
Media filter backwashes	17		1	
RO element chemical cleanings	3		0	

Units are mg/l unless otherwise specified.
TDS   = total dissolved solids
TSS   = total suspended solids
TOC   = total organic content
Tot Alk = total alkalinity

Army. The 3,000-gph ROWPU will be employed in the rear portions of combat areas. This system was designed to survive in a nuclear battlefield, as well as to withstand contamination by NBC warfare agents. Although the 3,000-gph ROWPU was designed for operations at temperature extremes, a winterization kit is used when the ambient air temperature is below 32°F.

The 3,000-gph ROWPU, as seen in Figure 12.3, is enclosed in an 8' × 8' × 20' container, with the high pressure pump mounted in front of the container. The electrical power needed to operate the ROWPU is provided by a standard military 60 kW diesel generator. Mobility of the system is enhanced by having all the components mounted on a standard military 22½-ton Army trailer. The 3,000-gph ROWPU, like the 600-gph ROWPU, is transportable by all modes of transportation.

As seen in Figure 12.4, the flow through the 3,000-gph ROWPU is quite similar to that of the 600-gph ROWPU. Added features in the pretreatment process are the cyclone separators at the raw water pump, which remove the larger suspended particles such as sand and heavy dirt, and the basket strainer to remove large particles that may damage the media filter. The twelve 8-inch elements are divided into two parallel streams, each with six RO elements in series. As with the 600-gph ROWPU, a NBC posttreatment is used with the 3,000-gph ROWPU when operating in an NBC environment.

Alarms at the control panel, set off by increases in differential pressure, indicate to the operator when to change cartridge filters, clean the basket strainer, or backwash the media filter. Other conditions that require media filter backwashing are specified in the operator's manual.

FIGURE 12.3.   3,000-gph ROWPU at a field site.

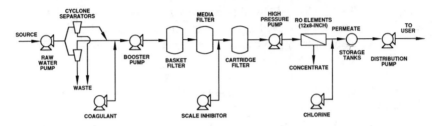

FIGURE 12.4.   Flow diagram—3,000-gph ROWPU (normal operation).

Chemical cleaning of the RO elements is conducted on a schedule based on hours of operation, feed-water condition, and the RO differential pressure. Element cleanings are also conducted when there is a decrease in permeate quality or quantity.

As this system is still in production, operational field data are not available. Table 12.5 summarizes the operational characteristics of the 3,000-gph ROWPU.

150,000-GPD ROWPU

The 150,000-gpd ROWPU was used as an interim measure until the 3,000-gph ROWPU could be designed and deployed. This system most closely resembles commercial RO plant designs. A "short-term fix," the 150,000 gpd ROWPU was not designed to satisfy all the unique military requirements. Although the 150,000-gpd ROWPU was designed for fixed-location operations on land (see Figure 12.5), in an interesting adaptation some of the 150,000-gpd ROWPUs

TABLE 12.5   Characteristics of the 3,000-gph ROWPU

Flow rates (at 77°F)	Feed permeate	101.0 gpm
	Fresh	53.0 gpm
	Brackish	43.0 gpm
	Seawater	33.5 gpm
Normal operating pressure	RO feed	≤900 psi
Operator actions	Cartridge filter change	ΔP > 10 psi
	Multimedia backwash	ΔP > 25 psi
Cartridge filter	10 each, 30-inch long, 5 μm (nominal)	
RO vessels and elements	4 Pressure vessels × 3 elements (8-inch)	
Chemicals	Coagulant	
	Antiscalant	
	Calcium hypochlorite	
Operating temperature range	−35 to 110°F	

FIGURE 12.5.   150,000-gpd ROWPU at a field site.

have been mounted on standard military barges to provide an off-shore desalination capability. There are two such designs—a 150,000-gpd and a 300,000-gpd system.

The usual operational scenario calls for the 150,000-gpd ROWPU to be set up on a beach, with the purified water being either stored near the site or pumped farther inland into a large storage complex. In the barge-mounted systems, the product water is chlorinated and then pumped to a shore-based storage complex.

Since the initial procurement of the system in 1981, units of this design have been used in several field exercises in the United States and overseas. The 150,000-gpd ROWPU has also been used to supply water to some fixed facilities on an emergency basis. Data on the raw water and permeate from one such operation are listed in Table 12.6. In a 22-day period, two ROWPUs were used

TABLE 12.6   150,000-gpd ROWPU—Operational Data

	Raw	Product
pH (pH units)	8.3	5.8
Temperature (°F)	—	98.0
Total dissolved solids	34,300.0	39.0
Total alkalinity (as $CaCO_3$)	125.0	3.0
Hardness (as $CaCO_3$)	6,875.0	8.0
Calcium	1,375.0	2.0
Chloride	20,500.0	16.0

Units are in mg/l unless otherwise specified.

to produce approximately 3.6 million gallons. It should be noted that in this operation each ROWPU was operated for about 10 hours each day.

The components of the 150,000-gpd ROWPU are mounted on eight skids. The standard site layout is shown in Figure 12.6, and the flow diagram is given in Figure 12.7. The complete set includes consumable supplies and spare parts, as well as replacement pumps and engines for independent operations at a remote site.

The 150,000-gpd ROWPU has less mobility than the other ROWPU systems. Material-handling equipment, such as forklifts, is needed to set up the system prior to operation. This system may also require site preparation. Once set up, the 150,000-gpd ROWPU is expected to operate from that location for long periods of time.

The water treatment process in this system is similar to that in other ROW-PUs. The booster pump, in addition to maintaining the flow through the multimedia filters and the cartridge filters, is also used for backwashing the media filters and chemical cleaning of the RO elements. In common with the other systems, cartridge filter changes and multimedia filter backwashing are performed in response to increases in pressure differentials. The backwashing procedure is interesting in that each multimedia filter is washed in sequence, the

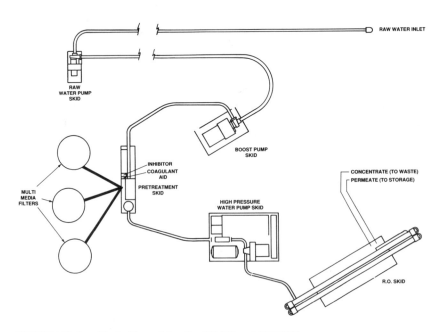

FIGURE 12.6.   Typical layout for the 150,000-gpd ROWPU.

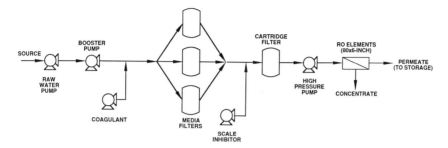

FIGURE 12.7.   Flow diagram—150,000-gpd ROWPU.

effluent from the other two being used for backwashing the third. The flow rates and length of backwashing are dependent on the water temperature. The RO elements have to be chemically cleaned when the pressure drop across the vessel increases by 10 psi over the starting value. The operational characteristics of the system are summarized in Table 12.7. In contrast to the other ROWPU designs, the 150,000-gpd ROWPU has no chlorination capability. The reason is that the output from this system goes into a standard storage and distribution system, which has chlorination capability.

TABLE 12.7   Characteristics of the 150,000-gpd ROWPU

Flow rates (at 77°F)	Feed	350 gpm
	Permeate	108 gpm
	Concentrate	242 gpm
Normal operating pressures	Booster pump	100 psi
	RO feed	≤820 psi
Operator actions	Cartridge filter change	$\Delta P \geq 12$ psi
	Multimedia backwash	$\Delta P \geq 35$ psi
	RO element cleaning increase	$\Delta P \geq 10$ psi
Cartridge filters	12 each, 40-in long, 10 μm (nominal)	
Multimedia filters	3 media filters with:	
	Flow rate: 10 gpm/sq ft	
	Media:	
	Anthracite	
	Silica sand	
	Fine garnet	
	Coarse garnet	
RO vessels and elements	16 pressure vessels × 5 elements (6-inch)	
Chemicals	Coagulant	
	Antiscalant	
Operating temperature range	−25 to 110°F air temperature	
	+32 to 110°F water temperature	

REVERSE OSMOSIS ELEMENTS

A thin film composite membrane was developed by private industry to meet the military's need. Part of the funding for the effort was provided by BELVOIR. The membrane was fabricated into 6-inch spiral wound elements to be used in military systems. To increase competition, additional sources for membranes were developed.

ROWPU systems use standardized seawater RO elements in two sizes. The 6-inch element is used with the 600-gph ROWPU and the 150,000-gpd ROWPU, whereas the 3,000-gph ROWPU uses an 8-inch element. The performance requirements for these elements are given in Table 12.8.

In military use, an element is expected to last 2,000 hours, with a 4,000-hr life as the desired goal. This relatively short life is due to the greater demands placed on the membrane. Varying and rigorous operating conditions, such as incomplete removal of suspended particles, biofouling, exposure to extremes of temperature, and the variety of raw water sources encountered by a typical ROWPU, all contribute to the reduced membrane life.

The military purchases RO elements from sources that have been previously qualified. The qualification process, which is extensive, verifies that a particular RO element can meet the requirements of military service. The qualification program for 6-inch elements consists of the following tests:

- A 200-hr Intermittent Operation Test
- Cold (32°F) Environment Storage Test
- Hot (120°F) Environment Storage Test
- High Temperature (120°F) Feed-Water Test
- NBC Warfare Agent Removal Evaluation

TABLE 12.8   Military RO Elements

	6-inch RO Element	8-inch RO Element
Dimensions (inches) (diameter × length)	6 × 40	8 × 40
Rejection	>99.0% Cl⁻	>98.7% Cl⁻
Flow (gpd)	1,850–2,550	6,000
Maximum operating pressure (psi)	1,000	1,000

Test Conditions:
  Synthetic seawater (19,000 ± 1,000 mg/l Cl⁻)
  pH = 7.0 ± 0.2
  Pressure = 800 psi
  Temperature = 25°C (77°F)

Currently there is only one supplier for the 8-inch element. A qualification program similar to that for the 6-inch element will be conducted for the 8-inch elements to develop additional sources.

CHEMICALS

Chemicals are used with ROWPUs for coagulation and scale prevention and to clean, disinfect, and preserve RO elements. These chemicals are listed in Table 12.9. The need for compatibility with different membranes, and effectiveness over a broader operational environment, makes the use and selection of chemicals more complex than in commercial practice. Of course, all chemicals selected for use have to meet health, safety, and environmental requirements.

ONGOING AND FUTURE EFFORTS

In response to a need to provide water treatment equipment for small (50 to 150 personnel) military units, BELVOIR is developing a small ROWPU. In addition to military applications, this system is expected to be used in support of disaster relief and other civic action operations. The design concept calls for a fresh water treatment system, with an add-on single-pass RO module for salt and brackish water applications. The small ROWPU will have a production rate between 85 and 250 gph.

Fouling of RO elements, especially in fresh and brackish waters, is a problem. The foulant is typically a complex of colloidal and organic material with

TABLE 12.9   ROWPU Chemicals

	600-gph ROWPU	3,000-gph ROWPU	150,000-gpd ROWPU
Coagulant	DADMAC	DADMAC	Hydrapol 50[3]
Antiscalant	SHMP	AC 1000[1]	Hydrapol 100[3]
RO cleaning	Citric acid	Citric acid	Hydrakleen 20[3]
	Surfactant	PA 111[2]	
		ATMP	
Water disinfectant	Calcium hypochlorite	Calcium hypochlorite	—
RO membrane pre-servative disinfectant	—	Sodium metabisulfite	—

DADMAC = Diallyldimethyl ammonium chloride
SHMP = Sodium hexametaphosphate
ATMP = Aminotri (methylene phosphonic acid)
[1]AC 1000 is a proprietary chemical sold by Aqua Chem, Inc., Milwaukee.
[2]PA 111 is a proprietary membrane cleaner sold by Water Systems Technical Service, Cave Creek, Arizona.
[3]Hydrapol 50, Hydrapol 100, and Hydrakleen 20 are proprietary chemicals sold by Hydranautics, Inc., San Diego.

some iron. There are several ongoing efforts to evaluate causes and remedies for element fouling.

In an effort to improve the performance of ROWPUs and extend the operational life of RO elements, extensive research and development efforts are being undertaken in several areas: RO elements, chemicals, and pretreatment. In conjunction with private industry, different membrane configurations and types of element construction are being evaluated. Extensive effort is being expended on various water treatment chemicals. Specifically, work has been initiated to determine improved alternates to chemicals currently in use. The new chemicals will be safer, more effective, and environmentally acceptable.

It is important to remember that although the technical goals and objectives such as (1) the elimination of fouling problems, (2) the development of light-weight pretreatment, and (3) the determination of improved chemicals are all important, the ultimate objective of the military water program is to ensure that deployed forces have the best means available to obtain purified water.

13

An Overview of RO
Concentrate Disposal Methods

Laura S. Andrews
Camp Dresser & McKee

Gerhardt M. Witt
Parsons Brinckerhoff Gore & Storrie

INTRODUCTION

Membrane technology is an essential application for numerous industries throughout the world. As such, reverse osmosis (RO) has become an integral and vital process in a wide range of industrial and domestic uses. However, disposal of concentrate generated from the process can be an obstacle to membrane water treatment applications. Concentrate disposal is directly associated with membrane applications and should receive careful consideration simultaneously with the development of the membrane facility or in-line processes.

RO applications produce a waste stream or residual referred to as concentrate, brine, reject, or by-product. For the purpose of this overview, the term *concentrate* will be used to refer to the side-stream.

The concentrate is a condensed form of the raw feed source water quality. As defined by the American Society for Testing and Materials (ASTM), *concentrate* is "the residual portion of an aqueous solution applied to a membrane." The concentrate quality and quantity are measurably a direct function of the membrane application, process design, system operation, and raw feed-water quality.

An emerging issue requiring detailed study and resolution by the membrane industry and regulators is the identification of environmentally acceptable methods of concentrate disposal. Through heightened public awareness and increased regulatory constraints, limitations have been placed on concentrate disposal mechanisms. In addition, the permitting and disposal of concentrate is becoming and will remain a more complex process for users of membrane systems because of stricter environmental laws and regulations.

For a membrane technology project to be implemented, a method for disposal of the concentrate must be available. Some process designs incorporate staging and/or other processes in series to achieve additional recovery, in which the concentrate of one stage is the feed source for another. This method can effectively reduce the volume of concentrate discharge. For purposes of discussion within this text, however, the concentrate is considered to exist in its final form as it exits the membrane treatment facility site.

The following are typical means of concentrate disposal.

- Surface water discharge (ocean, tidal waters, nonsaline waters)
- Deep well injection
- Spray irrigation
- Waste water treatment facilities
- Thermal evaporation
- Solar evaporation ponds
- Drain fields and boreholes

It is important to recognize concentrate disposal as an integral part in the overall membrane process design. Disposal alternatives must be considered on a site-specific basis. For example, a surface water discharge may not be a feasible alternative if the closest surface water is 50 miles away from the operation.

REGULATORY BACKGROUND

Historically, concentrate disposal has not been in the forefront of the regulatory agencies' concerns. With the progress of the Clean Water Act and the Safe Drinking Water Act, surface water disposal (direct and/or indirect) and well injection into groundwater aquifers have come under intensified scrutiny. As water quality regulations become more stringent, the level of effort and amount of lead time required for developing and obtaining acceptable disposal methods will significantly increase. Pursuing early regulatory involvement, by allowing for preliminary discussions and input from applicable authorities, will save time and money for the disposal project.

RO concentrate is classified by the U.S. Environmental Protection Agency (EPA) as a secondary industry process waste water. It is defined in EPA regulation 40 CFR 122.2 as "any water which, during manufacturing or processing, comes into direct contact with or results from the production or use of any raw material, intermediate product, finished product, by-product, or waste product." The industrial classification of concentrate developed as a carryover from the state of Florida's 1970s classification of concentrate (Conlon 1990) as an industrial waste. This classification has led to an industrial permitting process for concentrate disposal.

Public perception tends to be negative when the term *industrial waste* appears in a permit application. There is debate as to whether concentrate represents an industrial waste when it is the concentrate from a municipal water treatment facility. Because drinking water facilities using RO applications generally produce a concentrate stream of higher quality than an industrial facility's waste stream, many proponents oppose application of the industrial classification to municipal facilities.

SURFACE WATER DISPOSAL

Surface water discharges are common to many types of industry. RO systems are no different when it comes to using surface water sources for concentrate disposal. Surface water discharge of concentrate is one of the most common methods owing to the availability, cost-effectiveness, water quality, and mixing volume associated with a surface water receiving body. A surface water discharge can range from an ocean outfall to a drainage canal, as long as the concentrate can meet the criteria established for the particular receiving body of water.

For the most part, existing regulations do not specifically address RO concentrate surface water discharge. Rather, these regulations, whether federal, state or local, define water quality criteria and the allowed uses for the receiving surface water body. Water quality criteria, in regard to pollutant type and allowable concentrations, are typically defined by the state, which must apply the minimum established federal requirements.

Florida is one of the largest users of municipal membrane technology in the United States. This can be directly attributed to the need for producing potable water from limited and often poor water quality sources for an ever-growing population demand. Membrane technology is applied in coastal areas because of high concentrations of chlorides and sulfates in the groundwater. As of 1990, approximately 12 percent of the RO plants in Florida produce more than 1 million gallons per day (gpd) of drinking water. A great number of these facilities use surface water discharge as their means of concentrate disposal (DeGrove 1990).

Florida's high level of RO application provides a fundamental base to characterize surface water disposal requirements and issues. RO concentrate is classified by the Florida Department of Environmental Regulation (FDER) as an industrial waste. This classification requires RO facilities, regardless of their type of application, to obtain an industrial discharge permit from the state for disposal of the concentrate.

In cases of surface water disposal, the concentrate quality must be acceptable to the receiving body's water quality standards. Surface waters of Florida are categorized by defined surface water quality criteria, and each surface water classification has minimum quality standards that cannot be exceeded. Typically, RO concentrates have at least one constituent that exceeds acceptable levels. For instance, RO drinking water facilities probably have one or more of the following parameters that do not meet surface water quality standards:

- Hydrogen sulfide
- Dissolved oxygen
- pH
- Specific conductance
- Chlorides
- Radionuclides
- Fluoride
- Metals

If the concentration of a constituent cannot be reduced to acceptable levels by treatment before discharge, a mixing zone must be established. A mixing zone is a neutral boundary area where the discharge quality may temporarily exceed water quality standards while allowing for mixing. It must be demonstrated that the discharge will be in compliance by the end of the specified mixing zone. Mixing zone evaluations have become a major study effort, particularly in tidally influenced water bodies. Supporting information such as modeling results, pollutant loadings, mass balances, and mixing zone length (Andrews et al. 1991) calculations must accompany a mixing zone proposal. In many cases, this information will require extensive documentation in regard to receiving water quality background and flow characteristics.

Western states, such as California, apply a salt balance methodology on a regional basis in the evaluation of concentrate disposal. In these states, it is recognized that the RO process removes the dissolved salts from the source. An overall basin salt balance is evaluated for the discharge.

Discharge is usually made to tidally influenced water bodies where a dilution ratio can be established. For concentrate disposal, the surface water is typically a brackish or salt water system. Fresh water (DeGrove and Baker 1990) sources generally cannot support the varied concentrate quality, particularly of chloride concentrations. There are certain instances in which concentrate may be discharged to nonsaline surface water sources.

Toxicity testing of the concentrate to the receiving water and its ecosystem is frequently requested by the permitting authorities. Hydrogen sulfide and dissolved oxygen content are common substances that have a possible toxic impact.

It is often necessary for surface water discharges to comply with recently evolved antidegradation standards. In keeping with the Clean Water Act, the EPA established an antidegradation policy, defined in 40 CFR 131.12. Florida derived an antidegradation policy, effective October 1989, from this federal policy. The main points of the policy are as follows:

- Existing uses of the water body must be fully maintained.
- Water quality standards of the water body must not be violated.
- New or expanded discharges must meet public interest tests.

The maintenance of surface water quality and usage are nothing new to the industry. Although the public interest tests have increased the level of effort necessary to achieve acceptable surface water disposal, these tests do provide some flexibility for the acceptance of the concentrate disposal plan. The public interest tests are divided into two parts. The first part involves a balancing test, demonstrating a benefit to public health, safety, and welfare when weighed against possible adverse effects to the environment. The second part requires an options review. This review must demonstrate that other means of disposal or locations of discharge have been considered and are shown not to be economically and/or technologically reasonable.

Federal requirements for surface water discharge to waters of the state, direct or indirect, call for permitting through the National Pollutant Discharge Elimination System (NPDES) program. The EPA generally administers the NPDES program; however, some states have taken over and maintain the responsibility of the program. RO concentrate is classified as a secondary industry process waste water and therefore requires an industrial-type NPDES permit for surface water disposal. In those states that have been delegated the NPDES permitting program, the permit is administered from the state permitting authority.

In many cases, surface water disposal of concentrate presents itself as one of the most cost-effective methods available. However, disposal to a surface water requires in-depth studies and evaluations to ensure acceptability to the regulatory entities and to address environmental concerns and issues.

DEEP WELL INJECTION

Another method of concentrate disposal is the discharging of the concentrate underground through the means of deep well injection. When conditions exist that may not allow for the use of surface water discharges, such as receiving water body acceptability, source availability, or regulatory constraints, deep well injection can be explored as a viable alternative. Deep well injection is a technically feasible means of discharging the concentrate underground.

The injection of fluids into underground water sources has been in wide practice for many years. Deep well injection systems are generally used to dispose of concentrate, waste water effluent, and industrial and hazardous wastes. Deep wells are classified by type of use and injected fluids. RO concentrate is classified as an industrial waste and requires a Class I injection well for disposal. There are several RO facilities in Florida using this means of disposal for their concentrate.

The *EPA Rules for Underground Injection Control* (United States Environmental Protection Agency 1988a) offer a comprehensive description of regulations and permitting procedures. In some instances, primacy for the enforcement of these regulations is delegated to the state level, provided the state's regulations are as stringent as the EPA's. For instance, Florida's Department of Environmental Regulation handles the deep well injection permitting program (United States Environmental Protection Agency 1988b). The EPA rules for underground injection control provide for the involvement of a Technical Advisory Committee (TAC), at the discretion of the state's and/or EPA's director or regional administrator, to evaluate each permit application for an injection well system (Florida Department of Environmental Regulation 1985). The TAC is composed of representatives from the EPA, the state, United States Geologic Survey (USGS), and various regional and/or local authorities. The TAC is responsible for the evaluation of the adequacy of the test/injection well design and testing program prior to issuance of construction and operating permits. The review includes the feasibility, geologic conditions, injection system design, testing, construction, monitor well program, and operation.

The TAC and/or the state agency and EPA administrator is advised of all phases of construction and testing for the injection well. TAC concurrence with major construction decisions (regarding casing depths, monitoring zones, etc.) are essentially necessary. Upon the completion and testing of the injection well, the results are compiled and presented to the TAC for review. The TAC reviews the injection system in its entirety and recommends approval, approval with conditions, or disapproval to the permitting authority (Florida Department of Environmental Regulation, 1985). Like surface water disposal, deep well injection is evaluated on a case-by-case basis. The feasibility of deep well injection is extremely site specific—the many subsurface geologic and groundwater conditions are the determining factors. This method of disposal is feasible from regulatory and technical viewpoints provided certain subsurface conditions exist. Important considerations typically include the following:

- The area of study must be seismographically stable, where the injection horizon is structurally stable with no known faults in the area of influence and/or in the down gradient of the injection well.
- Highly permeable, areally extensive injection zones must be present for large quantities of concentrate to be disposed of at reasonable injection pressures.
- Injection zones must contain saline water with total dissolved solids (TDS) concentrations in excess of 10,000 mg/l.
- Injection into the receiving formation water should have no economic impact nor endanger mineral resources, nor affect gas storage or fresh water storage systems.
- Confining beds must be of sufficient thickness and impermeability to confine the concentrate to the injection interval.

- The concentrate quality should be compatible with the aquifer source.
- The injection horizon must be confined by lithologic units having very low vertical permeability.

The last requirement is to keep injected fluids from migrating and contaminating aquifers that function as underground sources for draining water supplies. The EPA defines aquifers with TDS concentrations less than 10,000 mg/l as underground drinking water sources. Upward leakage of injected (United States Environmental Protection Agency 1988b) fluids is an issue of paramount importance, which is receiving closer scrutiny.

Water quality of the receiving aquifer is reviewed closely. The RO concentrate is generally of higher quality than the receiving aquifer. Detailed testing, explorations, and modeling are conducted for the proposed site. The successful completion of the injection well system is greatly dependent on the reliability of the information collected during the geologic review of the area and data obtained from the drilling of the test/injection well.

Drilling of test/pilot holes and collection of romation cores are necessary to determine the monitor zones, confining beds, injections horizons, aquifer quality, and casing depths. Upon completion of the test/pilot hole, geophysical logs are conducted in the bore hole by means of logging equipment. Geophysical logging should, at a minimum, include the following:

Single point electric	Gamma ray
Temperature	Flow meter
Caliper	Dual induction
Borehole compensated sonic logs	

The design of the injection system is based on the flow rate of the concentrate. The flow rate can be adjusted by designing retention systems to average out the peaking factors. Casing size is based on the injection velocities and pressures for average daily, maximum day, and peak hour flow rates. Injection pressures are considered in the design of the pumping system. These pressures are a function of the friction loss in the well casing and the injection fluid viscosity. In addition, the density differential between the concentrate, the injection horizon fluid, and the bottom hole driving pressures (transmissivity) of the formation must be evaluated in the feasibility determination.

There are stricter design and construction requirements for injection wells handling concentrate than for those handling waste water effluent. This is due to the industrial classification. The basic difference in the construction involves the tubing and packer of the well. EPA regulations require the injection casing to have a minimum 0.500-inch wall thickness. A typical injection well design is shown in Figure 13.1. The injection velocity is generally not recommended to exceed 8 feet per second in the well casing operating under peak flow conditions.

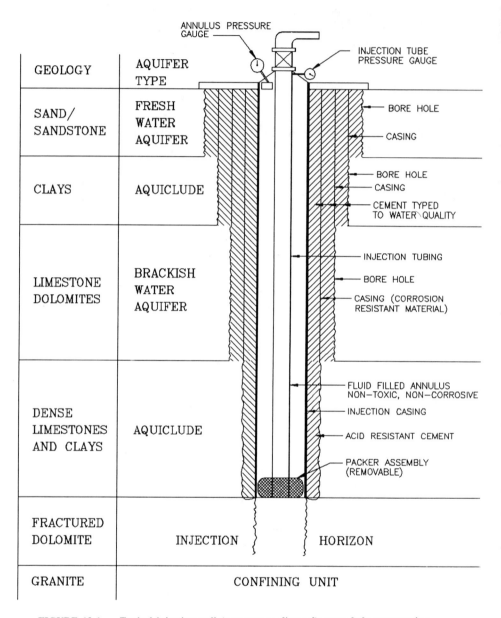

FIGURE 13.1.   Typical injection well (concentrate disposal): open hole construction.

The injection well system should have backflushing capabilities for remediation. Backflushing functions as a mechanism in the event of plugging relief under pressure. The backflushing performed routinely during low flow conditions enhances the longevity of the well and reduces the possibility of the formation becoming plugged.

Another consideration in the injection well design is a review of the RO concentrate quality. This is necessary from the standpoint of the compatibility with the casing type and material selection. Concentrate can be mildly corrosive to steel and can cause encrustation of the well's packer, casing, and formation. EPA regulations require the injection well to have a monitoring network system that can detect failures of the injection well and/or the confining sequences (Warner and Lahr 1977). It is possible that concentrate can migrate upward; therefore, employing a monitoring well network provides a warning system for the protection of the upper aquifers. Monitoring wells are typically located in aquifers above the injection zone to monitor water quality degradation. Operating permits often include a number of parameters to be tested regularly by sampling from the monitoring wells.

It is important to note that deep well injection systems are required to have a backup disposal method. The alternate method is essentially an entirely different means of disposal, either another injection well or some other permissible form of concentrate disposal (Witt and Ameno 1989). The levels of effort to study, design, and permit are basically the same. Deep well injection, although offering a viable alternative to surface water discharge disposal, can be quite costly. The use of an injection well for concentrate disposal may be the best alternative when no other acceptable means exists.

SPRAY IRRIGATION/LAND APPLICATION

Concentrate disposal can be accomplished by land application irrigation. Use of concentrate for irrigation purposes can offer a means of water conservation by offsetting the use of potable water, groundwater, or surface water for irrigation. Although this type of disposal method may appear beneficial, it is generally restricted by the concentrate quality. In most cases, RO concentrate quality, without the use of blending or extensive pretreatment, may not be acceptable for irrigation because of high concentrations of dissolved solids and chlorides.

Spray irrigation disposal is also used in limited applications. Such applications have been initially used as pilot study programs to assess possible environmental impacts, vegetation tolerance, and overall program feasibility.

The concentrate may be pretreated or blended with other water sources, such as treated waste water effluent, to provide a usable irrigation product. Concern with this method of disposal focuses primarily on the flora, fauna, and/or crop tolerance and impacts to groundwater and receiving surface water systems. The limiting constituents in the concentrate are usually TDS, chlorides, radionuclides, and metals. Issues to be addressed prior to implementation of a spray irrigation program are the effects of salinity, soil permeability, specific ion toxicity, and constituent elemental buildup (Edwards and Bowdoin 1988).

The possible impacts to the groundwater aquifers through percolation and to surface waters that may receive runoff must be evaluated. Prior to implementing a study of spray irrigation, the involvement of federal, state, and local regulatory agencies is important in establishing program feasibility and testing protocol.

Demand for irrigation water in relative proximity to the source is an important consideration. Wet weather management practices and storage have to be established. If wet weather storage can not be accomplished, a backup disposal method must be in place. With this type of disposal, a monitoring well network system and regular parameter testing can be expected to provide data on groundwater background quality and the impacts of the irrigation program.

OTHER DISPOSAL METHODS

There are a variety of other disposal methods that offer acceptable and viable alternatives for an RO facility. Use of an existing waste water treatment plant is a possibility for concentrate disposal. This alternative can eliminate much of the cost in studying, designing, permitting, and operating an independent concentrate disposal system. There are no direct federal regulations limiting the discharge of concentrate to a waste water facility. However, discharges must not cause a violation of the waste water facility's treatment capability, effluent disposal permits, or sludge disposal permits.

Federal regulations require discharges to sewer systems to comply with criteria established by the Nuclear Regulatory Commission (NRC) for radioactive materials (United States Environmental Protection Agency 1991). Furthermore, federal, state, and local industrial pretreatment requirements for waste water facilities must be reviewed.

Use of this method is strictly dependent on a waste water plant's capacity in terms of quantity and, more important, the plant's ability to treat the concentrate without upset to its own treatment regime. This requirement, coupled with local sewer ordinances, usually limits the feasibility of using a waste water treatment facility.

Some of the more innovative technologies for concentrate disposal are the applications of evaporation systems, solar ponds, and distillation systems. These types of disposal alternatives can be cost intensive. Evaporator systems have high energy demands. The concentrate residual may be increased to levels that are classified as hazardous. Solar systems can be land intensive and are dependent on climatic conditions for the sunlight necessary for evaporation.

Drain fields and boreholes offer still other means of disposal. Discharge is to the surficial aquifers. Drain fields or boreholes may be permitted in areas near shorelines where brackish water underlies the land. The concentrate water quality must meet the requirements for surficial groundwater aquifers (Conlon 1990). Use of these methods is dependent on appropriate soils and permeability. Monitoring systems are required to monitor the water quality.

ACKNOWLEDGMENT

The authors would like to thank William Conlon, Camp Dresser and McKee; Robert Carnahan, University of Southern Florida; and William Suratt, Richard Moore, and Marlene Hobel, Camp Dresser and McKee, for review of the manuscript and Jane Burns for drafting work.

References

Andrews, L.S., R.O. Moore, and C.T. Rose. 1991. "Permitting the discharge of reverse osmosis concentrate to a surface water." In *American Water Works Association Membrane Processes Proceedings,* 10–13 March, Orlando, Florida.

American Society for Testing and Materials. Standard definitions of terms relating to water, ASTM Standard Designation D1129–826. *Annual Book of ASTM Standards.* Philadelphia.

Conlon, W.J. 1990. "Disposal of concentrate from membrane process plants." In *Florida Engineering Society—Reverse Osmosis and other Membrane Technology,* June, Tampa, Florida.

DeGrove, B., and B. Baker. 1990. Discharge of reverse osmosis concentrates to surface waters: The regulatory process in Florida, Vol. 1. In *National Water Supply Improvement Association Biannual Conference Proceeding.*19–23 August, Orlando, Florida.

Edwards, E., and P. Bowdoin. 1988. "Irrigation with membrane plant concentrate: Fort Myers case study." 1988. In *Proceedings on Disposal of Concentrates from Brackish Water Desalting Plants.* National Water Supply Improvement Association and South Florida Water Management District. 18 November, Palm Beach Gardens, Florida.

Florida Department of Environmental Regulation. 1985. *Florida administrative code,* chaps. 17–28.011.

Warner, D.L., and J.H. Lahr. 1977. *An introduction to the technology of subsurface wastewater injection,* 600/2-77-240, Springfield, Va.: Environmental Protection Agency Technical Information Service.

Witt, G.M., and J.J. Ameno. 1989. City of Plantation Northwest Treatment Plant: R.O. Reject Injection Wall Preliminary Design (consultants report). Camp, Dresser and McKee Inc., Plantation, Florida.

United States Environmental Protection Agency. 1988a. *Permit regulations for the underground injection control program,* part 144.

United States Environmental Protection Agency. 1988b. *Regulation on state underground injection control programs,* part 147, subpart K-Florida 147.500.

United States Environmental Protection Agency Office of Drinking Water. 1991. *Suggested Guidelines for the Disposal of Drinking Water Treatment Wastes Containing Naturally Occurring Radionuclides.*

Index

Accounting rate of return, 78–79
Activated carbon, chlorine, 118
Aftergrowth, biocide, 197–198
Alkalinity
  carbon dioxide, 114
  water, 104–105
Aluminum
  solubility and pH, 110, 111
  water analysis, 145
Analysis, 139–162
Anion, 24
  polymer nodule, 57
Antidegradation policy
  Clean Water Act, 382–383
  Florida, 382–383
Antiscalant, 118–119
Aramid, 7
Aromatic polyamide membrane
  biofouling, 187
  casting solution composition
    polymer aggregate size, 56
    second pore radius, 56
    triangular diagram, 55
  organic solute separation, 50
  pore size
    average, 51
    distribution, 51
Aromatic polyamide polymer
  chlorination, 45
  structure, 45
Asymmetric aramid fiber, 1
Atomic absorption, foulant
  advantages, 238
  disadvantages, 238
Attenuated total reflectance analysis, foulant, 259–266
  calcium carbonate, 263–264
  cellulose triacetate membrane, 267
  clay, 265
  hydrocarbon-oil, 265
  limitations, 263
  organic fatty acid, 266
  principles, 262

Auger, foulant
  advantages, 238
  disadvantages, 238
Auger spectroscopy, foulant, 271–272

B-9 polymer, halogen substitution pattern, 44
Backscattered electron imaging, foulant, 246–247
Bacteria, foulant, 253–256
Barium, water analysis, 145
Barium sulfate, brackish water desalination, 278
Beer, reverse osmosis, 314–316
  concentration, 316
  low-alcohol, 314–316
Bench scale test, industrial effluent treatment, 285–288
  batch concentration mode, 287
  total recycle mode, 285
Benzanilide, 44
Beverage industry, reverse osmosis, 314–318
Bicarbonate, 24, 105
  water analysis, 146
Biocide, 194
  additional processes, 194–195
  aftergrowth, 197–198
  2,2-dibrom-3-nitrilopropionamide (Labozid), 196
  effectiveness, 194
  formaldehyde, 195–196
  hydrogen peroxide, 196
  hypochlorite, 195
  ozone, 196
  peracetic acid, 196
  regrowth, 197–198
Biofilm
  composition, 183–186
  development, 181–183
  growth, 181–183
  primary detachment, 181
  structure, 183–186
Biofoulant sanitization, 193–197
  effectiveness evaluation, 197–198
Biofouling, 163–201. *See also* Foulant; Fouling
  aromatic polyamide composite membrane, 187
  biofilm composition, 183–186

biofilm growth and development, 181–183
biofilm structure, 183–186
cases, 164–165
causes, 164–165
chlorination/dechlorination system, 195
cleaner, 196–197
conditioning film, 170–171
control, 186–198
culturing techniques, 188–189
fluid parameter influence, 175
fungus, 188
identification, 168–169
inhibition, 180–181
mechanistic aspects, 170–186
membrane surface influences, 176–177
microorganism, 172–174
microscopic techniques, 188–189
module autopsies, 183–186
mold, 188
monitoring, 191–193
as operational definition, 199–201
plateau phase, 199
polyamide membrane surface, 184, 185, 186
prediction, 187
pretreatment, 189–191
    sources, 164–165
prevention, 186–188
primary adhesion
    kinetics, 177–180
    models, 177–180
primary biofilm detachment, 181
primary microbial adhesion of microorganisms to
    membranes, 171–177
Pseudomona vesicularis, 173
Pseudomonas diminuta, 172, 173, 174, 175, 178,
    180, 181
Pseudomones fluorescens, 173, 174
reverse osmosis system
    effects, 165–168
    signs and symptoms, 165–166
risk factors, 187
schematic depiction, 170
silt density index, 189
Staphylococcus warneri, 172, 175, 180, 181
Boiler feed water, high purity water production,
    335–337
Borehole, concentrate disposal method, 388–389
Brackish water desalination, 275–282. *See also* De-
    salination
ancillary equipment, 276–277
barium sulfate, 278
biological control, 279
calcium carbonate, 278
calcium sulfate, 278
current trends, 280–282
defined, 275
ferrous ion, 279
forecast, 33
hydrogen sulfide, 279
magnesium carbonate, 278
module array configuration, 276
polyelectrolyte crystallization inhibitor, 278
precipitate prevention, 278–279

pretreatment, 277–278
silica, 279
sodium hexametaphosphate, 278
standard methods, 280
strontium sulfate, 278
Brine. *See* Concentrate
Bundle, defined, 8
By-product. *See* Concentrate

Calcium, 24, 105
    water analysis, 145
Calcium carbonate, 263–264
    brackish water desalination, 278
Calcium sulfate, brackish water desalination, 278
Calcium sulfate dihydrate, optical microscopy, 240,
    243, 244
Capital cost
    distillation, 93
    electrodialysis, 93–95
    evaporation, 93
    ion exchange, 90–92
        chemical costs, 92
        energy costs, 92
        insurance, 92
        labor costs, 92
        maintenance, 92
        replacement, 92
        resin attrition, 92
        taxes, 92
    reverse osmosis, 87–90
        chemical conditioning, 88–89
        electricity, 89
        maintenance, 90
        operating labor, 90
        polishing, 90
        repair, 90
        reverse osmosis membrane cleaning, 89–90
        reverse osmosis membrane replacement, 89–90
Carbon dioxide
    alkalinity, 114
    high purity water production, differential passage,
        347
    water, 105
    water analysis, 148
Carbonate, 24, 105
    water analysis, 147
Cardo membrane, inner skin layer transmission elec-
    tron micrographic image, 56–57
Cartridge, defined, 8
Cartridge filter analysis, water analysis, 156–158
Casting solution composition, aromatic polyamide
    membrane
    polymer aggregate size, 56
    second pore radius, 56
    triangular diagram, 55
Cation, 24
    polymer nodule, 57
Cellulose acetate, 1, 6–7
    salt rejection vs. membrane age, 119–120
    system flow rate vs. membrane age, 119, 121
    system flux decline, 121
Cellulose triacetate hollow fiber permeator, 2
Cereal processing, reverse osmosis, 318–322

Cereal processing *(continued)*
  by-product recovery, 318–319
  dilute sweet water concentration, 322
  reverse osmosis permeate and evaporator overhead polishing, 322
  steep water reverse osmosis evaporation, 322
  thin stillage treatment, 318–319, 320
Charged composite membrane, sodium chloride separation, 65
Cheese whey concentration, reverse osmosis, 301–305
  flow diagram, 304
  membranes, 303
Chemical recovery process, industrial effluent treatment, 290, 291
Chloride, 24, 105
  water analysis, 147
Chlorination, aromatic polyamide polymer, 45
Chlorination/dechlorination system, biofouling, 195
Chlorine
  activated carbon, 118
  sodium bisulfite, 118
Chlorine resistance, reverse osmosis membrane, long-term test, 47
Clay
  foulant, 254–256
  optical microscopy, 240, 241, 242, 244
Clean Water Act, antidegradation policy, 382–383
Cleaning, industrial effluent treatment, 284–285
Colloid 189/zinc chloride, 226
Colloidal fouling, 163
Compaction, reverse osmosis membrane, 5
Composite membrane. *See also* Specific type
  performance data, 39
  productivity, 38
Composite reverse osmosis membrane, classification, 38
Concentrate, defined, 379
Concentrate disposal method, 379–389
  borehole, 388–389
  deep well injection, 383–387
    open hole construction, 386
    requirements, 384–385
  distillation system, 388
  drain field, 388–389
  Environmental Protection Agency, 380–381
  evaporation system, 388
  land application, 387–388
  regulatory background, 380–381
  solar pond, 388
  spray irrigation, 387–388
  surface water disposal, 381–383
  waste water treatment plant, 388
Concentrate flow, pressure drop, different spaces, 63
Concentration polarization, 60
  reverse osmosis membrane, 5–6
Conditioning film, biofouling, 170–171
Conversion, reverse osmosis membrane, 4–5
Corn wet-milling, reverse osmosis, 320–322
Corn whey treatment, reverse osmosis, 319–320
Corrugated plate, flow, 63–64
Cost, water analysis, 159
Cylindrical pore, membrane surface
  solute concentration profile, 53
  water concentration profile, 52

Dairy industry, reverse osmosis, 301–308
Dairy waste water treatment, reverse osmosis, 307–308
Deep well injection
  concentrate disposal method, 383–387
    open hole construction, 386
    requirements, 384–385
  Environmental Protection Agency, 384
Deposit removal, fouling
  acids, 214–215
  alkalies, 215
  chelants, 216
  formulated products, 216–217
  techniques, 214–217
Desalination. *See also* Brackish water desalination; Seawater desalination; Specific type
  economics, 76–102
    methodology prerequisites, 76–77
  electrodialysis, 83–84
  energy consumption, 80–81
  ion exchange, 82–83
  phase change process, 84–87
    crystallization, 86–87
    flash evaporation, 86
    multiple effect evaporation, 84–85
    vapor recompression, 85–86
  processes, 81–87
  reverse osmosis, 81–82
  standard methods, 280
Desalination plant, by process, 26
Desalting process, selection guidelines, 25
2,2-Dibrom-3-nitrilopropionamide (Labozid), biocide, 196
Dispersant, 118–119
Distillation, 26–27
  capital cost, 93
  concentrate disposal method, 388
  operating cost, 93
  reverse osmosis, combination, 28–29
  seawater desalting, energy consumption, 27
Double-pass reverse osmosis, high purity water production, 347–349
Drain field, concentrate disposal method, 388–389
Dynamic membrane, 68–71
  reproducibility, 70

EDX spectrum, fouling, 220–221
Electrodialysis, 26
  capital cost, 93–95
  desalination, 83–84
  operating cost, 93–95
Electrolyte solute rejection, nanofiltration membrane, 40
Electronics industry, high purity water production, 337–338
  specification, 338
Electroplating, industrial effluent treatment, 291–294
Element, defined, 8
Elevated temperature
  HT-RO element, 41
  spiral reverse osmosis element, 41
Energy consumption, desalination, 80–81
Energy dispersive X-ray, foulant, 247–253, 258
  advantages, 238
  clay, 259

disadvantages, 238, 256–257
low atomic number, 248–250
low Z detector, 249–250
mapping, 250–253
Energy recovery, 28
Environmental Protection Agency
concentrate disposal method, 380–381
deep well injection, 384
EPA Rules for Underground Injection Control, 384
ESCA, foulant, 272–273
advantages, 238
disadvantages, 238
Evaporation
capital cost, 93
operating cost, 93
Evaporation system, concentrate disposal method,
388

Fat processing, reverse osmosis, 323–327
waste water treatment, 323–324
Feed stream bypass, spiral wound cartridge, 61
Feed water
analysis, 144–153
reverse osmosis system, 23–24
Ferrous ion, brackish water desalination, 279
Financial analysis, methods, 77–80
Florida
antidegradation policy, 382–383
reverse osmosis application, 381–382
surface water disposal, 381–382
Flow
corrugated plate, 63–64
estimating performance, 15
tubular space, 62, 63
Fluid flow dynamics, 61, 62
Fluoride, 105
water analysis, 147
Flux, reverse osmosis system, 125–126
Food processing, new developments, 31
Formaldehyde, biocide, 195–196
Foulant. *See also* Biofouling; Fouling
analytical identification techniques, 237–273
atomic absorption
advantages, 238
disadvantages, 238
attenuated total reflectance analysis, 259–266
calcium carbonate, 263–264
cellulose triacetate membrane, 267
clay, 265
hydrocarbon-oil, 265
limitations, 263
organic fatty acid, 266
principles, 262
auger
advantages, 238
disadvantages, 238
auger spectroscopy, 271–272
backscattered electron imaging, 246–247
bacteria, 253–256
clay, 254–256
energy-dispersive X-ray
advantages, 238
disadvantages, 238
energy dispersive X-ray analysis, 247–253, 258
clay, 259

disadvantages, 256–257
low atomic number, 248–250
low Z detector, 249–250
mapping, 250–253
ESCA, 272–273
advantages, 238
disadvantages, 238
frustrated multiple internal reflection, 259–266
industrial effluent treatment, 284–285
infrared spectroscopy, 257–271
advantages, 238
disadvantages, 238
optical microscopy, 239–240
advantages, 238
disadvantages, 238
scanning electron microscopy, 240–246
advantages, 238
bacteria, 257
blood cell, 260, 261
clay, 259
disadvantages, 238, 256–257
secondary electron imaging, 245–246
transmission spectroscopy, 266–271
gypsum powder, 271
X-ray fluorescence
advantages, 238
disadvantages, 238
Fouling, 58–60. *See also* Biofouling
deposit removal
acids, 214–215
alkalies, 215
chelants, 216
formulated products, 216–217
techniques, 214–217
EDX spectrum, 220–221
identification, 221–224
likely performance effects, 222
mechanistic aspects, 170–186
reverse osmosis membrane, 211
reverse osmosis system, 19–23, 220–224
scaling, 211, 212
scanning electron micrograph, 220–221
Free chlorine, water analysis, 147–148
Fruit juice processing, reverse osmosis, 308–311
Frustrated multiple internal reflection, foulant, 259–
266
FT–30 composite membrane, 7
Full-fit construction, spiral wound cartridge, 62–63
Fungus, biofouling, 188

Halogen substitution pattern, B–9 polymer, 44
Hardness, water, 104–105
High chlorine resistant membrane, materials development, 44–46
High-pressure pump, reverse osmosis system, 19
High purity water production, 334–362
applications, 335–340
boiler feed water, 335–337
carbon dioxide, differential passage, 347
double-pass reverse osmosis, 347–349
electronics industry, 337–338
specification, 338
field experience, 354–361
flow diagram, 355
initial operating results, 359, 360

High purity water production *(continued)*
  ionic removal, 341
  layout, 356–357
  market growth projections, 361–362
  market size projections, 361–362
  medical applications, 339–340
  membrane, 64–66
  membrane bypass, 344–346
  metal finishing, 339
  organics, removal, 342
  packaging, 340
  pharmaceutical applications, 339–340
  posttreatment, 349–354
    biological control, 351–352
    ion exchange polishing, 353–354
    ionic stabilization, 351
    particle stabilization, 349–350
    polishing treatment, 353–354
  pretreatment, 349–354
    biological control, 351–352
    ion exchange polishing, 353–354
    ionic stabilization, 351
    particle stabilization, 349–350
    polishing treatment, 353–354
  process design considerations, 354–356
  quality requirements, 335–340
  recontamination, 343
  reverse osmosis vs. ion exchange cost, 95–102
  silica
    differential passage, 347
    removal, 342
  start-up, 357–361
  steam generator water quality, 336
  system plot plan, 357
  system problems, 343–347
Hollow fiber, 11–12
  advantages, 12
  disadvantages, 12
  manufacturers, 32
  operation, 12
  standard rating conditions, 15
  structure, 11–12
Hollow fine fiber polyaramid, 122
HT-RO element
  elevated temperature, 41
  performance data, 41–42
Hydrogen peroxide, biocide, 196
Hydrogen sulfide
  brackish water desalination, 280
  water analysis, 147
Hydrous zirconium (IV)/polyacrylate membrane, 70
Hypochlorite, biocide, 195

Industrial effluent treatment, 282–296
  bench scale test, 285–288
    batch concentration mode, 287
    total recycle mode, 285
  characteristic reverse osmosis process configuration, 290–291
  chemical recovery process, 290, 291
  cleaning, 284–285
  electroplating, 291–294
  foulant, 284–285
  mine drainage, 296
  overview, 282–284
  pilot-plant tests, 288–290
  power station effluent, 294–295
  pretreatment, 284–285
    before disposal, 291, 292
  pulp and paper bleach effluent, 295–296
  reject and permeate total recycle, 290
  reverse osmosis examples, 291–296
  reverse osmosis process design, 284–285
  reverse osmosis process evaluation, 284–285
Infrared spectroscopy, foulant, 257–271
  advantages, 238
  disadvantages, 238
Instrumentation
  pretreatment, 127–132
  reverse osmosis system, 127–133
Internal rate of return, 79–80
Ion exchange, 24–26
  capital cost, 90–92
    chemical costs, 92
    energy costs, 92
    insurance, 92
    labor costs, 92
    maintenance, 92
    replacement, 92
    resin attrition, 92
    taxes, 92
  desalination, 82–83
  operating cost, 90–92
    chemical costs, 92
    energy costs, 92
    insurance, 92
    labor costs, 92
    maintenance, 92
    replacement, 92
    resin attrition, 92
    taxes, 92
  reverse osmosis, high purity water production comparison, 95–102
Ion separation, pH, 65, 66
Ionic removal, high purity water production, 341
Iron, 105
  water analysis, 146
Iron oxide, optical microscopy, 239–240, 243, 244
Iron-reducing bacteria, polyamide composite membrane, 212

Laminar boundary layer, mass balance equation, 58
Land application, concentrate disposal method, 387–388
Low-pressure composite membrane, silt fouling, 211, 213

Magnesium, 24, 105
  water analysis, 145
Magnesium carbonate, brackish water desalination, 278
Makeup ion exchange, 353
Manganese, 105
  water analysis, 146
Mass balance equation, laminar boundary layer, 58
Mass transfer coefficient, 59
Membrane, defined, 8
Metal finishing, high purity water production, 339
Microbial species, reverse osmosis membrane, 169

Microorganism
  biofouling, 172–174
  water, 169
Milk concentration, reverse osmosis, 305–307
  membranes, 303
Mine drainage, industrial effluent treatment, 296
Miscella distillation, reverse osmosis, 324–327
  membrane characteristics, 325
  recovery process, 326
Module design, 60–64
Mold, biofouling, 188
Monitoring, biofouling, 191–193
Mycobacteria, water, 169
Mydale crude permeate
  nickel, 67, 69
  vanadium, 67, 69

Nanofiltration membrane
  applications, 42
  electrolyte solute rejection, 40
National Pollutant Discharge Elimination System,
  surface water disposal, 383
Net present value, 79
Nickel, mydale crude permeate, 67, 69
Nitrate, 105
  water analysis, 147
Nodule, 47–48
Nodule aggregate, 48
Nonaqueous solution separation, 66–68
Nuclear plant water treatment, 70

Oil processing, reverse osmosis, 323–327
  waste water treatment, 323–324
Operating cost
  distillation, 93
  electrodialysis, 93–95
  evaporation, 93
  ion exchange, 90–92
    chemical costs, 92
    energy costs, 92
    insurance, 92
    labor costs, 92
    maintenance, 92
    replacement, 92
    resin attrition, 92
    taxes, 92
  reverse osmosis, 87–90
    chemical conditioning, 88–89
    electricity, 89
    maintenance, 90
    operating labor, 90
    polishing, 90
    repair, 90
    reverse osmosis membrane cleaning, 89–90
    reverse osmosis membrane replacement, 89–90
Operating pressure
  solute, preferential sorption, 54
  water, preferential sorption, 54
Optical microscopy
  calcium sulfate dihydrate, 240, 243, 244
  clay, 240, 241, 242, 244
  foulant, 239–240
    advantages, 238
    disadvantages, 238
  iron oxide, 239–240, 243, 244

polarized light illumination, 239–240
  silica, 240, 241, 242, 244
Orange juice, reverse osmosis, 308–311
Organic fouling, 163, 211, 213
Organic material, 101
  high purity water production, removal, 342
Organic solute separation
  aromatic polyamide membrane, 50
  PEC–1000 membrane, 43
Osmosis, 1
Osmotic pressure, van't Hoff equation, 3
Oxidizing agent, 114–115
Oxygen, water analysis, 148
Ozone, biocide, 196

Packaging, high purity water production, 340
Payback period, 77–78
PEC–1000 membrane, organic solute separation, 43
Peracetic acid, biocide, 196
Permasep B–9 permeator, 1
Permasep B–10 permeator, 1
Permeator, defined, 8
pH, aluminum solubility, 110, 111
pH
  ion separation, 65, 66
  water, 105
    large municipality, 140
  water analysis, 148
Phase change process, desalination, 84–87
  crystallization, 86–87
  flash evaporation, 86
  multiple effect evaporation, 84–85
  vapor recompression, 85–86
Phase inversion technique, membrane preparation, 47
Phosphate, 105
  water analysis, 146
Plate and frame device, 13–14
  advantages, 14
  disadvantages, 14
  operation, 14
  structure, 13–14
Polarized light illumination, optical microscopy,
  239–240
Polishing ion exchange, 353
Poly-m-phenylene-iso-(x)-co-tere(100 − x)-
  phthalamide copolymer, 49
Polyamide, 38
Polyamide composite membrane, iron-reducing bacteria, 212
Polyelectrolyte crystallization inhibitor, brackish
  water desalination, 278
Polymer aggregate, 48
Polymer nodule, 48
  anion, 57
  cation, 57
  interstitial domain, 48–49
  intramolecular chain segment distance, 57
  pore size, 54–55
  schematic representation, 48
  size, 56–57
Pore size
  aromatic polyamide membrane
    average, 51
    distribution, 51
  polymer nodule, 54–55

Pore size *(continued)*
  reverse osmosis membrane, control, 54–58
  small, 42–43
  solute, preferential sorption, 54
  synthetic membrane, 46
  water, preferential sorption, 54
Posttreatment
  high purity water production, 349–354
    biological control, 351–352
    ion exchange polishing, 353–354
    ionic stabilization, 351
    particle stabilization, 349–350
    polishing treatment, 353–354
  reverse osmosis system, 16
Potassium, 24, 105
  water analysis, 146
Power station effluent, industrial effluent treatment,
    294–295
Precipitate prevention, brackish water desalination,
    278–279
Preferential sorption-capillary flow mechanism, 50
Pressure
  reverse osmosis membrane, 5
  reverse osmosis system, 125–126
Pressure drop, concentrate flow, different spaces, 63
Pretreatment
  biofouling, 189–191
    sources, 164–165
  brackish water desalination, 277–278
  high purity water production, 349–354
    biological control, 351–352
    ion exchange polishing, 353–354
    ionic stabilization, 351
    particle stabilization, 349–350
    polishing treatment, 353–354
  industrial effluent treatment, 284–285
    before disposal, 291, 292
  instrumentation, 131–136
  methods, 20–22
  new developments, 30
  record keeping, 131–136
  reverse osmosis system, 19–23, 108–119
    antiscalant, 116
    calcium carbonate, 113–116
    cartridge filter, 112
    clarification, 109
    dispersant, 116
    medial filter, 109–112
    organics, 117–118
    oxidizing agent, 118–119
    scale control, 112–119
    silica, 116–117
    silt reduction, 108–109
    suspended solids, 108–109
    ultrafilter, 111
  seawater reverse osmosis, 27–28
  techniques, 23
Pseudomona vesicularis, biofouling, 173
Pseudomonas diminuta, biofouling, 172, 173, 174,
    175, 178, 180, 181
Pseudomones fluorescens, biofouling, 173, 174
PTA/PTB (Lutonol M410 from BASF), 226
Pulp and paper bleach effluent, industrial effluent
    treatment, 295–296

Recontamination, high purity water production, 343
Record keeping
  pretreatment, 131–136
  reverse osmosis system, 131–137
Recovery
  reverse osmosis membrane, 4–5
  reverse osmosis system, 126–131
Regrowth, biocide, 197–198
Reject. *See* Concentrate
Rejuvenation, 224–225, 226
  procedure, 227
Rejuvenation agents, 226
Reproducibility, dynamic membrane, 70
Reverse osmosis
  applications, 24–29, 33
  basic equations, 2–4
  beer, 314–316
    concentration, 316
    low-alcohol, 314–316
  beverage industry, 314–318
  capital cost, 87–90
    chemical conditioning, 88–89
    electricity, 89
    maintenance, 90
    operating labor, 90
    polishing, 90
    repair, 90
    reverse osmosis membrane cleaning, 89–90
    reverse osmosis membrane replacement, 89–90
  cereal processing, 318–322
    by-product recovery, 318–319
    dilute sweet water concentration, 322
    reverse osmosis permeate and evaporator over-
      head polishing, 322
    steep water reverse osmosis evaporation, 322
    thin stillage treatment, 318–319, 320
  cheese whey concentration, 301–305
    flow diagram, 304
    membranes, 303
  corn wet-milling, 320–322
  corn whey treatment, 319–320
  current industry state, 32–33
  dairy industry, 301–308
  dairy waste water treatment, 307–308
  desalination, 81–82
  distillation, combination, 28–29
  fat processing, 323–327
    waste water treatment, 323–324
  flow diagram, 8
  food industry applications, 300–327
  fruit juice processing, 308–311
  future trends, 37–72
  industrial applications, 33
  ion exchange, high purity water production com-
    parison, 95–102
  milk concentration, 305–307
    membranes, 303
  miscella distillation, 324–327
    membrane characteristics, 325
    recovery process, 326
  oil processing, 323–327
    waste water treatment, 323–324
  operating cost, 87–90
    chemical conditioning, 88–89

electricity, 89
maintenance, 90
operating labor, 90
polishing, 90
repair, 90
reverse osmosis membrane cleaning, 89–90
reverse osmosis membrane replacement, 89–90
orange juice, 308–311
recent developments, 29–31
salt whey desalting, 307
seawater desalting, energy consumption, 27
soft drink industry, 318
sugar industry, 327
system components, 17
tea product, 327
theory, 2–6
U.S. military applications, 364–378
  3,000-GPD, 369
  150,000-GPD, 372
  600-GPH, 367–369, 370
  background, 364–365
  chemicals, 377
  design considerations, 365–366
  future development, 377–378
  reverse osmosis elements, 376–377
vegetable juice processing, 312–314
vegetable protein, 323
vs. other desalting processes, 24–27
wine processing, 316–318
Reverse osmosis inorganic membrane, 42
Reverse osmosis membrane
applications, 2, 64–67
array, 126–131
bypass, high purity water production, 344–346
chlorine resistance, long-term test, 47
commercial era, 1
compaction, 5
concentration polarization, 5–6
configurations, 8–14
conversion, 4–5
cylindrical pore
  solute concentration profile, 53
  water concentration profile, 52
defined, 8
design considerations, 16–17
device performance, 14–15
devices, 8–14
factors affecting performance, 4–6
failure problems, 211
fouling, 211
ideal membrane characteristics, 6
manufacturers, 31–32
materials development, 6–8, 29–30
  recent, 37–44
microbial species, 169
overview, 1–2
pore, origin, 46–50
pore size
  control, 54–58
  small, 42–43
pressure, 5
recent developments, 37–44
recovery, 4–5
selection, 119–125

standard rating conditions, 14–15
state of the art, 1–33
sterilization, techniques, 225–228
temperature, 5
transport, 50–54
ultrapure water production, 64–66
Reverse osmosis membrane autopsy, water analysis,
    158
destructive phase, 158
nondestructive phase, 158
Reverse osmosis membrane cleaner
customized, 217
generic, 217
selection criteria, 217–219
Reverse osmosis membrane cleaning, 210–234
case histories, 228–234
system schematic, 224
Reverse osmosis system
biofouling
  effects, 165–168
  signs and symptoms, 165–166
chemical problem evaluation, 220
construction materials, 108
design considerations, 16–24, 104–138
design parameters, 125–131
end use, 16
feed water, 23–24
flux, 125–126
fouling, 19–23, 220–224
high-pressure pump, 19
instrument operation verification, 219
instrumentation, 127–133
maintenance, 18
mechanical problem evaluation, 220
membrane array, 126–131
operating data review, 219
operation, 18
posttreatment, 16
pressure, 125–126
pretreatment, 19–23, 108–119
  antiscalant, 116
  calcium carbonate, 113–116
  cartridge filter, 109–112
  clarification, 109
  dispersant, 116
  medial filter, 109–112
  organics, 117–118
  oxidizing agent, 118–119
  scale control, 112–119
  silica, 116–117
  silt reduction, 108–109
  suspended solids, 108–109
  ultrafilter, 111
procedure, 223
record keeping, 131–137
recovery, 126–131
scaling, 19–23
sterilization, techniques, 225–228
temperature, 125–126
troubleshooting, 219–224
water analysis design criteria comparison, 220
Reverse Osmosis Water Purification Unit system,
    364–378
  3,000-GPD, 369

Reverse Osmosis Water Purification Unit system
(*continued*)
600-GPH, 367–369, 370
background, 364–365
chemicals, 377
design considerations, 365–366
future development, 377–378
reverse osmosis elements, 376–377
water, 366–367

Salt passage flow, estimating performance, 15
Salt transport, 3–4
equation, 3–4
Salt whey desalting, reverse osmosis, 307
Sampling, water analysis, 141–143
holding time, 144
point selection, 142
preservation, 144
sample-collection containers, 141–142, 144, 161
Scaling, 163
fouling, 211, 212
reverse osmosis system, 19–23
Scanning electron microscopy
foulant, 240–246
advantages, 238
bacteria, 257
blood cell, 260, 261
clay, 259
disadvantages, 238, 256–257
secondary electron imaging, 245–246
Seawater desalination, 27–29, 40. *See also* Desalination
distillation, energy consumption, 27
forecast, 33
reverse osmosis, energy consumption, 27
Seawater reverse osmosis, pretreatment, 27–28
Secondary electron imaging
foulant, 245–246
scanning electron microscopy, 245–246
Shell, defined, 8
Silica, 116–117
brackish water desalination, 279
high purity water production
differential passage, 347
removal, 342
optical microscopy, 240, 241, 242, 244
temperature, 116, 117
water analysis, 146
Silt density index, 106, 107
apparatus, 107
biofouling, 189
water
analysis, 149–153
apparatus, 135, 137
bias, 137
calculations, 136–137
precision, 137
procedures, 135–136
reference documents, 134
report, 137
scope, 134
significance, 135
standard test method, 133–138
terminology, 134
use, 135

Silt fouling, low-pressure composite membrane, 211, 213
SiO_2, 105
Sodium, 24, 105
water analysis, 145
Sodium bisulfite, chlorine, 118
Sodium chloride separation, charged composite membrane, 65
Sodium hexametaphosphate, brackish water desalination, 278
Soft drink industry, reverse osmosis, 318
Solar pond, concentrate disposal method, 388
Solute
operating pressure, preferential sorption, 54
pore size, preferential sorption, 54
Solute mass transfer, 60
Spiral reverse osmosis element, elevated temperature, 41
Spiral wound cartridge, 9–11, 60–63
advantages, 10–11
disadvantages, 11
feed stream bypass, 61
full-fit construction, 62–63
improvement, 60–62
manufacturers, 31
operation, 10
standard rating conditions, 15
structure, 9–10
tubular spiral, 62–63
Spiral wound cellulose acetate, 119–122
construction, 122
selection, 119–122
Spiral-wound polyamide thin film composite membrane, 2
Spray irrigation, concentrate disposal method, 387–388
Standard rating conditions
hollow fiber, 15
reverse osmosis membrane, 14–15
spiral wound cartridge, 15
Staphylococcus warneri, biofouling, 172, 175, 180, 181
Steam generator water quality, high purity water production, 336
Sterilization
reverse osmosis membrane, techniques, 225–228
reverse osmosis system, techniques, 225–228
Strontium, water analysis, 145
Strontium sulfate, brackish water desalination, 278
Sugar industry, reverse osmosis, 327
Sulfate, 24, 105
water analysis, 147
Surface force-pore flow mechanism, 50
Surface water disposal
concentrate disposal method, 381–383
Florida, 381–382
National Pollutant Discharge Elimination System, 383
Synthetic membrane, pore size, 46
System analysis, water analysis, 153–156

Tea product, reverse osmosis, 327
Temperature
reverse osmosis membrane, 5
reverse osmosis system, 125–126

silica, 116, 117
    water analysis, 148
TFC membrane, 7
Thin film composite, 7–8, 123–125
    membrane characteristics, 124
    recommended operating ranges, 124
Total dissolved solids, water analysis, 104–105, 147
Total organic carbon, water analysis, 148
Transmission spectroscopy, foulant, 266–271
    gypsum powder, 271
Tubular configuration, 12–13
    advantages, 13
    disadvantages, 13
    manufacturer, 32
    operation, 13
    structure, 12–13
Tubular spiral, spiral wound cartridge, 62–63
Turbidity, water analysis, 105, 148–149
Turbulence, membrane surface, 63–64

Ultrapure water production. *See* High purity water
    production

Vanadium, mydale crude permeate, 67, 69
Van't Hoff equation, osmotic pressure, 3
Vegetable juice processing, reverse osmosis, 312–
    314
Vegetable protein, reverse osmosis, 323
Vinac/Gelva, 226

Waste treatment, new developments, 30
Waste water treatment plant, concentrate disposal
    method, 388
Water
    alkalinity, 104–105
    carbon dioxide, 105
    characteristics, 104–107
    hardness, 104–105
    microorganism, 169
    Mycobacteria, 169
    operating pressure, preferential sorption, 54
    pH, 105
        large municipality, 140
    pore size, preferential sorption, 54
    Reverse Osmosis Water Purification Unit system,
        366–367
    silt density index
        apparatus, 134, 137
        bias, 137
        calculations, 136–137
        precision, 137
        procedures, 135–136
        reference documents, 134
        report, 137
        scope, 134
        significance, 135
        standard test method, 133–138

    terminology, 134
    use, 135
    source investigation, 140–141
    supplies, 104–107
    testing, 106–107
    total dissolved solid, 104–105
    turbidity, 105
Water analysis, 106–107
    aluminum, 145
    barium, 145
    bicarbonate, 146
    calcium, 145
    carbon dioxide, 148
    carbonate, 147
    cartridge filter analysis, 156–158
    chloride, 147
    cost, 159
    feed-water analysis, 144–153
    fluoride, 147
    free chlorine, 147–148
    hydrogen sulfide, 147
    iron, 146
    magnesium, 145
    manganese, 146
    membrane autopsy, 158
        destructive phase, 158
        nondestructive phase, 158
    nitrate, 147
    oxygen, 148
    pH, 148
    phosphate, 146
    potassium, 146
    sampling, 141–143
        holding time, 144
        point selection, 142
        preservation, 144
        sample-collection containers, 141–142, 144, 161
    silica, 146
    silt density index, 149–153
    sodium, 145
    strontium, 145
    sulfate, 147
    system analysis, 153–156
    temperature, 148
    total dissolved solids, 147
    total organic carbon, 148
    turbidity, 148–149
    types, 143–158
Water sample, 140
Water sample kit, 143
Water transport, 2–3
    equation, 2–3
Wine processing, reverse osmosis, 316–318

X-ray fluorescence, foulant
    advantages, 238
    disadvantages, 238